22 95/100
7/15

Herbert B. Sachse

Semiconducting Temperature Sensors and their Applications

Herbert B. Sachse

Semiconducting Temperature Sensors and their Applications

A Wiley-Interscience Publication

JOHN WILEY & SONS, INC.

New York / London / Sydney / Toronto

RECID = 7120-1

Library of Congress Cataloging in Publication Data:

Sachse, Herbert, 1903-

Semiconducting temperature sensors and their application.

"A Wiley-Interscience publication."

Includes bibliographical references and index.

1. Thermistors. 2. Thermometers and thermometry. 3. Semiconductors. I. Title.

TK7871.98.S23 621.3815′2 75-2017

ISBN 0-471-74835-8

Printed in the United States of America

10 9 8 7 6 5 4 3 2 1

Preface

The author of this book has been working in the field of temperature-sensitive resistors since the early 1930s. At that time he measured the electronic conductivity of a large number of oxides and arrived at conclusions concerning the electronic structure of their metal constituent and their stoichiometry. Although these conclusions had to remain qualitative, considering the numerous side effects in polycrystalline and powder materials (their defect structure and surface states), they still have a degree of validity; in some ways they presaged the present duality of *n* and *p* semiconductors.

His later involvement in the development of practical types of semiconducting oxide resistors in the Central Laboratory of Siemens and Halske expanded into a large-scale production, terminated only by the disruption of industrial activities in Germany after World War II. After the author joined the Research Staff of Keystone Carbon Company in St. Marys, Pennsylvania and acted as its Research Director, a part of his duties again brought him into contact with thermistors. One of his contributions during this time was the development of reliable and reproducible temperature sensors for the cryogenic range from 3 K up to 200 K, thus connecting the cryogenic range with that of regular thermistors. The author does not intend in the present book to dwell on a large collection of composition formulas or processing data, many of them still rather vivid in his mind. In avoiding this, not only could more space be devoted to basic concepts, but also the principle of propriety has been served. In treating the basic properties of materials, some freedom was taken to delve into the present theoretical models and to report some interesting experimental data. Since this excursion had to be short, ample references are given to satisfy the reader's intellectual appetite. Another liberty was taken to mention certain types of nonmetallic sensors competitive with semiconductors. Although they could act as semiconductors, their dielectric and pyroelectric properties are used, and it is of interest to compare their usefulness as sensors.

In describing applications some harsh decisions were necessary to avoid "overeating": In some cases some basic investigations of earlier years received more attention, but recent work has also been quoted to the limit of available space.

The patent literature is amply recorded here, though without claim to completeness. The critical reader familiar with this field will recognize that many patents have only a small degree of inventive novelty and others with more original ideas apparently have not made much impact on the commercial field. This might help prospective inventors to concentrate their efforts on real novelties.

In some cases novel production steps are mentioned that have not yet passed the rigid tests of reliability and economy. The author felt obliged to mention such possibilities since a new method, practiced by the right mind and in skilled hands, can prosper where other people have failed. The author, of course, has his own reservations with regard to the success of some innovations (for example, new contacting methods), but he also feels that progress should not be discouraged by the so-called expert experience.

The author wishes to express his gratitude to those who helped to prepare the manuscript of this book. Foremost among them is his wife Gerda, who not only willingly accepted the restrictions to our private life imposed by this work, but also made vital contributions in collecting the selected literature, preparing the indexes together with her daughter Renate Sachse, and finally typing the manuscript.

Professor C. Wagner of Göttingen, Germany has through many years stimulated the author's interest in problems of solid-states chemistry and was so gracious to read a few chapters. The author also thanks Dr. F. Brickwedde of Penn State University, Professor Bruno Klinger of Colorado State University, Dr. Francis Shull, George Hall, formerly with Kodak and A. D. Little Company, respectively, and his son, Wolfgang Sachse, assistant professor at Cornell University for critical reading of major parts of the text. Finally the author wishes to express his appreciation and gratitude to the staff members of John Wiley and Sons who have made great efforts to prepare this book for publication.

Herbert B. Sachse
P.O. Box 206
St. Marys, Pa. 15857
U.S.A.

September, 1974

Contents

Herbert B. Sachse

Semiconducting Temperature Sensors and their Applications

1

Introduction

For many years platinum was the only material used in resistance thermometry. Its corrosion resistance and its high melting point made it attractive for a wide range of applications. It could be purified to such a degree that it attained the status as secondary temperature standard with a temperature coefficient of 0.0392% K^{-} at 300 K. However, its high price together with its low resistivity of 9.8 $\mu\Omega$ cm at 273 K were serious drawbacks in many applications. In order to obtain a sufficiently high ratio of thermometer to lead resistance, coils had to be wound in a strain-free manner, which increased production costs beyond those of the high cost of the material, also increasing mass and volume of the sensor. Furthermore, the small temperature coefficient required the use of precision bridges. When platinum was replaced by nickel as suggested by Grant and Hickes,[1]* only the material costs were reduced while the other drawbacks remained. It was therefore quite natural that the advent of electrical instrumentation in industrial process control, which often implies the measurement and control of temperature, stimulated the development of novel resistive temperature sensors. They are made not only of less expensive materials but also with much higher temperature coefficients as well, or with a wide range of resistivity. This trend coincided with the demand for resistors in telecommunication circuits that could be continuously regulated, either by external heating or by self-heating.

This was the situation forty years ago when the present concept of the thermistor was born. At first this component had no name of its own and was simply called a regulating resistor. In Germany in the early 1930s the name *Heissleiter* (hot conductor) was coined, while in the United States the device was called a *thermistor*.

* All references are listed at the end of the book, pp. 334–360.

Control and communication engineers were soon intrigued by the possibility of using the horizontal range of the voltage-versus-current characteristic to stabilize the input or output voltages in circuits over a rather wide current range. Automatic amplitude control gained importance with the rapid extension of long-distance telephone systems. The remote operation of electrical systems for unmanned amplifier relay stations was the next application. A defective amplifier could be replaced by activating a reserve amplifier using a remotely heated thermistor as switch.

Under normal ground conditions the operation of electrical networks did not require temperature compensation of the resistivity of copper wire in coils and cables. However, with the rapid development of high-flying aircraft, civilian and particularly military electronics had to be reliable to low temperatures existing in high altitudes. By 1940 it was generally specified that components and assemblies had to function reliably between temperatures of about 220 and 375 K.

The temperature coefficient of copper wire is approximately 0.4% K^{-1}. Therefore for a temperature difference of 150 K, a resistance change of 60% occurs whenever copper wire is used in electric circuits, as in relay or generator coils. Since the voltage output of batteries is also temperature dependent, a self-regulating device was necessary to make aircraft systems reliable.

The first commercial thermistors were made of UO_2 in Germany in 1932 by Osram. This was soon replaced by $MgTiO_3$ (spinel), which was reduced with H_2. These thermistors retained the trade name "Urdox" even after their production from UO_2 was abandoned. Resistivity and temperature dependence of these units could be widely varied in a controlled manner by the degree of reduction and the admixture of MgO in excess of the stoichiometric amount.

In 1936 Philips followed with another type of semiconducting resistor made from silicon or ferrosilicon sintered together with inorganic binders and sold under the name Startotube. These never obtained as wide a distribution as Urdox units. Both products, Urdox and Starto, required a protective atmosphere to prevent oxidation that would cause large resistance increases during operation. Therefore they were sealed into bulbs filled with H_2, N_2, or Ar, which made them less suitable for temperature measurements because of the high heat resistance between the sensor and the environment.

The use of UO_2 was revived a few years later by Siemens in Germany for the production of voltage self-regulating thermistors. They were capable of stabilizing ac voltages of 2 or 6 V within $\pm 10\%$ over a current ratio of 20 (between 0.1 and 2 mA or 0.4 and 8 mA). For smaller current ratios the voltage stabilization could be improved to $\pm 1\%$. Their small input was most useful for nonstationary equipment with limited power supply. Their maximal resistance ratios of cold to hot were 30 and ~100, respectively. Temperature measurement and compensation required thermistors with faster response and better thermal contact with the environment. For this purpose during the latter 1930s Siemens developed thermistors of CuO that were stable in air up to 475 K. Great scientific and technological efforts were necessary to produce these thermistors with a wide range of resistivities, only a small variation in temperature sensitivity, and good stability.[2]

During World War II the production of these thermistors, though steadily increasing, was mainly for military applications; only a small fraction reached the commercial market. After the war their production was not resumed.

Bell Telephone Laboratories was confronted by the same problem as European companies working in the telecommunication field, namely, to compensate resistance changes in transmission lines caused by temperature variations. They also developed compensating resistance elements, as described in 1940 by Pearson.[3]

A few patents were granted to Bell Telephone Laboratories for silver sulfide (Ag_2S) resistors with a high negative temperature coefficient of resistivity[4–6]; silver sulfide can be considered the first semiconductor. It was discovered by Faraday in 1834. Its technical application is hampered by its β modification existing below 452 K, where less than 20% of the conductivity is electronic, the balance being ionic.[7]

Ionic semiconductors change their composition and properties under current flow, especially with direct current, and therefore have never been seriously considered as thermistor materials. The most serious drawback, however, is that the material changes its resistivity in air by sulfur losses.

Further research at Bell Telephone Laboratories concentrated on transition metal oxide systems. (See Table 15 in Section 3.5.1 and the patent list, pp. 318–329).

Philips reentered the field 1942 with a number of oxide systems in which iron oxides of different valence state were the conductive component.

Between 1950 and the present, the art of controlling the resistivity and its temperature dependence in transition metal oxide systems has been advanced by thermistor manufacturers, together with improved cycling and long-range stability. Many new applications in the chemical and medical-biological field required interchangeability and miniaturization. These needs were met by the development of a variety of bead types in glazed or glass-encapsulated form. After 1955 the missile and space technology industries stimulated the development of special cryogenic thermistors, since conventional types would become insulators at temperatures below 100 K.

The rapidly expanding technology of pure and doped Ge and Si promoted the development of reproducible and sensitive thermometers and bolometers reaching down to a few degrees Kelvin, thus providing an excellent substitute for carbon resistors previously used as thermometers in this range. The temperature sensitivity of nonohmic devices such as diodes and transistors made them eligible as cryogenic thermometers, and superconductive transitions deserved attention for applications where only a narrow temperature range was of interest. A few paragraphs on dielectric and pyroelectric materials are included in this book to offer users of temperature sensors a wide scope of options for their applications. It seemed advisable to use only one temperature scale (Kelvin), even if it meant some inconvenience for readers interested in medical, biological, and chemical applications. A few concessions were made in the chapter dealing with the manufacturing processes, especially sintering, where the use of the Fahrenheit scale is still common. In these cases the Fahrenheit value is added in parentheses. Pressures are still expressed in torrs since the new unit pascal (= 7.5 m Torr) is less well known. The electrical quantities are expressed in mks units (ampere, volt, ohm, volt per meter, joule). Exceptions are made if theoretical considerations of another author are quoted.

2

Fundamentals of semiconductors

2.1. MATERIALS WITH NEGATIVE TEMPERATURE COEFFICIENT OF RESISTIVITY (NTC MATERIALS)

Semiconduction from the dawn of its discovery by Faraday until 20 years ago was synonymous not only with certain ranges of resistivity (approximately 10^{-2} to 10^{9} Ω cm), but more so with the phenomenon of a large negative temperature coefficient (TC) of resistivity. Fundamental experimental and theoretical studies between 1930 and the early 1950s led to a rather good qualitative understanding and in many cases to a quantitative concept; these were condensed in a number of classical books on semiconductor physics and parts of textbooks.[8–10]

The explosion in semiconductor activity after 1955 was accompanied by a large output of books dealing not only with special aspects but also treating the fundamentals. This book does not propose to give a full up-to-date bibliography. Therefore only a few additional sources are mentioned for each language region.[11–16]

2.1.1. Intrinsic Semiconductors

The ideal model of a semiconducting crystal of an element (Ge, Si, diamond) is shown by the band scheme in Figure 1. At temperatures approaching 0 K all electrons are bound in the valence band. With rising temperature, electrons in increasing numbers are thermally excited to an energy level E_g, which makes them mobile, and they form the conduction band. Since the electron concentration in the conduction band increases with temperature, semiconductors always have a negative temperature coefficient of resistivity as long as this model can be applied. The removal of electrons from the valence band creates an equivalent number of electron deficiencies (holes),

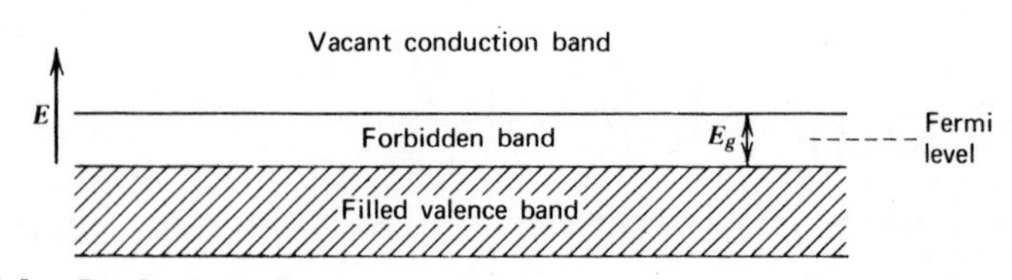

FIGURE 1. Band scheme for intrinsic conductivity. From Kittel, *Introduction to Solid State Physics,* Wiley, New York, 1956, p. 348.

which can be treated formally as positive current carriers. The concentration of electrons in the conduction band is determined in first approximation by the Boltzmann distribution law, which correlates the equilibrium ratio of electrons in the valence and conduction band with the energy difference between these states:

$$\frac{n_e}{n_v} = k \exp\left(-\frac{E_g}{kT}\right) \tag{1}$$

This equilibrium is dynamic, since free electrons always may recombine with holes and other electrons will be excited to the conduction band, following statistical thermal fluctuations. In equilibrium, the thermal excitation rate is equal to the recombination rate, which is

$$\frac{dn}{dt} = -k_1 n_e n_h \tag{2}$$

Since $n_e = n_h$,

$$\frac{dn}{dt} = -k_1 n_e^2$$

The minus sign indicates the disappearance of electrons. The excitation rate, which is the number of electrons reaching the conduction band per time unit, is given by

$$\frac{dn}{dt} = k_2(N - n_e) \exp\left(-\frac{E_g}{kT}\right) \tag{3}$$

where N is the total number of excitable valence electrons. In equilibrium both rates are equal and $n_e \ll N$; therefore

$$k_1 n_e^2 = k_2 N \exp\left(\frac{E_g}{kT}\right) \quad \text{or} \quad n_e = k_3 \exp\left(-\frac{E_g}{2kT}\right)$$

The energy gap E_g is a characteristic material constant; it depends somewhat on the temperature.

The mobility of the electrons μ_e in the conduction band and of the corresponding holes in the valence band μ_h are normally different. Therefore the conductivity of the semiconductors is defined in first approximation by

$$\sigma = e(n_e\mu_e + n_h\mu_h) \tag{4}$$

Only for a very pure semiconductor material are electrons and holes found in equivalent numbers, that is, $n_e = n_h$.

It is well known that the Boltzmann distribution is not applicable to the distribution of electrons in solids and that the Fermi–Dirac distribution comes nearer to reality. In solids, discrete atomic levels broaden into bands of closely spaced energy levels because of the atomic interactions. Conduction electrons occupy different energy levels because of the Pauli principle.[17–19]

Furthermore, the motion of electrons in response to an electrical field is different in a metal or semiconductor from such motion in free space. The same applies to electron deficiencies (holes). This can formally be expressed by substituting for the mass of a free electron m_0 (9.107×10^{-28} g) an effective mass m^* that modifies the acceleration produced by an electrical field:

$$a_e = \frac{eE}{m_e} \quad \text{or} \quad a_h = \frac{eE}{m_h} \tag{5}$$

By application of Fermi–Dirac statistics and using the effective masses of electrons and holes, the conductivity of an intrinsic semiconductor ($n_e = n_h$) can be written as

$$\sigma = 2e\left(\frac{2\pi kT}{h^2}\right)^{3/2}(m_e{}^*m_h{}^*)^{3/4}(\mu_e + \mu_h)\exp\left(-\frac{E}{2kT}\right) \tag{6}$$

In practical applications the resistivity of semiconductors in dependence on temperature is written as

$$\rho = A \exp\frac{B}{T} \tag{7}$$

or

$$\sigma = \frac{1}{A}\exp\left(\frac{-B}{T}\right) \tag{8}$$

In comparing Equations 6 and 7 it can be seen that $B = -Eg/2K$ and A is a factor proportional to $T^{-3/2}$ and inversely proportional to the mobility of the current carriers.

The mobilities of electrons and holes in the intrinsic region have their own temperature dependence. In very pure materials only lattice scattering by collisions between lattice and electron waves occur. Since impurity centers can exist even in nominally pure materials, scattering of electron waves and mobility impediments can happen even at very low temperatures where lattice vibrations are minimized.

2.1.2. Extrinsic Semiconductors

In most practical applications, semiconductors are not intrinsic ($n_e \neq n_h$). Impurities are added in small concentrations to influence the electrical properties. They act either as donors or acceptors of electrons.

In general, an impurity added to an elemental semiconductor is electrically active if its valence is different from that of the atoms forming the host lattice. Substituting a P, As, or Sb atom for a Ge or Si atom will leave an extra valence electron, which can act as conduction carrier. On the other hand, substitution by a B, Al, Ga, or In atom creates an electron deficiency, and holes conduct the current. The energies necessary to ionize donors or acceptors are much smaller than the energy gap of the intrinsic semiconductors. Therefore the exponential term, which determines the temperature dependence of the conductivity (or resistivity), becomes much smaller. In other words, the temperature coefficient of the resistivity of doped materials is much smaller than that of pure materials.

While the energy gaps of intrinsic semiconductors are of the order of 0.1 to ~1 eV, the ionization energies of donors or acceptors are between 0.01 and 0.1 eV, depending on the dopant atom and also its concentration. These low values have been explained by assuming the Bohr model of the hydrogen atom as example of a donor atom from which an electron can dissociate. The ionization energy of the free hydrogen atom is

$$E = -\frac{e^4 m_0}{8\epsilon_0^2 h n^2}$$

where ϵ_0 is the dielectric constant of the vacuum, e is the electron charge, and m_0 is the mass of the free electron. If n, the principal quantum number, is 1, the ionization energy of the hydrogen atom is 13.6 eV. For the case of a donor atom in the Ge or Si lattice, the dielectric constants are 15.8 and 11.7, respectively. The effective mass of the "free" valence electron is also different ($0.12m_0$ in Ge and $0.25m_0$ in Si). This would reduce the ionization energy E_d of a donor by a factor $0.12/15.8^2$ in Ge and $0.25/11.7^2$ in Si, resulting in values of 0.0065 and 0.025 eV, respectively. For acceptor atoms a similar approach can be made. The observed donor and acceptor ionization energies are in most cases slightly larger than the calculated values since the chosen estimates ignore anisotropy effects. The values are in most cases comparable to kT at room temperature (0.026 eV). Therefore nearly complete dissociation can be expected, because $E_d/kT \approx 1$. The band scheme for impurity semiconductors is shown in Figure 2. If only donors

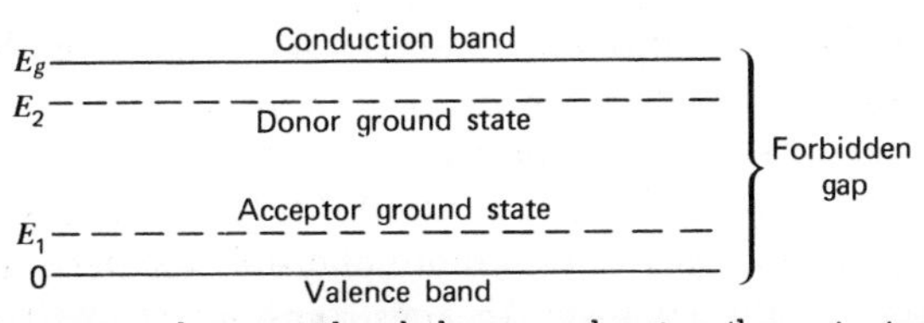

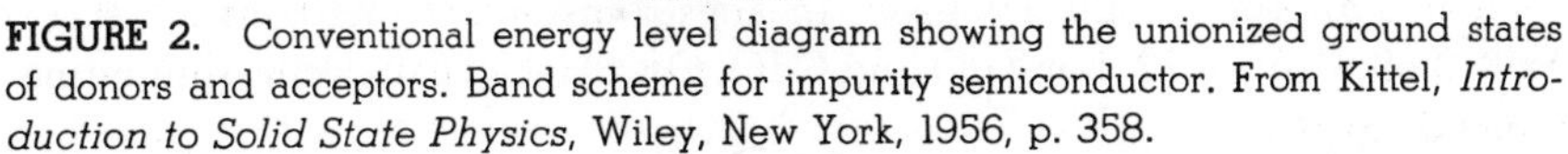

FIGURE 2. Conventional energy level diagram showing the unionized ground states of donors and acceptors. Band scheme for impurity semiconductor. From Kittel, *Introduction to Solid State Physics*, Wiley, New York, 1956, p. 358.

were present, the conduction carriers would be electrons, in the case of acceptors holes. In reality both types of carriers eoexist, since thermal activation of electrons in the valence band goes parallel with dissociation of donors. Furthermore, it is very difficult to exclude other small impurities that can act as acceptors (or vice versa). If the majority of the carriers are electrons, the conduction is n-type with holes as minority carrier. In the opposite case, the conduction is p-type.

Electrical neutrality requires that the sum of the conduction electrons and the negatively charged acceptors is equal to the number of holes and donors with one positive charge

$$n + N_a^- = p + N_d^+.$$

If the donor concentration is large and $N_d \gg N_a$, the concentration of electrons is given by

$$n = (2N_d)^{1/2} (2\pi k T m_e/h^2)^{3/4} \exp(-E_d/2kT)$$

At low temperature the donors are only partially ionized and an increasing fraction of electrons is needed to fill up acceptors, thus changing the exponential to $-E_d/kT$. Scattering of electrons by neutral and ionized impurity atoms adds to the effect of lattice scattering on the mobilities.[20] If all donors or acceptors are ionized, the conductivity is calculated in straightforward manner as

$$\sigma_n = e\mu_e N_d \quad \text{if} \quad n_e \gg p_h$$

$$\sigma_p = e\mu_h N_a \quad \text{if} \quad n_e \ll p_h$$

So far only the case of elemental semiconductors has been treated, which would apply to Ge, Si, or doped diamond temperature sensors. The majority of commercial semiconducting sensors consists of compounds, mainly oxides or complex oxidic systems such as spinels or perovskites. In these cases the stoichiometry of the compounds plays a decisive role. Drastic increases of the electronic conductivity have been found in the systems CuI_{1+x},[21] NiO_{1+x}[22], and Cu_2O_{1+x}[23] with $x < 0.02$. Wagner and Schottky[24] have shown that nonstoichiometry in inorganic compounds is not an exception, but a general phenomenon based on the defect structure of solids.

However, instead of claiming the existence of O- excess in cubic close-packed NaCl structures of the oxides FeO, MnO, NiO, CoO, NbO, and VO, the assumption of cation vacancies is more plausible for sterical reasons (see Section 3.3). Ionic radii are given in angstroms as 0^{2-}, 1.40; Mn^{2+}, 0.80; Ni^{2+}, 0.69; Co^{2+}, 0.72. For ZnO and CdO, O vacancies can occur, simulating a metal excess.

Since all metal oxides are ionic compounds, their anions have always the charge -2, while their cations can have charges between $+1$ and $+6$. In nondefect ideal crystals the ratio of cation to oxygen ions is determined by the principle of electroneutrality. In real crystals, point defects may also be charged, for instance, metal vacancies $V_{\bar{m}}^{+1}$, $V_{\bar{m}}^{+2}$ . . . or oxygen vacancies $V_{\bar{0}}^{-1}$ or $V_{\bar{0}}^{-2}$.

The semiconducting transition metal oxides have not only attracted extensive theoretical interest; they have also gained importance in practical applications. The energy levels of their 3*d* electrons can split up in the crystal field, depending on the symmetry of the lattice and the presence of other cations. The importance of 3*d* electrons in these oxides for their electronic conductivity had been recognized before the formulation of the band model for semiconductors had been proposed by Mott and Wilson. Now the localized electron model has regained attention, after the band model had entirely failed to describe the electrical properties of a number of transition metal oxides. Elementary band theory of Bloch and Wilson predicts MnO and NiO to be metallic, while in reality they are good insulators. The reasons for the spectacular failure of the band theory have been discussed at length by a number of authors,[25–29] however, these discussions have no relevance to the substance of this book. It is sufficient to say that most pure transition oxides in a nearly stoichiometric state have a high resistivity and large energy gap, although it is rather difficult to give absolute data. Not only minute nonstoichiometric deviations, but also very small concentrations of foreign ions with different valence can influence the electrical data (see pp. 67–83). Finally, in polycrystalline oxides surface states, also on grain boundaries, can appreciably influence resistivity.

Literature data on the electrical resistivity of oxides should be considered with great caution even for those data measured with single crystals. This applies even more to polycrystalline samples, especially if they had been sintered under poorly defined O_2 partial pressure, which is known to affect the defect equilibrium in the oxide. The resistivity differences observed in transition oxides sintered in air or oxygen are an example for these effects. In complex oxide phases such as Ni(Co)-Mn spinels, the sites on which the ions are located can contribute to the electric properties.

More details on the semiconductivity in oxidic materials with negative temperature coefficient of resistivity are given in Section 3.5.1.

2.2. MATERIALS WITH POSITIVE TEMPERATURE COEFFICIENT OF RESISTIVITY (PTC MATERIALS)

Until about 20 years ago, semiconductivity was synonymous with positive temperature coefficient of conductivity, therefore the opposite trend for resistivity. In 1955, N. V. Philips Gloeilampenfabrieken (Netherlands) was granted a German patent (No. 929,350, priority 21 May 1952) for production of semiconducting materials with high positive temperature coefficient (PTC) of resistivity. Its principle is based on the discovery that doped $BaTiO_3$, sintered at 1600–1680 K in air or other atmospheres with greater than 0.05 torr O_2 partial pressure, has a positive temperature coefficient of resistivity (up to 20% K^{-1}) and a much lower resistivity than undoped $BaTiO_3$ sintered under the same conditions. Dopants can be used as well as trivalent elements such as Bi, Yttrium, or other rare-earth elements or Sb and W. The positive resistance characteristic exists only over a certain temperature range and can be shifted to lower temperature by partial substitution of Sr for Ba. Since the drastic reduction of resistivity was caused by ions (for instance, La^{3+} or Nb^{5+}) differing in valence from that of Ba^{2+} and Ti^{4+}, the effect of controlled valence was obvious.[30] The possibility of shifting the positive characteristic to lower temperature by Sr substitution was a sufficient hint to link the PTC effect to the ferroelectric properties of the titanate host material. The Curie temperature $Ba_{(1-x)}\,Sr_xTiO_3$ is shifting to lower temperatures with increasing x.

Heywang[31] made the first approach toward developing a model for PTC semiconductors. The ionization energy for impurity centers in a doped homogeneous semiconductor depends on the dielectric constant of the undoped material. With Mott and Gurney's concept[32] of a quasihydrogen model for an electron bound to an impurity center, the activation energies of p- and n-doped Ge have been explained. Their calculation is based on the relation

$$E_a = \frac{E_H}{\epsilon^2}\,\frac{m_{\text{eff}}}{m_0}$$

where E_a is the activation energy of the impurity center, E_H is the dissociation energy of the H atom, and m_{eff} and m_0 are the effective and free-electron mass and ϵ is the relative dielectric constant.

In most doped semiconductors, E_a is constant, but not in ferroelectric materials, which exhibit over a wide temperature range, preferentially

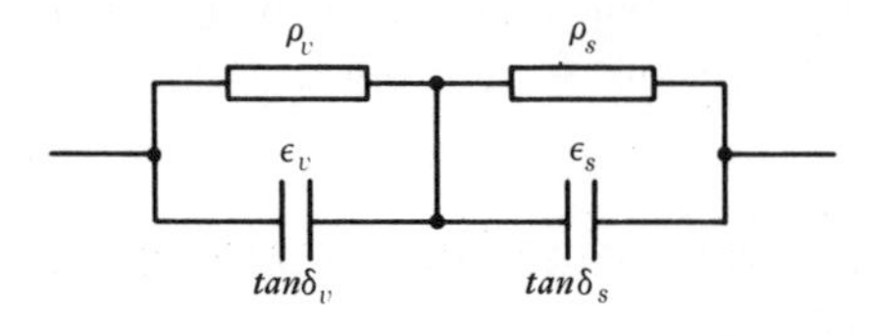

FIGURE 3. Equivalent circuit for PTC grains. From *J. Am. Ceram. Soc.*, **47**, 486 (1964).

below and above their Curie temperature, a strong temperature dependence of ϵ. If the dielectric constant determines the temperature dependence of the activation energy (and therefore also the resistivity), its value at high fields must be taken into consideration. Heywang developed this qualitative model into a well-founded theory that has in principle remained valid, at least for the temperature region above the Curie temperature, except for some quantitative modification.

A second important feature had to be introduced into this concept. Heywang[33] and coworkers showed that the steep increase of resistivity is not a property of the bulk material but is caused by barrier layers at grain boundaries within a semiconducting PTC body. These barrier layers result from the interaction of surface states acting as acceptors with conduction electrons formed by dissociating donors (dopant ions.).

The height of the potential barrier and the width of the space charge (or carrier depletion zone) have been calculated by measurements of the complex impedance of $BaTiO_3$ sinter bodies doped with 0.1 wt % Sb between 0.1 and 1000 KHz. With an equivalent circuit for PTC grains as shown in Figure 3, the resistivity data for the "interior" of the semiconducting grain and the intergranular zones were found at 90, 300, and 470 K; these values are given in Table 1.

The individual barrier layer has approximately a thickness of one-fiftieth of the grain diameter. Therefore 1–10% of the thickness of a sintered specimen is present as sequence of boundary layers with high resistivity. The

TABLE 1 Resistivity (ohm cm)

	90 K	300 K	470 K	Units
Grain interior	10^4–10^5	10–100	10 –100	ohm cm
Grain boundary zone	10^5–10^6	10–100	10^5–10^6	ohm cm
Potential barrier	0.55	0.08	0.55	eV
Conductivity region		Transition		
	NTC	NTC PTC	PTC	

potential barriers in the boundary zones, given in Table 1, determine the temperature dependence of resistivity and the voltage dependence of the material. Using these and other data, a doping concentration (10^{19} cm^{-3}) and a donor activation energy (0.15 eV) in his model of interior barriers, Heywang was able to calculate the positive resistance characteristics and the voltage dependence of resistivity quantitatively (Figure 4).

The surprising fact that the controlled valency doping produces low resistivity of pure or Sr-substituted $BaTiO_3$ only within a very limited doping concentration range has stimulated studies to explain this limitation. For example, with 0.1–4 at. % La or Nd and 0.25 at. % Sb corresponding to dopant concentrations of $\sim 10^{18}$–10^{19} cm^{-3}, a resistance minimum is reached.

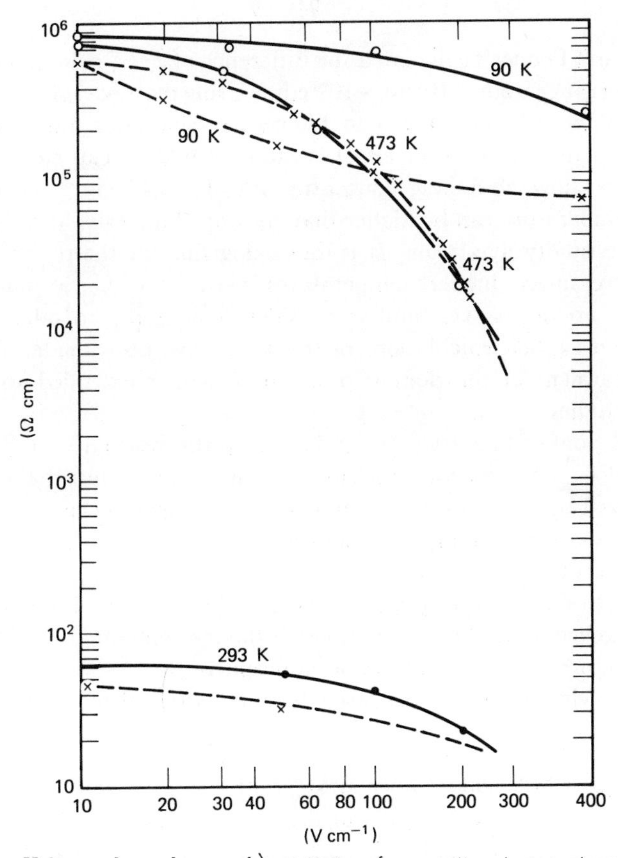

FIGURE 4. Voltage dependence of resistivity of a positive temperature coefficient thermistor at various temperatures. Solid curves, measured values; dashed curves, theoretical values. From *J. Am. Ceram. Soc.*, **47**, 486 (1964).

For doping concentrations greater than 1.5 at. % the same systems became nearly insulating again. The resistivity of the PTC materials is in the first approximation determined by the resistivity of the barrier zone ρ_B:

$$\rho_B = \alpha \rho v \, e^{\phi/kT}$$

The factor α depends on the grain size and the depth of the space charge zone and does not vary much for identical preparation conditions.

The crucial parameter is the barrier potential ϕ, which not only depends on the effective dielectric constant but also on the doping concentration. However, ϕ increases above the Curie temperature with $1/\epsilon$, and ϵ decreases according to the Curie law:

$$\epsilon = \frac{C}{T - \theta}$$

Heywang and Fenner[34] calculated the difference of ϕ/kT cold and ϕ/kT hot for a doping range from $\sim 10^{17}$ to $\sim 10^{20}$ cm^{-3}. Using the model of Heywang[33a,b] they found the relations shown in Figure 5a. Since this difference enters into the exponential term, its effect is rather strong. It can clearly be seen that for very low Sb-doping concentrations of $\sim 10^{18}$ cm^{-3} the resistivity at room temperature can be higher than for doped material at temperatures near the resistivity maximum. It is interesting that for the resistivity maximum, increasingly higher temperatures above the Curie temperature ($\sim$393 K) are necessary. Similar considerations apply to other dopants, although no explicit calculations of this kind have been made. This more formal treatment of the doping problem was later extended to physicochemical studies.

By 1964 Jonker had tried to explain why the resistivity of $BaTiO_3$ as function of dopant concentration goes through a minimum that is visually characterized by the transition from white for undoped $BaTiO_3$ to dark blue at the resistivity minimum and followed by reversal to nearly white at a doping concentration of $\sim$0.1 at. %.[35]

In order to simplify the approach, the law of mass action is first applied to the heterogeneous equilibrium between a binary compound $M^{m+}X^{m-}$ with defect structure and the surrounding atmosphere.

The following notations are used for the lattice defect concentrations per cm^3.[36]

M_M: M^{m+} ion on normal M^{m+} site.
F_M^+: Foreign ion with valence $(m + 1)^+$ on normal M^{m+} site.
V_M^0: Vacancy resulting from M-atom missing at M^{m+} site.
V_M^{a-}: Vacancy V_M after attraction of a electrons.
n: Number of electrons per cubic centimeter.

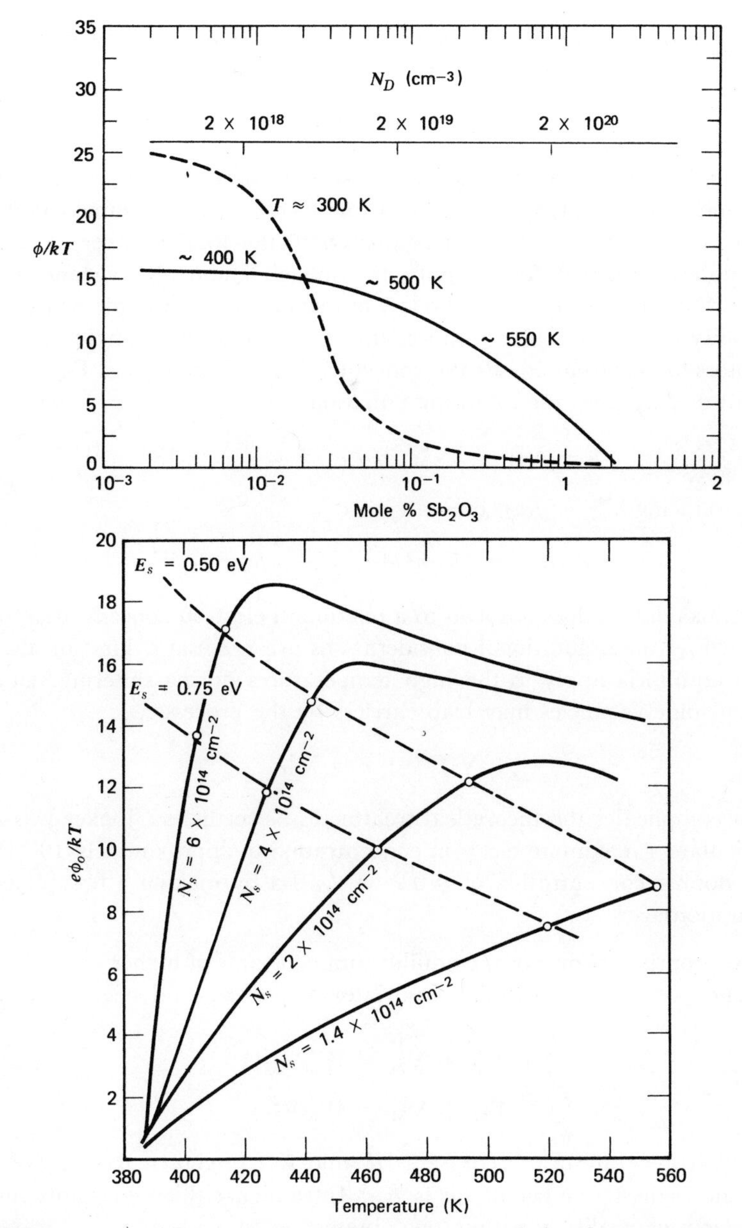

FIGURE 5. *(a)* (ϕ/kT)/max and (ϕ/kT) as function of the doping concentration. Numbers on solid line n/D define the excess temperatures above temperatures for the resistivity maximum. From *Zeitschr. angew. Phys.*, **18**, 316–318 (1965). Redrawn with computed data. *(b)* $e\phi/kT$ as function of the temperature for different surface states characterized by their density N_s in cm^{-2} and their distance E_s from the conduction band. From *Mat. Res. Bull.*, **2**, 406 (1967). Courtesy of the author.

K_1 to K_a represent equilibrium constants for the reactions between vacancies and electrons according to

$$M_M + ae = V_M^{a-} + M_{gas\ (neutral)}$$

For equal preparation conditions, $V_M^{a-} = K_a n^a$ ($a = 1, 2, 3, \ldots, a$) since a vacancy V_M^0 can attract more than one electron. It is assumed that the defect concentrations are small compared to the total number of cations per cubic centimeter. To simplify the model, anion (X) vacancies and holes are neglected. Applying the law of mass action and implementing the neutrality condition, which requires that the sum of all vacancies acting as acceptors for electrons equals the concentration of foreign ions, F_M^+, used as dopant at M_M sites, the following equation results:

$$n + V_M^- + 2V_M^{2-} + \ldots + a\,V_M^{a-} = F_M^+$$

By introducing $V_M^{a-} = Kan^a$ one can write

$$n + K_1 n + 2K_2 \cdot n^2 + \cdots + aK_a n^a = F_M^+$$

Since this relation does not lead to a maximum electron concentration for a certain F_M^+ value, additional considerations are necessary. First of all, the defect equilibria apply to the high temperatures during sintering. During cooling some vacancies may trap carriers by the process

$$V_M^{(a-1)} + e = V_M^{a-}$$

which complicates the theoretical treatment. Nevertheless, Jonker was able to calculate a maximum electron concentration of approximately 10^{19} cm^{-3} for a doping concentration of $\approx$0.2 wt. % La by making a few plausible assumptions:

1. Appropriate choice of the equilibrium constants of higher order.
2. The possibility of forming associates of positive and negative centers such as

$$F_M^+ + V_M^- = [F_M^+ V_M^-]^0$$

$$F_M^+ + V_M^{2-} = [F_M^+ V_M^{2-}]^-$$

for which the equilibrium constants K_5 and K_6 are chosen. For tetravalent Ti^{4+} the highest number of Ka is K_4.) Introducing these constants into a new electroneutrality equation, one obtains

$$n + [V_M^-] + 2[V_M^{2-}] + [(F_M^+ V_M^{2-})^-] = [F_M^+]$$

or

$$(K_1 + 1)\,n + 2K_2 n^2 + K_6 K_2 [F_M^+]\,n^2 = F_M^+$$

Although these rather simplifying considerations have been able to explain the existence of a resistivity minimum at a certain dopant concentration, they have not explained the possibility of the increase of resistivity with temperature. The assumption of a grain boundary model to explain the positive resistance characteristics remained inevitable.

In addition to Heywang's impedance measurements at frequencies of 10–10^6 Hz, there is sufficient other experimental evidence to indicate that the resistance increase near and above the Curie temperature is a boundary effect that is also influenced by nonequilibrium oxidation states in this layer. This was clearly demonstrated by sintering of Y-doped $BaTiO_3$ at 1670 K in an atmosphere with O_2- partial pressures between about 0 and 230 torr with Ar as balance against 1 atm. With 4 hr sinter time, the resistance ratio of hot to cold varied from about 0 to greater than 800, the "cold" resistance increasing much less.[37] It is obvious that variations of the O_2- partial pressure not only during sintering but also during the cooling process, together with the cooling rate, offer possibilities for modifying the properties in PTC material even without substantial changes in the formula. This principle has been incorporated into U.S. Patent 2,981,699, April 25, 1961 (Patent list p. 320).

The large effect of the cooling rate and of short-time annealings in the temperature range 970–1520 K can depend strongly on the doping concentration. This was shown for $BaTiO_3$ doped with 0.2–1 at. % Gd.[38] Only slowly cooled samples had the characteristic resistivity minimum at a certain doping concentration (~0.2 at. %). Quenched samples initially showed no PTC effect, until after $1\frac{1}{2}$ hr annealing in air at $\approx$1520 K.

The dramatic boost of the hot-to-cold resistance ratio and the maximal attainable TC of $BaTiO_3$, $Ba_{0.80}Sr_{0.20}TiO_3$, and $Ba_{0.94}Pb_{0.06}TiO_3$, all doped with La, Sr, or Nb, by treatment with F-, Cl-, or Br-containing inorganic or organic compounds at temperatures greater than 1000 K is obviously connected with the change of acceptor levels in the grain boundaries.[39]

Jonker[40] has discussed the theoretical aspects of this treatment. He calculated the exponential factor exp $e\phi_0/kT$ determining the grain boundary resistance as function of the temperature T for combinations of the bulk electron concentration N_0, the density of surface states N_s, and the distance E_s of their energy level to the conduction band. As shown in Figure 5b, at constant E_s and increasing N_s the maximum of $e\phi_0/kT$ increases and shifts to lower temperatures, which means a steeper PTC characteristic. On the other hand, increasing E_s at constant N_s has no influence on the initial slope but raises the maximum and shifts it to higher temperatures. Treatment with O_2 increases N_s, while halogen treatments seem to increase both N_s and E_s. An effort to correlate the experimental results to Pauling's sequence of electronegativity of the adsorbed atoms fails. Seebeck[41] measurements have shown that the bulk electron concentration remains nearly unchanged if all donors are dissociated. Most crucial was the observation that the PTC-effect was practically absent in a single crystal of $BaTiO_3$ doped with 0.5 milli at. % mole samarium, but appeared if polycrystalline sinter bodies were made from it.[42] The crystal was crushed to powder, which was molded at 2450 kg/cm^2 isostatically in a polystyrene vial to avoid contamination. After sintering 1 hr in O_2 at 1670 K, a resistance jump of more than 100 was observed above the Curie temperature (398 K). The fact that $BaTiO_3$ single crystals doped with 0.1 wt % Nb show a sharp resistivity jump of a factor $\sim$10 at the Curie temperature (besides a similar distinct drop by a factor $\sim$5 at a lower transition temperature of $\sim$280 K) is interesting, but it does not carry enough weight to invalidate the barrier model. Just below and above the Curie temperature of 395 K the crystal has NTC characteristics with activation energy of 0.15 and 0.20 eV.[43]

A noise anomaly in semiconducting $Ba(Sr)TiO_3$ doped with 0.3 at. % La supports the model of a "grain" structure with barrier layers. It is known that the current noise expressed as the mean square fractional resistance fluctuation is given by $(\Delta R)^2/(R) = \alpha\Delta f/Nf$ where N is the number of free current carriers, Δf and f are the bandwidth and frequency, and α is a constant that for homogeneous metals and semiconductors is of the order of 10^{-3}. For semiconducting $Ba(Sr)TiO_3$, α was found to be $\sim$800; it can be considered as a measure of the internal electrical structure with high- and low-resistivity zones.[44]

The work function of La-doped $BaTiO_3$ and commercial Siemens P 350-C 15 PTC thermistors measured with the Kelvin capacitor method using a rotating Pt reference electrode increased between 300 and 400 K by about 0.5 eV, which is of the same order of magnitude as the barrier potential assumed by Heywang. The reversal to lower values above the Curie temperature and the presence of large thermal hysteresis effects limit the relevance of these measurements to a qualitative support of the barrier-layer model. However, the side effect observed during gas adsorption can be a meaningful aid for understanding drifts in stability of PTC units.[45]

For the low-resistance region below the Curie temperature, Tonker pointed out another boundary effect caused by the permanent polarization of ferroelectric domains.

In a grain boundary between two crystallites of undoped $BaTiO_3$, domains of different direction of permanent polarization will meet, resulting in strain of the surface layers. For doped semiconductive material, a pattern of surface layers with opposite polarity can develop that either enhances or reduces the contact resistance, in the latter case shunting the depletion zone of the first case.

Recently Kulwicki and Purdes[46] have added a new approach to Tonker's domain concept. They calculated the potential profiles of depletion layers for materials with high and low permittivity at 001-001 interfacial barriers in semiconducting $BaTiO_3$ and concluded that the internal stresses below the Curie temperature T_c drop sharply at T_c with the result that the electron concentration also increases, owing to lattice distortion. This would revive an earlier model[47] proposed by Peria and coworkers that was neglected by the obvious success of Heywang's theory.

However, in principle the authors accept the models of Heywang and Jonker. They try to refine them by taking into account the component of the existing permanent polarization below T_c directed toward the grain boundary using the thermodynamic concept of Devonshire's theory of ferroelectricity.[48] The temperature dependence of the spontaneous polarization Ps below T_c is responsible for the resistance increase in this range.

Visual evidence for the existence of barrier layers has been shown by optical[49] and electron microscopy.[50] In the latter case a strong field of the order of several kV cm^{-1} at the grain boundaries exists if a field of >30 V cm^{-1} is applied across the specimen. It deflects secondary electrons emitted after ion bombardment in an emission electron microscope, thus making the barrier zones visible as dark areas that follow field reversals. With more experimental proofs for the physical reality of barrier layers, increased probing into their nature and behavior was undertaken. The initial model was based on surface states at the interfaces without going in details of their nature, except that they were acceptors. Empirical data based on variations of the PTC effects with the impurity profile of the raw materials and the striking influence of milling conditions[51] led to the following conclusions:

1. It is of crucial importance whether the Sb dopant goes to Ba or Ti sites. In the first case, semiconductivity is observed, in the second case a small quantity of an insulating phase $Ba(TiSb)O_3$ is formed at the grain boundaries. Its lower melting point produces a fine-grained material with lower voltage dependence, higher resistivity, and a drastic increase of a PTC effect. This could be the result of higher acceptor activation energy in the surface, possibly by the presence of Sb^{5+} in $Ba(TiSb)O_3$.

2. The observation that the resistance increase tends to become smaller with decreasing impurity content of the raw materials, shrinking to less than a factor 10 for ultrapure materials, suggested that in addition to the controlled-valence doping, certain impurities preferentially contribute to the PTC effect. The usual method of tracing down the effective agents is the doping of ultrapure materials with well-defined concentrations of specific additives. CuO and Fe_2O_3 have been found to be especially effective in

controlling the cold resistivity and the resistance ratio hot to cold.[52] In each case optimal concentrations have been found, for instance ~30 ppm for CuO and ~50 ppm for Fe_2O_3. The great sensitivity of the PTC characteristic with respect to small concentration differences seems to indicate that these act mainly in the barrier layer, since this is also the site of the PTC mechanism. Considering the fact that these layers cover grains of 10–25 μm, the absolute concentration values necessary for grain boundary doping appear reasonable. Brauer has investigated these effects and proposed their explanation with a controlled reduction of the electron concentration by the second dopants Cu and Fe, which, acting as acceptors, contribute to make the contact between adjacent grains to *n-p-n* junctions. Direct proof for the localized action of Cu has been given by microsonde pictures.[53]

The doping problem continued to stimulate research with different perspectives:

1. To obtain more understanding for the limited doping range suitable for semiconduction.

2. To broaden the aspects of wider modifications of resistance–temperature relation.

1. Following earlier work correlating lattice constant and resistivity minimum[54] in Sb-doped $BaTiO_3$, Schmelz developed a model for this system that takes into account the defect structure of the system and the possible lattice site to be occupied by the dopant.[55] Excess TiO_2 results in increased concentration of Ba vacancies, which can be filled by the trivalent dopant, thus making the material semiconductive. After all Ba vacancies are filled, the Sb-dopant substitutes on Ti^{4+} sites forming a compound $Ba(Ti_{1-x}Sb_x)O_3$ in which Sb has to be present with a ratio of $Sb^{3+}/Sb^{5+} = 1$ to ensure electroneutrality. This material is not conductive. X-diffraction data were supplemented by neutron-diffraction measurements to determine the sites at which Sb was incorporated into the lattice.[56] Heywang has pointed out[57] that pairing of Sb^{3+} and Sb^{5+} ions leads to mutual attraction which acts similarly to an external pressure, which is known to lower the Curie temperature as in Sr substitution.

In the case of doping with trivalent rare-earth elements, substitution at Ti^{4+} sites is not probable. The resistivity minimum could coincide with the exhaustion of the Ba^{2+} vacancies.

Electrical and paramagnetic resonance absorption measurements suggest that the doping characteristic is related to a change from a simple substitutional role of the trivalent ion to a more complex solid solution with an important influence of the ionic radius of the dopant atom.[694]

GRAVIMETRIC DOPING STUDIES. A novel approach to the doping problem in $BaTiO_3$ is based on chemical experiments and calculation.[58] Starting from the straightforward consideration that introduction of any doping atom must not only meet the condition of electroneutrality in the lattice but also requires an equivalent of oxygen according to its valence as ion, the oxygen uptake when doping with La and Nb has been measured by gravimetric analysis. Doping was extended up to 20 at. % La and 2 at. % Nb to increase the accuracy of the method. For comparison the "natural" nonstoichiometry of pure undoped $BaTiO_3$ was determined to be ~0.07 at. % at 1273 K, corresponding to a formula $BaTiO_{2.998}$ in good agreement with a titrometric redox method.[59]

It has been observed that at low oxygen pressure[60] the free-electron concentration, that is the reducing power of $BaTiO_3$, is proportional to the La^{3+} doping concentration. Heating in oxidizing atmosphere must result in O_2 uptake. To exemplify this idea, the formula of $BaTiO_3$ with 10 at. % La-doping is given in the reduced or oxidized state:

$$\text{Reduced:} \qquad (Ba_{0.90}{}^{2+}La_{0.10}{}^{3+}Ti_{0.90}{}^{+4}Ti_{0.10}{}^{+3})O_3{}^{2-}$$

$$\text{Oxidized:} \qquad Ba_{0.90}{}^{2+}La_{0.10}{}^{3+}Ti_4{}^{+4}O_{3.05}{}^{2-}$$

This shows an oxygen uptake of 50 milliatoms per mole.

The following experimental procedure was applied: The donor-doped $BaTiO_3$ was prepared by a wet process that did not require filtration of a precipitate and was therefore free of potential material losses with nonstoichiometry as a consequence. A solution of Ba-Ti-citrate complex in ethylene glycol was heated to evaporate the solvent and to form a rigid polymer. After calcination at 970–1170 K the composition was molded at 3500 kg cm^{-2} (25 tsi), sintered in air at 1670–1770 K, and then exposed at 1330 K either to CO or O_2 atmosphere. If a specimen previously heated in CO is brought into O_2, an oxygen absorption takes place which is reversible by returning to the CO atmosphere. The O_2 absorption was found to increase linearly by 0.034% per at. % La^{3+} doping. Model experiments with dopant ions of higher valence charge increased this effect correspondingly.

Efforts to explain how the surplus of oxygen would be accommodated had to consider the respective ionic radii. It could be shown that the excess oxygen is not accumulated at grain boundaries and does not result in formation of a new phase with higher oxygen content. This leaves an option on several defect structures: either random point defect distribution or the formation of an extended well-organized defect structure (Magneli phase). The experimental x-ray data support a compromise of both possibilities for La-doped $BaTiO_3$, while for nonstoichiometric $TiO_2 - n$ ($n \geqq 0.05$) the second option prevails. The oxygen demand for oxidation of doped TiO_2 between 10^{-15} and 1 at O_2 is 0.09 wt % 0 per at. % Nb^{5+}.

Information on the doping effects with different M^{3+} ions ($M^{3+} = Y^{3+}$, La^{3+}, Ce^{3+}, Pr^{3+}, Nd^{3+}, Pm^{3+}, Sm^{3+}, Eu^{3+}, Gd^{3+}, Tb^{3+}, Dy^{3+}, Ho^{3+}, Er^{3+}, Tm^{3+}, Yb^{3+}, Lu^{3+}) is of considerable interest. It is generally assumed that the differences among the M^{3+} are small, and empirical data seem to support this assumption.

Some relations between the ionic radii (Table 2) of the rare-earth dopants La, Ce, Pr, Sm, Gd, Er, and Yb in pure $BaTiO_3$, the optimal dopant concentration to obtain a resistivity minimum, and the temperature region of steepest PTC characteristic were found. At an ionic radius of 0.97 Å the optimal doping concentration was 0.04 at. % (corresponding to Y, Tb, and Sm). For 0.45 at. % Er the maximum of the temperature coefficient $\Delta R/R \times K^{-1}$ of 8 coincides with the minimum of the resistivity.[61] Figure 6 illustrates some of these effects.

2. It was known that the various phase transitions of $BaTiO_3$ below the Curie temperature, for instance from tetragonal to orthorhombic at 278 K and then to rhombohedral at 183 K, also influence the activation energy of semiconduction after doping. On the other hand the transition points between the different phases can be shifted by Zr substitution of Ti.[62]

It was therefore of considerable interest to study the influence of phase shifting on the resistivity characteristics of doped $BaTiO_3$. This was done

TABLE 2 Ionic radii (in 10^{-8} cm) in sequence of their size.

Ionic Charge	Ion[a]								
3+	Sb 0.76	Lu 0.85	Yb 0.86	Tm 0.87	Er 0.89	Ho 0.91	Dy 0.92	Y 0.92	Tb 0.93
	Gd 0.97	Sm 1.00	Nd 1.04	Pr 1.06	Pm 1.06	Ce 1.07	La 1.14	Eu 1.15	
2+	Sr 1.12	Pb 1.20	Ba 1.34						
4+	Si 0.42	Ti 0.68	Zr 0.79						
5+	Sb 0.62	Nb 0.69							

[a] The value of the radius is given directly below the name of the element.

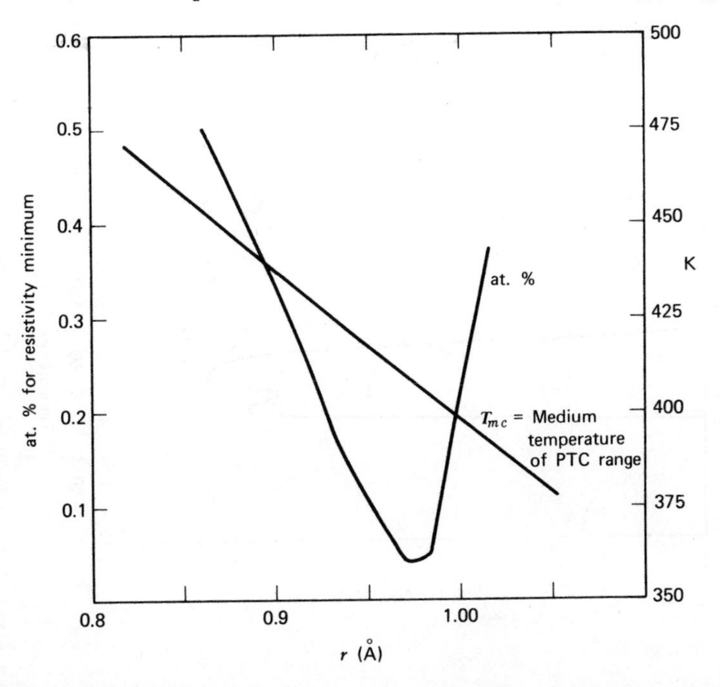

FIGURE 6. Optimal doping concentration and medium temperature of the PTC range as a function of the ionic radius of the dopant ion. From *Istvest. Akad. Nauk. SSSR,* **Ser. Fiz. 31**, 1821–1823 (1967).

with the system (see Figures 7 and 8) $Ba,_{0.99}$ $Y_{0.01}$, $Ti_{0.99}$, $Zr_{0.01}$, O_3.[63] The room-temperature resistivity increased drastically with the Zr content and also with decreasing oxygen partial pressure during 2 hr sintering at 1573 K. This opens up a wide range for variations of the entire NTC-PTC resistivity–temperature characteristics. Some of the most interesting features of this system are:

1. Temperature coefficients from −2 to −16% K^{-1} in the NTC region and +2 to +17% K^{-1} in the PTC region (Figure 7).

2. Narrowing the interval between the NTC and PTC branch of the characteristics in some cases to 50 K and less, which might be attractive for some applications (Figure 8).

Shifting of the PTC Characteristics by Substitution. As early as 1956 Sauer and Flaschen[64] applied the principle of partial Ba substitution by Sr to shift the range of steepest resistivity increase to lower temperatures, taking into account the well-known fact that the ferroelectric Curie temperature of $BaTiO_3$ also shifts in the same way. Pb substitution has the

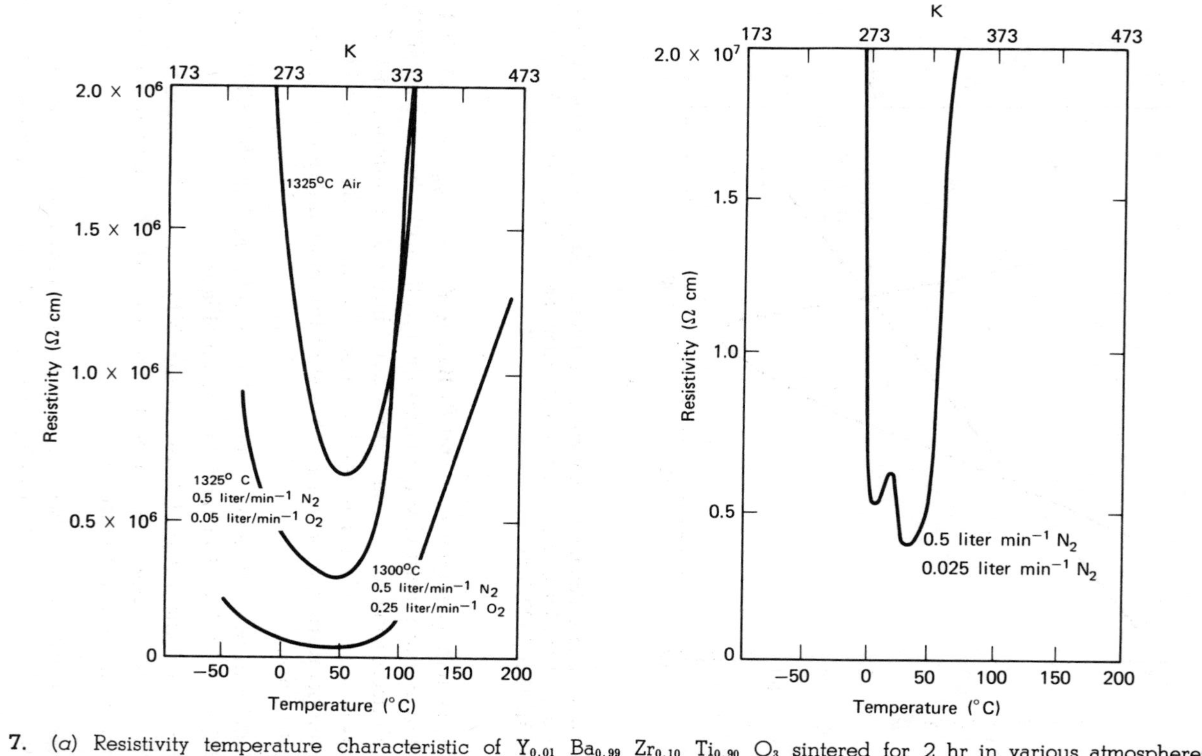

FIGURE 7. (*a*) Resistivity temperature characteristic of $Y_{0.01}$ $Ba_{0.99}$ $Zr_{0.10}$ $Ti_{0.90}$ O_3 sintered for 2 hr in various atmospheres. (*b*) Resistivity temperature characteristic of $Y_{0.01}$ $Ba_{0.99}$ $Zr_{0.20}$ $Ti_{0.80}$ O_3 sintered at 1300°C for 2 hr in N_2 with 5 vol % O_2. Courtesy *Japan J. Appl. Phys.*, **10**, 421–426 (1971).

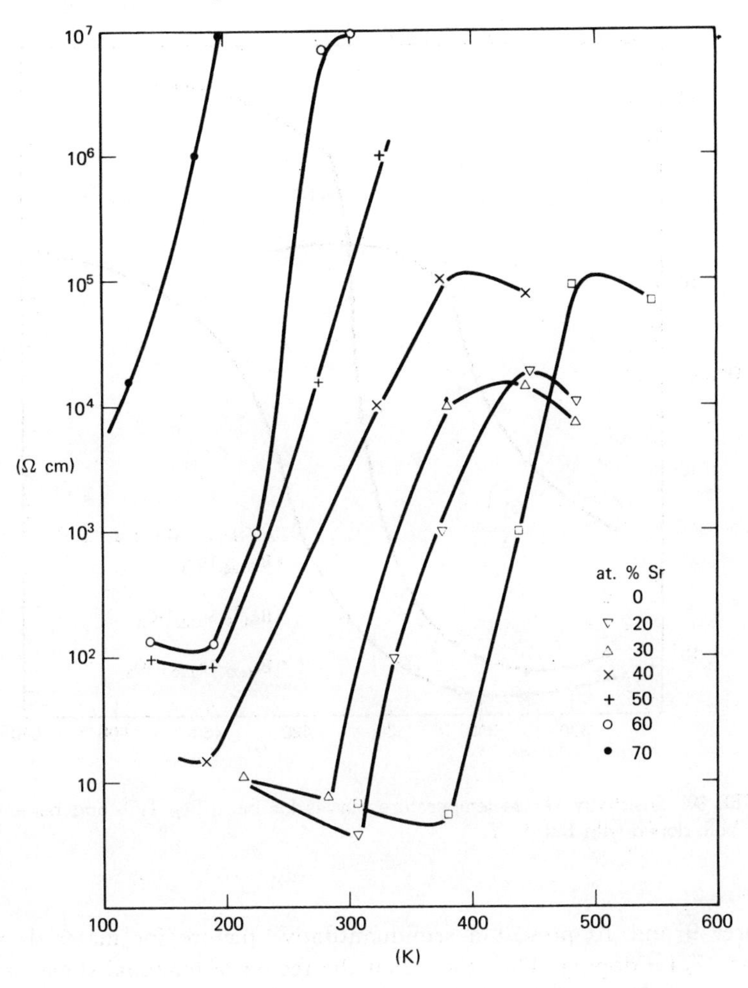

FIGURE 8. Resistivity versus temperature curves of $Ba_{1-x}Sr_xTiO_3$ doped with 0.3 at. % La (x = 20–70 at. %).

opposite effect. Systematic investigations of the influence of Sr and Pb substitution up to 20 at. % in doped semiconductive $BaTiO_3$ have been reported by Ichikawa and Carlson.[65]

With increasing degree of substitution, the resistivity at 300 K increases several orders of magnitude for Sr, and decreases, but less drastically, for Pb. It is not practical to give absolute quantitative data for this effect without additional specifications, since dopant concentration, deviations from stoichiometry [ratio of Ba(Sr) to Ti] and sinter conditions have an influence.

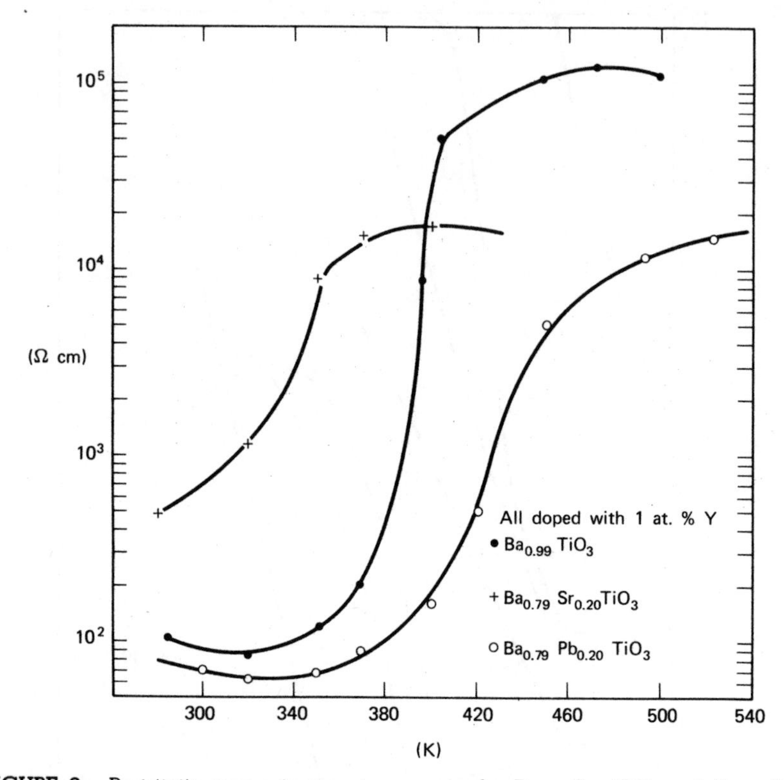

FIGURE 9. Resistivity versus temperature curves for $Ba_{0.79}$ Sr_{20} TiO_3 and $Ba_{0.79}$ $Pb_{0.20}$ TiO_3, both doped with 1 at. % Y.

Figures 9 and 10 present a semiquantitative picture for materials with 0.3 at. % La doping. They show that the region of maximal slope can be shifted from 200 to about 600 K. The maximal resistance ratio between the lower and upper inflection points changes with the substitution; with increasing Pb content it tends to become smaller. While Sr substitution does not present fabrication difficulties, Pb-containing compositions can lose PbO by evaporation during sintering and also react with ceramic boat materials. Sintering in closed nonreactive containers and hot molding (pressure sintering) minimize these undesirable effects.[66–69]

For the system $Ba_{1-x}Sr_xTiO_3$ with 0.3 at. % La doping, the maximal TC together with the medium temperature of the PTC range are given as function of x in Table 3. The room-temperature resistivity for higher x values increases rapidly, because it is much above the inflection minimum.

Comparative studies of various rare-earth dopants for $PbTiO_3$ led to the conclusion that Yb and Lu are useful: With 1 at. % Lu or Yb, a minimum

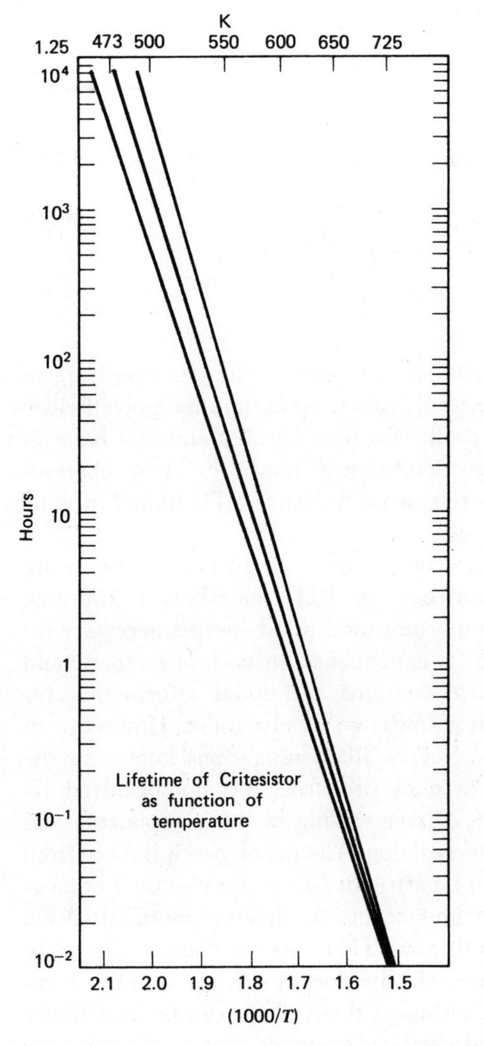

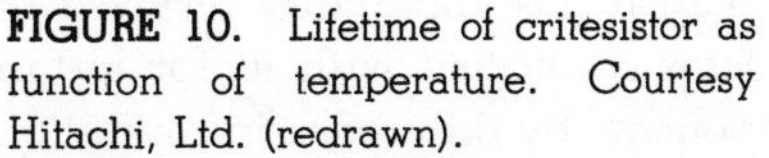

FIGURE 10. Lifetime of critesistor as function of temperature. Courtesy Hitachi, Ltd. (redrawn).

for the resistivity at 593 K was obtained (about 500 Ω cm), compared to about 10^7 Ω cm for undoped material sintered in PbO vapor. From 600 K up to the Curie temperature (760 K), ρ increases from about 3 Ω cm at the resistance minimum to about 200 Ω cm at 825 K.[70]

By impregnation of $BaTiO_3$, doped with 0.2 at. % La and sintered to a density of about 90% of the theoretical value with $BiCl_3$, a two-step PTC effect can be produced with an additional resistance jump of a factor of 10 at the lower transition point of about 278 K. Apparently, along the grain boundary Bi doping takes place. The effect is strongly field dependent.[71]

TABLE 3 $(La_{0.003}\ Ba_x\ Sr_{0.997-x})TiO_3$

x	1-0.003	0.9	0.8	0.7	0.6	0.5
ohm cm (293 K)	200	200	70	200	1500	$>10^7$
α max % K^{-1}	21	8	7	5	4	11
at ~ K	400	385	370	360	325	260

Studies on the PTC characteristics of composite materials consisting of ultrafine natural graphite homogeneously dispersed in paraffin–polyethylene mixtures have shown a large PT coefficient between 320 and 360 K, with large hysteresis effects and voltage dependence, however. These observations obviously prompted the news release on "plastic" PTC units[72] in some technical newspapers.

The material presented in this section cannot do justice to the entire experimental and theoretical knowledge on PTC materials. Within the scope of this book only the most important facts and concepts necessary for understanding the PTC effect and its dependence on various factors could be mentioned. The cited literature contains additional information for those who wish to gain a broader knowledge of this topic. However, an additional point should be made: Nearly all investigations known in the literature have been made with systems prepared and homogenized by solid-state reactions, generally dry or wet milling of the components and subsequent calcination to promote reaction. Chemical precipitation from aqueous or other solutions seems to be attractive to attain maximal homogenization and high reactivity,[73,74] and seems to deserve more attention than it has apparently received until now. There are, of course, also problems connected with it, for instance selective loss of dopant or other components by difference in solubility, although these effects can be kept under control by well-known physical and chemical methods.

2.3. MATERIALS WITH SEMICONDUCTOR-TO-METAL TRANSITION

2.3.1. Semiconducting Temperature Sensors with Sudden Large Resistivity Drop

Foex and his coworkers[75–77] have observed that the resistivity of V_2O_3 dropped suddenly below 173 K by a factor of 10^5. This transition was reversible in slow cooling or heating. Below the transition point, log ρ was proportional to $1/T$, indicating semiconduction, while it was only slightly temperature dependent above that point. Morin[78] has confirmed this effect and also found it in VO_2 at 340 K and in VO at 126 K. It was also found, though in a less spectacular form, in Ti_2O_3 at 450 K and in NbO_2 at 1070 K as well as in nonstoichiometric compostiions in the system Nb-O.[79] This also applies to nonstoichiometric compositions between $VO_{1.75}$ and $VO_{2.16}$.[80]

These experimental data stimulated a number of theoretical studies and related experimental investigations.[81,27] It is beyond the scope of this book to discuss the merits of different theoretical models. Goodenough considered the effect of strong cation-cation interactions in ionic solids on their magnetic and electrical properties and crystalline distortions. In the model of Adler et al., an energy gap caused either crystalline distortion or antiferromagnetic ordering decreases with increasing temperature, resulting in the disappearance of the gap at a critical temperature ("catastrophic collapse"). The interesting consequences of this model is that it predicts a transition temperature of 143 K, very near to the experimental value of 152 K.

Besides theoretical work the semiconductor-to-metal transition spawned during the recent 10 years many additional experimental investigations with single crystals, whiskers, and films of VO_2.[82–85]

The effect of doping with various impurities.[86,87] and of mixed systems with other oxides was also studied, including the behavior of ceramic compositions of 50 vol. % VO_2 in "low melting phase," used as a binder to promote sintering, which had a high resistivity of 125 Ω cm and an activation energy of 0.23 eV.[88]

Similar bonding with a flux found its expression in a Japanese patent.[89] Starting in 1964 a number of Japanese papers by Watanabe and coworkers[90–93] revealed the growing interest in materials having this transition.

The commercial effect of this activity became visible in 1965 by worldwide marketing of a new electronic component using this transition by Hitachi Ltd., Tokyo, Japan; it was offered under the trade name Critesistor with nine different specifications, which had in common a transition temperature of 341 K as observed for VO_2. However, they were made from a binary or ternary oxide system with V oxide as the major component in combination with oxides of Mg, Ca, Ba, Pb, and P, B, Si, the latter obviously helping to produce a flux. After sintering of the compositions in reducing atmosphere (N_2), they are quenched from above 1173 K. Any reheating to temperatures lower than $\sim$900 K would adversely affect their temperature-resistance characteristics.

For normal operating temperatures, this degrading effect is very slow. The life time of these units in air is a linear function of the reciprocal absolute

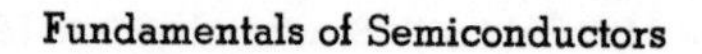

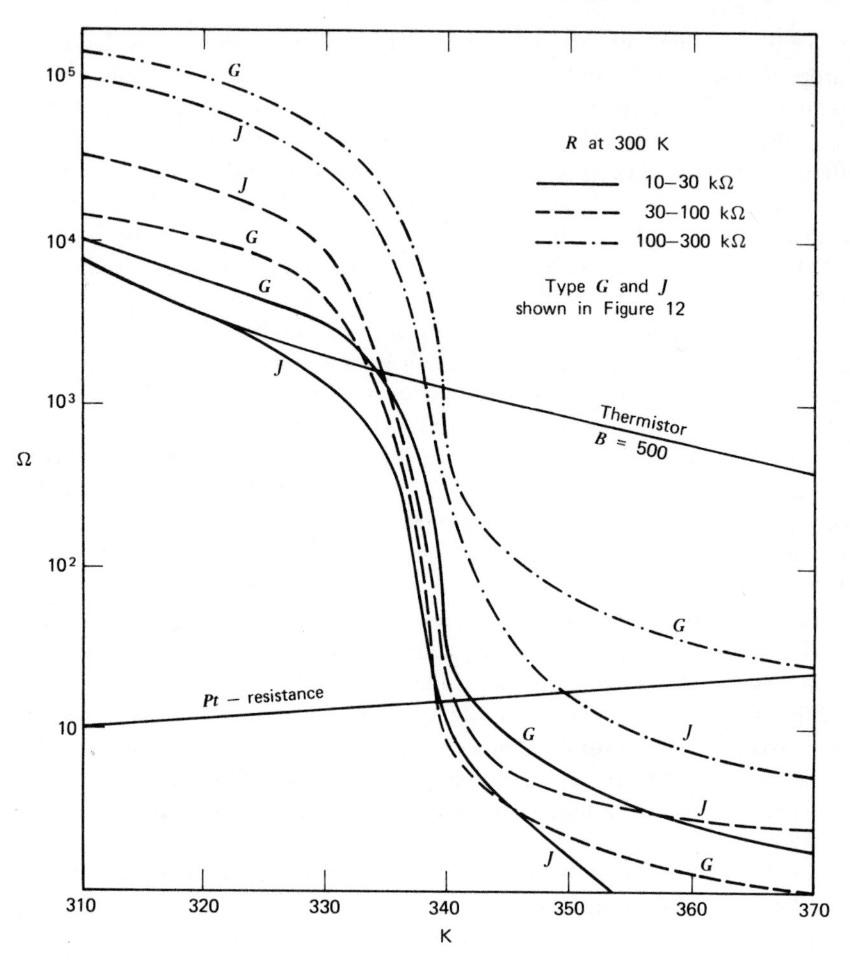

FIGURE 11. Resistance versus temperature curves of Pt, commercial thermistors, and critesistors.

temperature as shown in Figure 10. In N_2 atmosphere it is higher and estimated to be 100 years at 483 K for a binary oxide system and 50 years at 473 K for a ternary oxide system.

Tests by switching the units through the transition temperature during 1 year and 10^5 cycles also were satisfactory. Figure 11 shows a number of switching curves in comparison to the temperature characteristics of thermistors or Pt resistance thermometers.

Critesistors are also made as beads either with parallel or axial leads, always extending into opposite direction. Each of the basic design types as

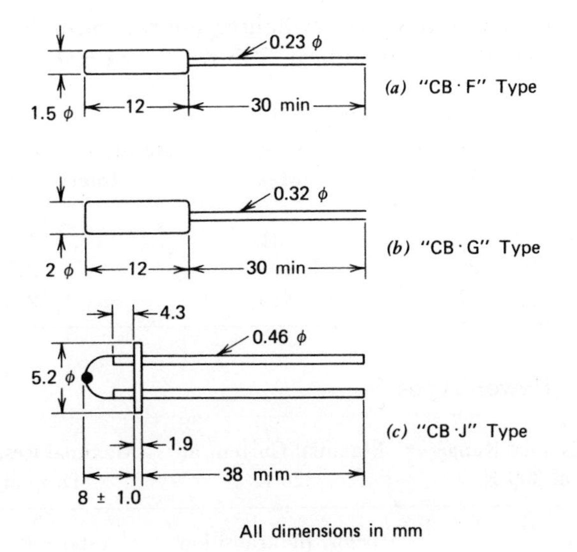

FIGURE 12. Types of critesistors. Courtesy Hitachi, Ltd., Pamphlet 1965.

shown in Figure 12 is available in three resistance ranges resulting in nine types (Table 4). The resistance ratio within the transition interval 341±1 K is always 2.5, indicating impure material. For a larger temperature interval much higher ratios are found (Figure 12). The temperature coefficient of the Critesistor in the transition region is 30 times larger than for normal thermistors with 3–4% K^{-1} temperature coefficient. The voltage-current characteristics are dramatically changed in the 313–443 K region as shown in Figure 14. (Temperatures at curves in °C).

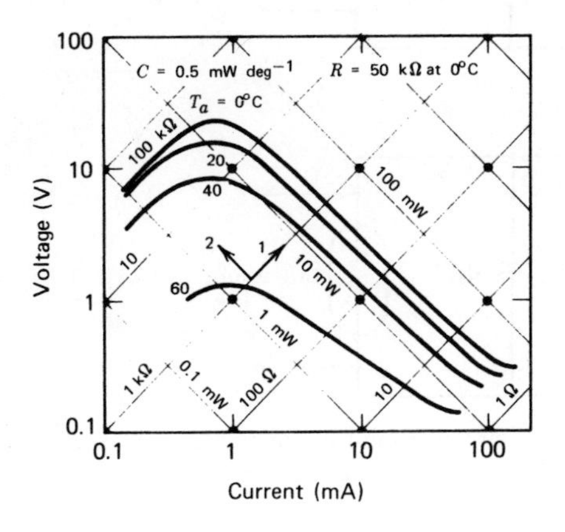

FIGURE 13. Voltage–current characteristics of self-heated critesistors. Courtesy Hitachi, Ltd., Pamphlet 1965.

TABLE 4a Hitachi Critesistor: switching temperature, 341 K; maximal operation temperature, 395 K. (Types made in 1965.)

Type	Resistance at 298 K (kΩ) Min.	Max.	Resistance Ratio in Transition Interval 340–342 K
CB 46 F,G,J	9	31	2.5
CB 56 F,G,J	29	103	2.5
CB 66 F,G,J	97	310	2.5

TABLE 4b Newer Types

Type	Resistance Range at 300 K	Maximal Current at 328 K (below the transition temperature) at 10 V	30 V	Maximal Residual Voltage Drop at 353 K (above the transition temperature) at 40 mA	120 mA
CG 66	100– 250 KΩ	1 mA	—	—	1 V
CG 86	900–3000 KΩ	—	0.5 mA	3 V	—

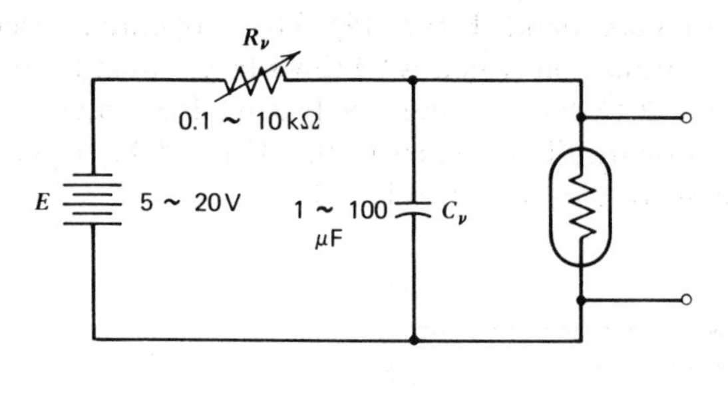

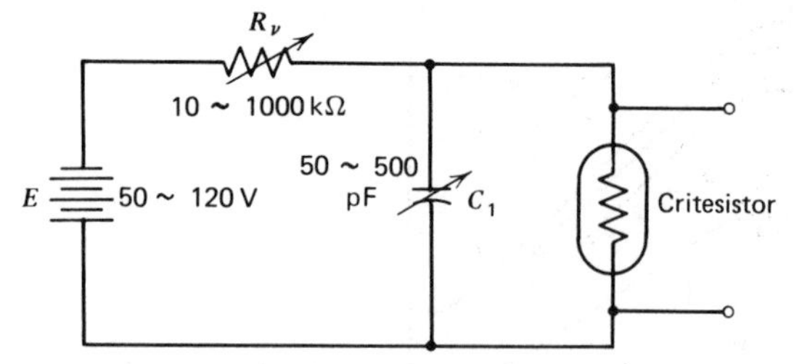

FIGURE 14. (*a*) Low-frequency oscillator circuit. (*b*) High-frequency oscillator circuit. From Hitachi, Ltd., (redrawn).

Typical applications for this component can be divided into two groups:

1. Temperature sensing and control near the critical temperature where resistance ratios up to 1000 permit switching operations for overheat protection in circuits and fire alarms in rooms.
2. Low- and high-frequency oscillation circuits using indirectly or self-heated critesistors (Figure 15).

Under certain conditions of series resistance Rv and voltage supply, the Critesistor Type b operates with a discontinuous (Figure 15) characteristic near the critical temperature as an oscillator with an oscillating period proportional to its heat capacity (Figure 16). Wide-band oscillations between 30 KHz and 10 MHz and input power of only 3 mW can be obtained by driving the critesistor with an AC current.

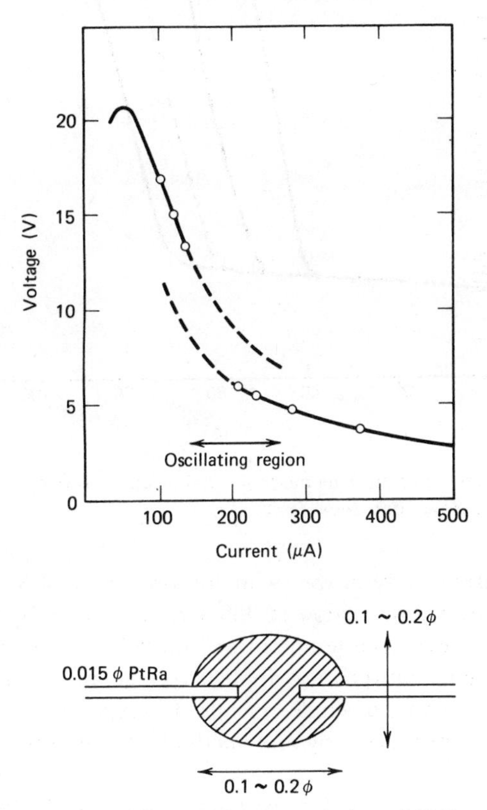

FIGURE 15. Discontinuity in voltage versus current characteristic of a critesistor near the critical temperature. Courtesy Hitachi, Ltd., Pamphlet 1965.

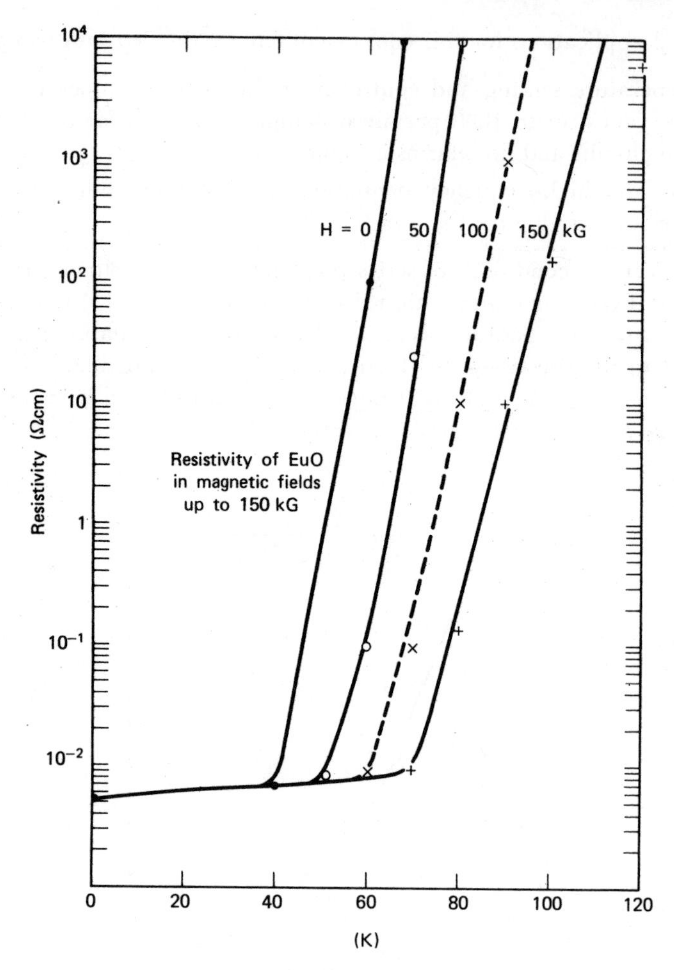

FIGURE 16. Resistivity of EuO in magnetic fields up to 150 kG. Courtesy American Institute of Physics, New York, New York.

A comprehensive study of the switching properties of VO_2 single crystals in the thermal transition region of 338.5 K was made by Guntersdorfer.[94] He measured the resistance temperature and voltage current characteristics to establish relations between the applied voltage and the resulting thermal process during switching of single crystals of maximum dimensions of 2.0 × 0.05 × 0.03 cm produced by the method of Kitahiro, Watanabe, and Sasaki.[95]

The resistivity drop at 338.5 K was approximately 40,000 starting with a resistivity in the semiconductive state of about 10 Ω cm. Self-heating of any conductor is always determined by a differential equation representing two terms: (1) the internal heating process, and (2) the heat dissipation to the environment. The resulting temperature increase ΔT is inversely proportional to the heat capacity mc of the conductor. Therefore:

$$\Delta T = \frac{1}{mc}\int_0^t \frac{V_2}{R_T\,dt} - \alpha \int_0^t \Delta T\,dt$$

where c is the heat capacity (electrical energy units, W sec g^{-1}, K^{-1}), V is the applied voltage (volts), R is the resistance (ohms), and α is the cooling constant (sec^{-1}), not identical with the conventional dissipation constant in W K^{-1}.

Differentiation after the time results in

$$\Delta T = \frac{1}{mc}\left(\frac{V^2}{R_T}\right) - \alpha\Delta T$$

For an intrinsic semiconductor R must be substituted for by

$$R_T = R_0 \exp \frac{B}{T}$$

The resistance at room temperature T_1 is $R_1 = R_0 \exp (B/T)$; after heating to a temperature $T_2 = T_1 + \Delta T$ the resistance of the semiconductor decreases to

$$R_{T_2} = R_{T_1} \exp - \left(\frac{\Delta TB}{T_1(T_1 + \Delta T)}\right)$$

For $\Delta T \ll T_1$

$$R_{T_2} = R_1 \exp\left(\frac{-\Delta TB}{T_1^2}\right)$$

Substituting this resistance in the differentiated second equation gives

$$\Delta T = \frac{V^2}{mcR_1} \exp\left(\frac{\Delta TB}{T_1^2}\right) - \alpha\Delta T$$

The quantitative solution of this equation leads to three regimes depending on the applied voltage and the ratio of semiconductor and series resistance:

1. Low voltage: Self-heating saturates at a plateau where low-voltage internally generated power equilibrates with the dissipation of heat.
2. Critical voltage: A temporary plateau is reached slightly below the critical transition temperature. After a slow increase, the unit switches and the temperature increases rapidly again.
3. High voltage: the temperature increases continuously from the start and pushes the unit through the switching region. Using known data for the

heat capacity of crystals 0.25 cm long with $\sim 1 \times 10^{-4}$ cm^2 cross section, the resistance value of the VO_2 crystals at room temperature of 85 kΩ and a series resistance of 10 kΩ, their response time is about 24 msec with an applied voltage of 19.2 V, which permits the attainment of the critical transition temperature with an energy input of 3.7 mJ. This input decreases linearly with increasing temperature, approximating zero in the vicinity of the transition temperature. The enthalpy change during the transition is very small. Much faster response has been attained by imbedding the crystals into a heat sink, which has to be an electrical insulator. A mixture of V_2O_3 and $SrCO_3$ (V/Sr = 1/3), molten between two Pt leads has a resistivity ratio 10^5 and a switching time of 5.10^{-8} sec, which is surprisingly small for thermal processes.

2.3.2. Influence of Stoichiometric Deviations

The activation energy in the semiconductive state and the transition temperature do not exhibit a monotonous trend between $VO_{1.5}(V_2O_3)$ and VO_2.[82] This was found with sintered specimens, and therefore would be less reliable (though good agreement exists with Morins' single-crystal data). Table 5 gives transition temperatures for the magnetic susceptibility and electrical conductivity of these sintered specimens.

TABLE 5

Composition VO_x	Transition Temperature (K)	
x	Magnetic	Electrical
1.50	168	152
1.67	235 ± 3	—
1.75	130 ± 3	136
1.80	162	—
2.00	345 ± 3	353
2.17	154	156

For $x = 1.67$ and 1.80 the electrical analog of the semiconductor-to-metal transition is missing. Investigations with Si-, Al-, Cr-, Ti-, Mn, Fe-, Co-, and Nb-doped single crystals of VO_2 produced by programmed decomposition of V_2O_5 in a N_2 stream up to 1620°K and slow cooling to room temperature had results given in Table 6.

TABLE 6

Addition (at. %)	Transition Temperature (K)	Resistivity below the Transition Temperature (Ω cm)	Resistivity Jump
−0	340	3	4,000
0.28 Al	342	100	1,400
0.05 Si	338	10	30,000
0.82 Ti	336	10	400
5.17 Cr	340	100	10,000
0.015 Mn	339	2×10^{-1}	1,000
0.15 Fe	338	8×10^{-2}	250
0.035 Co	338	2×10^{-2}	100
1.1 Nb	320	3×10^{-2}	100

Stoichiometry considerations might also enter in studies on the impurity effects on the VO_2 transition.[86] This has been clearly shown by reactive sputtering of VO_2 at controlled O_2 partial pressures between 0.5 and 2.5 mTorr. Transition temperature and resistivity ratio go through a maximum for stoichiometric composition.[686]

The switching ability of VO_2 stimulated the preparation and study of thin-film switching elements. Evaporation of V in 5.10^{-5} torr O_2 on glass substrates preheated to about 770 K produced 1000 Å thick films devoid of the resistivity jump.[96] Subsequent annealing at 770 K in O_2, even at less than $\frac{1}{5}$ at. % restored the jump, although only with a resistance ratio of 10^2. With multilayer films, each 1000Å, the ratio could be boosted to 10^3. Films with a large jump showed a resistance hysteresis of 5 K and those with small jump, 15 K in their R vs T curve. The fatigue cracking observed after repetitive cycling of bulk samples was not observed in these films.

Following a suggestion by Bongers and Enz[97] to use VO_2 crystals as bistable resistors with a stable high-current state and a stable low-current state at constant applied voltage with switching times of 100 μsec, two- and four-pole thin-film switching elements have been made with linear dimensions of 0.1×0.03 cm^2 using only single layers on 0.1-cm-thick glass substrates. The four-pole elements have a sandwich structure consisting of a single layer VO_2 film, covered by an insulating SiO film and Ni-Cr film used as an external heater to switch the VO_2. With 100 mW dissipated in the heated film, the resistance of VO_2 film dropped from 50 to 0.55 kΩ, with no detectable change after 30,000 heating cycles. The relatively large time constant of 0.2 sec prompted efforts to make 10×10-μm two-pole elements from VO_2 films by photoetching. They were frequency independent up to 500 Hz; however, their resistance jump decreased from 25 under static conditions to 20 at 1000 Hz at a point level of 1 mW.

Reactively sputtered VO_2 films of 670 ± 25 Å thickness with controlled O_2 pressure in the plasma obviously are very promising, being probably better defined in their stoichiometry. The resistance jump at 336 K is nearly 10^4 and the activation energy in the semiconductive state 0.28 eV.[98] As substrate, randomly oriented sapphire preheated to ≈700 K during and for 30 min after sputtering was used. The influence of amorphous mono- and polycrystalline substrates of TiO_2 and Al_2O_3 for different film thickness was further investigated by Hensler.[99]

While films on glass and glazed ceramic had small grain size, and a ratio of $\approx 10^2$ with broad transition, a ratio of 3.5×10^3 was obtained with larger grained films on sapphire and rutile.

VO_2 films 0.1–1 μm thich have been produced by pyrolytic decomposition of Vanadium acetylacetonate. Starting from about 10 Ω cm, their resistivity drops about three orders of magnitude at 333 K. Their switching time decreases from 10 μsec at 2 V to 1 μsec at 10 V for an electrode gap of 1 μm.[100]

V_2O_3 undergoes a semiconductor-to-metal transition at about 163 K with a resistivity jump of 10^6. Single crystals grown by flame fusion technique* showed a linear depression of their transition temperature with a coefficient $dT/dp = (4.1 \pm 0.3)\ 10^{-3}$ K bar^{-1} up to 15 K bar, resulting in a shift of about 55 K in this pressure range. This might be of interest for pressure measurements in this cryogenic range.[101]

An insulator-to-metal transition of inverted sequence occurs in the ferromagnetic *n*-type semiconductor EuO at $T_c = 69$ K. The resistivity jump is strongly influenced by the stoichiometry and impurity content of the specimens. The resistivity is about 10^{-3} Ω cm at 42 K and about 10^4 Ω cm at 300 K for samples with an excess of Eu. The large resistivity difference of greater than 10^6 within a temperature difference ΔT of less than 100 K seems to be attractive for temperature measurements. However, it is also dependent on the magnetic field, as shown in Figure 16. Furthermore, a rigid control of the stoichiometry must be accomplished to obtain reproducible sensors.[102]

* Supplied from Linde Co., Speedway, Indiana.

3

Sensor materials and their basic properties

3.1. PRESENT SENSOR MATERIALS

For elemental sensor materials such as Ge and Si that are often used in monocrystalline form, the basic properties can be derived from their impurity profile, which determines their conductivity and conduction type. Somewhat more complex are the conditions for semiconductor compounds such as GaAs, where deviations from stoichiometry can inject another factor.

Most commercial sensors are of polycrystalline oxidic nature, often nonstoichiometric and in most cases prepared by reacting of several oxides with each other. In these cases new phases are formed, such as spinels as in $NiMn_2O_4$ or perovskites ($BaTiO_3$). The solid-state chemistry of these oxidic systems is very complex for the following reasons:

1. The reactivity of the components can differ widely not only because of submicron particles size distribution but also because of large differences in defect concentration, caused by various preparation processes of component materials.
2. Atmosphere: The gas atmosphere and also its small contaminations (H_2O, organic vapors) can influence the reaction rate.
3. The reaction product (sinter body) is porous and has in most cases a considerably lower density than the crystal. It is often misleading to derive the electrical properties of the sinter bodies from those of single crystals with the same composition applying the naive concept of a foam or Swiss cheese, for which the density can be calculated with a simple mixing rule. On the contrary, these oxide compounds can be composed of grains with a concentration gradient of either of the components and with surface layers differing in their electrical properties from the interior.

An oxidizing atmosphere during reaction and sintering tends to shift the conductivity in such a way that compared to the interior of a grain the surface is less *n*- or more *p*-type, depending on the semiconductor type. The opposite occurs in a reducing atmosphere. Sometimes conductivity of surface and interior can be of different type, resulting in an inversion layer of $\approx 10^{-4}$ cm.

4. The grain or crystallite growth can differ considerably if a number of different phases are formed as consequences of nonstoichiometric ratios of the oxide components. The points (1) to (4) are valid even for equilibrium conditions, that is, for long reaction periods approaching infinity. In reality thermistor materials in many cases do not represent equilibria, since the actual reaction (sinter) time is much shorter for economical reasons.

5. Thermistor compositions often are not only made of the basic semiconducting oxide phases, but also contain additions, which can have a multiple purpose.

a. They directly control the resistivity of the oxide on the atomic level by producing electrons or holes in a charge exchange with the cations of the lattice (controlled valence).

b. They indirectly control the resistivity of the sinter units by accelerating or retarding the sinter rate. Numerous examples of such effects have been found, investigated, and explained by the theory of C. Wagner[103] stressing the importance of lattice vacancies for the reactivity in solid phases.

A classical example is the commercially important sintering of MgO bricks, which can be accelerated by small additions of Fe_2O_3. The presence of Fe^{3+} ions in MgO leads to the Mg vacant cation sites, according to the formula $[Mg^{2+}_{1-3x}\ Fe^{3+}_{2x}\ V_x]$ O; those vacant sites promote diffusion and sintering at high temperatures. The reduced solubility of Fe^{3+} in MgO at low temperatures, resulting in segregation of $MgFe_2O_4$, does not matter since the promoting effect had been consummated at higher temperatures. In ZnO other trivalent ions such as Al^{3+}, Cr^{3+}, Ga^{3+}, In^{3+}, and La^{3+} retard sintering, at least at temperatures below 1250 K. Above this temperature they are either inert or even act as promoters (Fe^{3+}).[104]

Extensive experimental and theoretical treatment of such phenomena has been given by Hauffe.[105] For more details see Section 5.3.

c. Certain additions have the unsophisticated purpose of increasing the resistivity by diluting the conductive oxide with insulating oxide phases such as Al_2O_3, bentonite, and kaolin, or to improve shrinkage and cohesion during sintering (glass addition acting as flux). In some cases these ingredients can have multiple effects simultaneously: valence control, sinter promotion, and flux bonding, though to very different degrees, and it is not easy to separate them from each other.

Some patents specify empirical additives which might not only be "black art," but might be based on their combined beneficial effects on the electrical properties.

The other parts of this chapter are devoted to the chemical, thermal, structural, and electrical properties of materials for semiconducting temperature sensors. Only properties relevant to their production, stability, and function are treated.

3.2. CHEMICAL PROPERTIES

3.2.1. Temperature Range of Chemical Stability in Different Atmospheres

The O_2 dissociation pressure of several oxides used in commercial thermistors in the temperature range used for sintering is shown in Figure 17. The horizontal lines correspond to commonly used O_2 concentrations in the sinter atmosphere. It is clearly visible that Mn_3O_4, Cu_2O, and PbO only are stable below ~2300°F (1530 K), besides NiO, CoO, and Fe_3O_4, which are not included in the figure, since their dissociation pressures even at 1600 K (2410°F) are far below 10^{-4} torr. The curves for Co_3O_4 and CuO, indicating dissociation in air at 1175 K (1660°F) and 1315 K (~1900°F), respectively, reveal not only the possibility of strong reabsorption of O_2, if the cooling is slow enough, but also explain the resistivity difference between quenched and slowly cooled units in chemical terms. More critical for reabsorption of O_2 is the temperature range below the equilibrium temperatures ~1170 K (~1650°F) for Co_3O_4 and ~1315 K (~1900°F) for CuO.

Controlled annealing at these temperatures permits adjustment of the electrical properties.

Even for much lower temperatures where in air no dissociation of oxides can be expected, the knowledge of their dissociation equilibrium pressures as functions of temperature is of considerable practical interest. Figure 18 gives information on the effect of reducing agents at operating temperatures on the chemical (and therefore also) electrical stability. Organic vapors or traces of CO can be present in sealed probes used at elevated temperatures or a gasket containing organic matter can be in direct contact with the thermistor.

Water vapor is also formed by decomposition of organic material in presence of oxygen donors. In this case the thermistor material may behave as if the partial pressure of O_2 is reduced by orders of magnitude. An estimate of the O_2 partial pressure in a gas mixture CO + H_2O leads at 600, 800, and 1000 K to values of 10^{-37}, 10^{-25}, and 10^{-18} torr, respectively, well below the

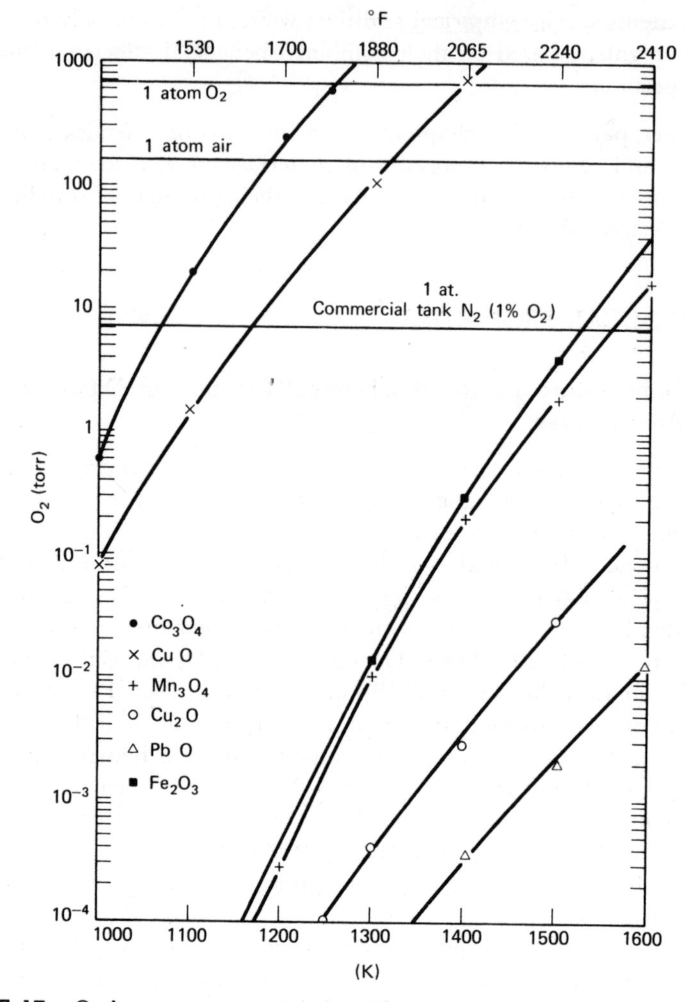

FIGURE 17. O_2 dissociation pressure of oxides between 1000 and 1600 K.

dissociation pressures of NiO, CoO, Co_3O_4, and Fe_3O_4 in this temperature range. In other words, these oxides could oxidize organic materials not only when present as vapors but also as solids in contact with the oxide. This is important when selecting protective coatings for thermistors operating at temperatures >400 K, where the reaction rate can become detectable over long time periods. Oxygen loss in a nonstoichiometric transition metal oxide is often not well reversible, especially at temperatures below 600 K. The assumption of classical equilibria between well-defined oxide phases is

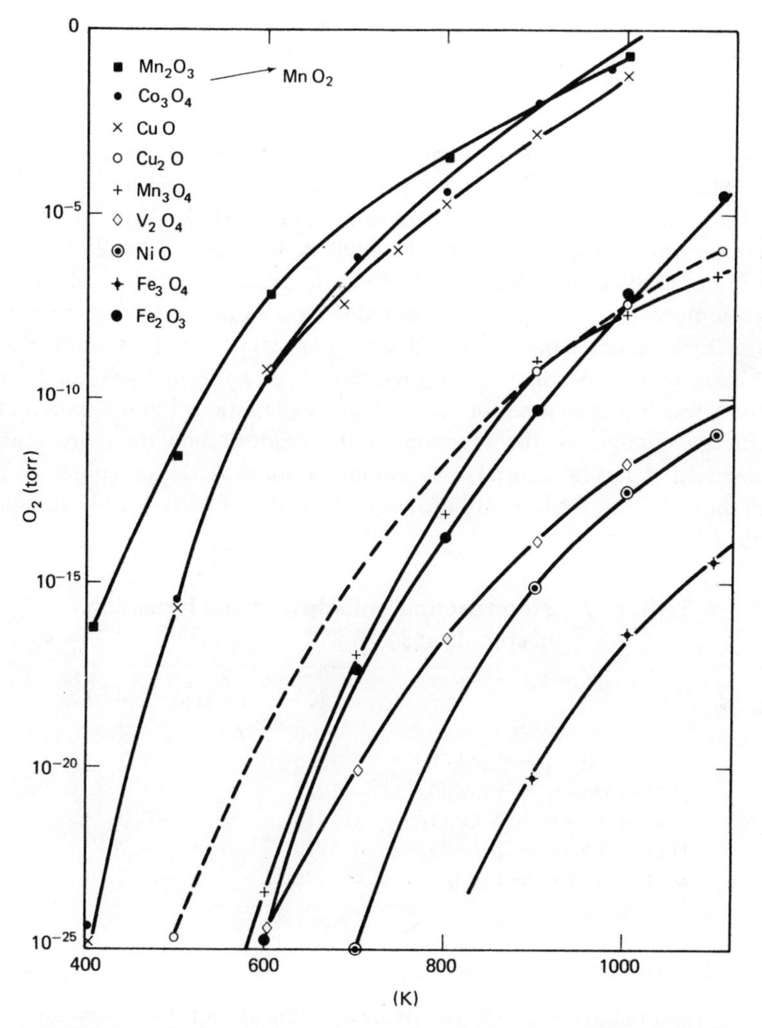

FIGURE 18. O_2 dissociation pressure of oxides between 400 and 1120 K.

of course an oversimplification and gives only a crude picture. In reality the influence of O_2 partial pressures on the defect concentration in the oxide phases determines their electrical properties. Basic studies of this type date back into the 1930s[106] and a comprehensive survey was given by Hauffe.[107]

In mechanical mixtures, the transition oxide with the highest dissociation pressure at a given temperature would determine the equilibrium for the entire system. However, thermistor compositions are not mechanical mixtures, but reaction products. Therefore their O_2 dissociation pressure is in general smaller than that of the pure components. Explicit data for binary and ternary oxide systems, even *thermodynamic data* to calculate the *equilibrium pressures*, are scarce. One well-investigated system is $NiAl_2O_4$,[108] which can be considered as a model case where only one of the components contributes to the observed dissociation pressure, at least at temperatures below 1400 K. It was found that the O_2 pressure over $NiAl_2O_4$ ($NiO-Al_2O_3$) over the temperature range from 1000 to 1300 K is about $1\frac{1}{2}$ to $2\frac{1}{2}$ orders of magnitude smaller than for pure NiO; this difference shrinks slightly at higher temperature. It appears reasonable to assume that other combinations or new phases (spinels) would show depressions of the same order of magnitude or even smaller if the dissociation pressure of the second component is much higher than for Al_2O_3. Exact calculations require data on the free-energy change by the reaction of the oxides; such data are scarce. Schmalzried[109] has measured and compiled some data for spinel phases which indicate that the free enthalpy is of the order of $-(6 \pm 2)$ kcal $mole^{-1}$ (Table 7).

TABLE 7 Free reaction enthalpy for the formation of spinels at 1273 K.

	K	kcal $mole^{-1}$
$NiO + Al_2O_3 = NiAl_2O_4$	1273	−5.2
$CoO + Al_2O_3 = CoAl_2O_4$	1273	−4.9
$NiO + Cr_2O_3 = NiCr_2O_4$	1273	−6.0
$FeO + Cr_2O_3 = FeCr_2O_4$	1273	−8.2
$FeO + Fe_2O_3 = Fe_3O_4$	1273	−7.5

3.2.2. Hygroscopicity of Oxides

The electrical behavior of sintered oxide systems at high humidity can be of great influence on the stability in life tests made with nonencapsulated units in normal atmosphere. This is drastically evident in systems containing a strongly basic oxide component such as BaO or SrO or even alkali oxides. These oxides have a strong tendency to react with water vapor (and also with CO_2, if present). However, other oxides present in thermistor compositions have also a certain hydroscopicity, differing not only from oxide to oxide, but also with their origin from certain compounds. These differences are probably caused by the submicron particle size (determining the active surface in $m^2\ g^{-1}$) of the oxide before molding and sintering, but they are leveled somewhat by the sinter process at temperature above 1200 K.

However, thermistor compositions are at first handled as powders. Therefore these hygroscopic data could be of interest, since they may influence the flow of powders into a mold of an automatic press with high repetition rate.* The hygroscopicity of Cr_2O_3, NiO, CoO, and ZnO prepared from carbonate, nitrate, oxalate, acetate, or hydroxide between 10% and 99% humidity has been found to differ considerably with the acid radical of the mother compound, being a maximum for the oxide made from carbonate, and a minimum for nitrate oxide.[110] Different concentrations of alkali traces could be the cause. This confirms the general experience found in earlier literature. The maximal percentage of water picked up at 99% humidity ranges from 30% for CoO and 32.5% for NiO to 9% for ZnO if made from carbonates. For the same oxides made from nitrates the corresponding values were 3.2%, 2.7%, and 0.7%.

3.3. MISCIBILITY OF BINARY SYSTEMS AND PHASE EQUILIBRIUM IN MULTICOMPONENT SYSTEMS

Nearly all oxide compositions used for NTC or PTC thermistors start with mechanically blended mixtures. To understand the electrical properties of the reacted sinter bodies, a knowledge of the phase equilibria is desirable, though in some cases the sinter time is insufficient to attain equilibrium. Table 8 presents some data for compacted oxide mixtures that are permitted to react with each other in solid state until complete equilibrium is reached.

For cubic oxides of divalent metals and rhombohedral oxides of trivalent metals, complete miscibility is probable if the difference of ionic radii is <12%. Between 12% and 29% difference, partial miscibility exists with the dissolved quantity inversely proportional to the percent difference.

Influence of Li Doping on Lattice Parameters of MnO, NiO, and CoO. In all cases the lattice constant decreases nearly linear with the Li^+ concentration. No new phases were formed. Within the common doping range of 2 at. %, the lattice contraction is less than 0.5%.[112]

The most common oxide system for thermistors, based on the oxides of Mn, Ni, and Co, can form a new crystal phase known as *spinel*. Its prototype is the mineral spinel $MgAl_2O_4$ (Figure 19) in which the Mg^{2+} ions occupy tetrahedral (A) sites, each surrounded by four oxygen ions and the Al^{3+} ions occupy octahedral (B) sites each surrounded by six oxygen ions (normal spinel). The unit cell of AB_2O_4 contains 32 O^{2-} ions forming a cubic close packing with 64 tetrahedral and 32 octahedral interstices; 8 of the first and 16

* For the effect of water adsorption on contact resistance see pp. 162–163.

TABLE 8[a]

System	Miscibility	a (Å)
CoO-NiO	complete above 1070 K	
	below T = 1050 K, 2-phase region	4.250–4.179
CoO-MnO	complete	4.250–4.442
CoO-MgO	complete above 1070 K	
	below T = 1050 K, 2-phase region	
CoO-CuO	up to 25 CoO complete for >25 new phase $Cu^{2+} Co^{2+} O_2^{2-}$ [111]	
CoO-ZnO	>30 CoO complete, green color, hexagonal	
	>30 ZnO complete, rose color, cubic	
	Between 30 and 70 CoO mixture of 2 phases	
NiO-MnO	complete	4.179–4.442
NiO-CuO	very low	
NiO-ZnO	very low	
CuO-ZnO	none	
FeO-MnO	miscibility gap dependent on metal deficiency[687]	
$NiO-Mn_2O_3$[152]	0–3 NiO, tetragonal	5.75 (c/a = 164)
	3–13 NiO, 2 phases	
	tetragonal	5.749–5.732
	spinel	8.458–8.428
	13–70 NiO, spinel	8.428–8.355
	70–90 NiO, 2 phases	
	spinel	8.35
	face centered cubic (fcc)	
	NiO	4.178
	90–100 NiO, (fcc) NiO	4.178
	with MnO in solid solution	

[a] All data in mol. %.

of the second are occupied by cations. The classical investigations by the Philips researchers have shown that[113–115]

1. Spinel structure can be expected for a large number of oxide systems containing divalent and trivalent metals.
2. The distribution of divalent and trivalent metal ions can vary from normal (Me^{3+} in octahedral position) to inverse (Me^{3+} in tetrahedral position) swapping site with Me^{2+} ions that go into octahedral position (inverse spinel).
3. Mixed types can exist in which only partial exchange of Me^{2+} and Me^{3+} occurs.

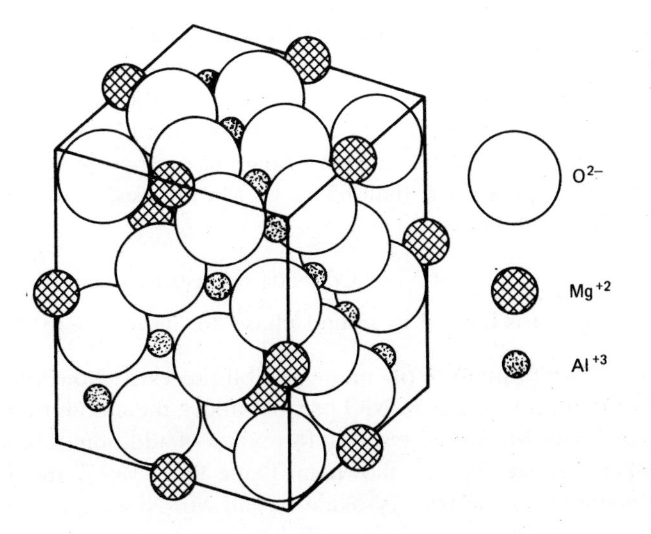

FIGURE 19. Spinel structure. From Kittel, *Introduction to Solid State Physics,* Wiley, New York, 1956, p. 446.

Certain cations prefer octahedral coordination and therefore determine whether normal or inverse spinels are formed, examples, Ni^{2+} and Fe^{2+} in $Mn^{2+}[Ni^{2+}\ Mn^{4+}]\ O_4^{2-}$ and $Fe^{3+}[Fe^{2+}\ Fe^{3+}]\ O_4^{2-}$.

It was later found that the spinel structures are much more abundant in composite oxide systems and they are not restricted to divalent and trivalent ions, for examples,

$$Ge^{4+}Ni^{2+}O_4^{2-} \qquad Ag_2^{1+}Mo^{6+}O_4^{2-}$$

$$Si^{4+}Co^{2+}O_4^{2-} \qquad Na_2^{1+}W^{6+}O_4^{2-}$$

This concept had a great impact on the understanding of the magnetic and electrical properties of these types of mixed oxide system and led to a new approach with the well-chosen name molecular engineering, especially successful in the ferrite field.

Since the conduction mechanism in semiconducting spinels of the transition elements depends on the distribution and valence of the different metal ions in the crystal lattice, a direct method for investigating the valence state of certain ions, for instance of Mn, is desirable in addition to the conventional x-ray or neutron diffraction. The x-ray K-absorption spectra of Mn in MnO, Mn_3O_4, MnO_2, and Co-Mn spinel led to the conclusion that di- and tetravalent Mn ions are coexistent in the normal and inverse spinels. This method could aid the understanding of differences in electric properties of oxide samples of equal composition, but slightly different heat treatment.[116]

In binary thermistor compositions of Ni-Mn or Co-Mn oxides, complete formation of spinels $NiMn_2O_4$ and $CoMn_2O_4$ requires always an atomic ratio 2 of Mn/Ni or Mn/Co corresponding to weight ratios of 1.873 and 1.864 for metal contents (~65 wt % Mn of 100 for both Ni and Co). Thermistor engineering requires that a broad range of compositions meet specifications. What happens if less than 65 at. % Mn are used in a composition? There are two possibilities:

1. (trivial): The surplus of Ni or Co oxide coexists as second phase:
2. Another spinel is formed as second phase, for instance $Co^{2+}[Co_2^{3+}]\,O_4^{2-}$.

In ternary compositions, still more possibilities exist; however, the rule still applies that any surplus of NiO or CoO above the amount necessary to form a spinel with Mn would result in formation of additional spinel phases, of normal or inverse type as shown in Table 9. X-ray[117] and neutron[118] diffraction studies of the ternary oxide system Mn-Ni-Co have been made and led to the following conclusions (Table 9):

1. Ni ions prefer *B* sites; Co ions prefer *A* sites.
2. The inverted Ni-Mn spinel is cubic, the normal Co-Mn spinel is tetragonal ($c/a = 1.13$).
3. Ni^{2+} ions at *B* sites can cause the formation of an equivalent number of Mn^{4+} ions to accomplish charge stoichiometry (composition of 67 at. % Mn).
4. If less than 25% of the *B* sites are occupied by Ni ions, the tetragonal distortion of the lattice starts. Spinel formation is incomplete after 16 hr heating of the blended oxide mixture to 1170 K (1650°F). Additional heating for 3 min to 1523 K (2280°F) followed by 1 hr annealing at 1123 (1565°F) completed spinel formation and eliminated the excess diffraction lines caused by unreacted NiO, CoO, Co_3O_4, Mn_2O_3, or Mn_3O_4.
5. Since technical sinter conditions are marginal in temperature and time to those necessary to attain complete spinel formation, the existence of unreacted small amounts of unreacted components would be always possible even for compositions with a component ratio near the theoretical one to produce a single spinel phase.
6. The O_2 partial pressure in the sinter atmosphere has to be large enough to obtain the stoichiometrically necessary oxygen for spinel formation. It has been found that annealing in air or oxygen and slow cooling produced the same structure. This indicates that between $\frac{1}{5}$ and 1 at atmospheric O_2 pressure, the stoichiometry of spinels is practically sustained, although the number of defects might change. Sintering at temperatures above 1370 K (>2000°F) and quenching to temperatures below 500 K is another story.

TABLE 9

	Formula		
at. % Mn	A Site	B Site	Structure
>67	Mn	$Mn_{3/2}Ni_{1/2}O_4$	cubic
	$Co_{1/2}Mn_{1/2}$	Mn_2O_4	tetragonal
67	Mn(trivalent)	$MnNiO_4$ Mn(trivalent)	cubic
(65 wt %)	Mn(divalent	$MnNiO_4$ Mn(tetravalent)	
	Co	Mn_2O_4	tetragonal
	$Mn_{1/4}Co_{3/4}$	$Mn_{7/4}Ni_{1/4}O_4$	tetragonal
	$Co_{1/2}Mn_{1/2}$	$Mn_{3/2}Ni_{1/2}O_4$	cubic
<67	$Co_{1/2}Mn_{1/2}$	$MnNi_{1/2}O_4$	cubic
	Co	$Mn_{3/2}Ni_{1/2}O_4$	cubic

or more generalized (at % A, B, C, and D)

For at. % Mn			
>67	Mn_A	$[Mn_B\ Ni_C\ Co_D]O_4$	
	$A + B \geqq 2$	$C + D \leqq 1$	
<67	$Mn_A\ Ni_C$	$[Mn_B\ Co_D]O_4$	
	$A + B \leqq 2$	$C + D \geqq 1$	

In this case considerable resistivity differences are possible between air and O_2-sintered specimens, and small stoichiometric O deficiencies have been determined.

7. The coexistence of identical metal ions in different valence states at identical lattice sites favors conduction by electron exchange and results in lower resistance values. It has been predicted that substitution of $\frac{1}{3}$ on Mn^{3+} of B sites by Ni^{2+} reduces the resistivity to a minimum by producing an equivalent concentration of Mn^{4+}.

Ni-Mn-oxide systems with more Mn than the stoichiometric amount in $Mn(NiMn)\ O_4$ have been investigated by the MIT-Lincoln Laboratory.[119]

The entire phase diagram for $(Ni_{1-x}\ Mn_{2+x})\ O_4$, varying from 0 to 1, was determined by heating the well mixed components NiO and Mn_2O_3 to 1370 K ($\approx$2000°F) for $x < 0.4$ and to 1480 K ($\approx$2190°F) for $x > 0.4$ and subsequent quenching. Slow cooling resulted in increased absorption of O_2 for x values greater than 0.5. In this composition, coexistence of cubic $Mn(NiMn)\ O_4$-spinel and tetragonal Mn_3O_4 must be assumed. A similar study with the system $Co_3O_4 - Mn_3O_4$ at 1273 K under equilibrium conditions (>40 hr sinter time) has led to the following conclusions[120]:

1. Mn_3O_4, a cubic spinel phase, is formed with a lattice constant increasing from 8.07 to 8.27 Å, starting from Co_3O_4. For less than 3 mole % Mn_3O_4 a second cubic phase with NaCl structure coexists, identified as CoO by its lattice constant of 4.27 Å. Between 43 and 63 mole % the cubic and tetragonal spinel phases exist; above 63% only the latter exists, increasing its tetragonal ratio with the Mn concentration.
2. In air the CoO phase coexists with the cubic spinel up to 30 mole % Mn_3O_4, the conditions above 50% remaining nearly the same as in O_2.
3. It is easy to understand that in Ar the CoO phase extends up to much higher Mn_3O_4 concentrations. Co_3O_4 is completely dissociated at 1273 K in Ar. Obviously Mn^{2+} ions enter the CoO lattice, since the lattice constant increased (see also Table 8).

The Co^{3+} ions needed for the formation of the cubic spinel are nonexistent; therefore only the tetragonal spinel can be formed, requiring only Co^{2+} ions.

The Ar sintering is not applicable to the conditions of the usual thermistor sintering; however, it might be of interest, if by freak accidents uncontrolled amounts of reducing agents contaminate the sinter atmosphere.

Cu-containing thermistor compositions, often used to obtain lower resistivities, also form a spinel $CuMn_2O_4$. Conditions in this case are more complicated due to the O_2 loss of CuO above 1200 K, where solid-state reactions to form the spinel phase are fast enough. Starting with Cu_2O at temperature below 1170 K,[121] formation of the spinel $CuMn_2O_4$ was observed with Cu^+ on *A* sites and the *B* sites occupied by Mn^{3+} and Mn^{4+}, to compensate for the lower charge of the Cu^+ ions.[122]

Since this investigation was made with a large Cu surplus (factor 3), CuO and Cu_2O were always coexistent with the spinel. A more systematic study of the Cu-Mn-oxide system under conditions similar to thermistor production with sinter temperatures 1270–1470 K (1830–2200°F), mainly concerned with its electrical properties, considered $CuMn_2O_4$ as the main reaction product and either Mn_3O_4 or CuO_x (x varying with temperature, but always >1) as unreacted by-products. There seems to be some evidence that the spinel $CuMn_2O_4$ can form a solid solution with Mn_3O_4.[122]

The spatial distribution of the existing phases has been examined by microscopy and related to the cooling rate. As a general rule, rapid cooling favors the existence of solid solutions in the range Mn_3O_4–$CuMn_2O_4$ (0–33 at. % or 0–37 wt %) based on Mn + Cu = 100; slow cooling leads to segregation of less conductive zones. There has been some controversy over whether $CuMn_2O_4$ is indeed a normal spinel.[123] Magnetic studies[124] support the assumption of Cu^{1+} ions at *A* sites, which would require either the presence of two Cu^+ or one Mn^{4+} for electroneutrality. X-ray investigations seem to favor a mixed spinel $Cu_x^{2+} Mn_{1-x}^{2+} [Cu_{1-x}^{2+} Mn_{1-x}^{4+} Mn_{2x}^{3+}] O_4^{2-}$.[125,126]

Systems with Iron Oxides

FeO_{1+x}.[127,128] FeO_{1+x} has NaCl (fcc) structure with a = 4.31 Å. Vacancies in the lattice are double the number of cation vacancies. Homogeneous phase with stoichiometric O excess resulting in initial resistivity increase up to 6 at. % O followed by decrease toward the resistivity of Fe_3O_4.

Fe_3O_4. <119 K orthorhombic (0.05% distortion); a = 8.41 Å; >119 K cubic spinel, inverse.

Fe_2O_3. Only of interest doped with TiO_2.

α-Fe_2O_3. Corundum structure with rhombohedral symmetry.

The latter is of interest mainly because of its ability to form inverse spinel phases with divalent metal oxides such as MnO, NiO, and CoO normally called ferrites. These ferrites have not yet found commercial applications as thermistors; however, their electrical properties have been explored.[129–131]

Fe_3O_4-Mn_3O_4. This system could have considerable interest for thermistor compositions. It would cover a wide resistivity range from 0.1 to 10^7 Ω cm at 300 K and activation energies from ~0.03 to ~1 eV. However, it requires operation in a sinter atmosphere with controlled O_2 pressure. Otherwise Fe_3O_4 would oxidize to Fe_2O_3, which even at 1300 K has only a O_2 dissociation pressure of ~10^{-2} torr, and would therefore be the stable compound in air. The optimal O_2 pressure during sintering of ≈10^{-2} torr. would inhibit oxidation of Fe_3O_4 and dissociation of Mn_3O_4. At 300 K, Fe_3O_4 and Mn_3O_4 are completely miscible as spinels. Up to ~61 at. % Mn, the cubic spinel phase exists. With >61 at. % Mn, the volume of the unit cell increases regularly and the axial ratio increases from 1 to 1.15 at 100 at. % Mn. Between 753 and 1423 K, miscibility is limited for this tetragonal range.

Heating of homogeneous tetragonal phases with >61 at. % Mn, above 750 K, produces a cubic phase in equilibrium with a tetragonal phase, the latter decreasing in percentage until again a single cubic phase is obtained at higher temperatures (>1423 K).[132]

The semiconducting properties of the mixed crystals Fe_3O_4-$MnFe_2O_4$-Mn_2FeO_4 were investigated by measurements of the electrical conductivities and the Seebeck voltages on single crystals and polycrystalline materials. The experimental data are interpreted by assuming that the conductivity arises exclusively from ferric and ferrous ions on octahedral sites according to the Verwey hopping mechanism (1947) and that the reaction $Mn^{2+} + Fe^{3+} = Mn^{3+} + Fe^{2+}$ is endothermic, involving an energy of approximately 0.30 eV.[130]

3.4. THERMAL PROPERTIES

3.4.1. Specific Heat

The response time of a thermal sensor is determined by its thermal properties, especially its heat capacity (mass and specific heat). At and above the ambient temperature, specific-heat data can be taken from the literature or, if unavailable, estimated with good approximation by adding the atomic heats of its components using the empirical value for oxygen in oxides of 4.75 cal/g (Kopp–Neumann rule, 1864). While temperature dependence of the specific heat of solids above room temperature is rather small, it becomes rather strong at temperatures below ambient. According to Debye the specific heat of solid metals can be expressed by the following function:

$$C_v = 9Nk\left(\frac{T}{\Theta}\right)^3 \int_0^{x_m} \frac{e^x x^4 \, dx}{(e^x - 1)^2}$$

where k is the Boltzmann constant, N is the number of harmonic oscillators in three dimensions, x is the hr/kT, V is the atomic vibration frequency, and $\Theta = hv_{\max}/k$. Knowledge of the characteristic Debye temperature of a solid permits calculation of its specific heat as function of temperature.

For practical use this has been simplified by modification of the Debye formula to

$$C_v = 3RD\left(\frac{\Theta}{T}\right)$$

and tabulation of the Debye function $D(\Theta/T)$. Eucken[133] has developed a monogram to determine C_v for any combinations of T and Θ. Since the factor $3R$ is valid for metals, it must be substituted for oxides by their specific heat at high temperature.[134]

For a number of semiconducting materials the following Debye temperatures are known:

Diamond	Si	Ge	GaAs
391 K	658 K	366 K	345 K

At very low temperatures the temperature dependence of C_v can be approximated by

$$C_v = 234Nk\left(\frac{T}{\Theta}\right)^3$$

This function is plotted and normalized in Figure 20. Empirical specific-heat data are given in Table 10, which does not reflect certain sharp maxima in the specific heats MnO and V_2O_3 at 118 and 173 K, respectively. Tables 10 and 11 give experimental data between 100 and 1500 K.

3.4.2. Heat Conductivity

Another parameter determining the response time of temperature sensors is heat conductivity. The Wiedemann–Franz law correlating the electrical (σ) to the thermal conductivity (K) by the Lorenz number can hardly be applied for most of the semiconductors used in sensors, since the electronic contri-

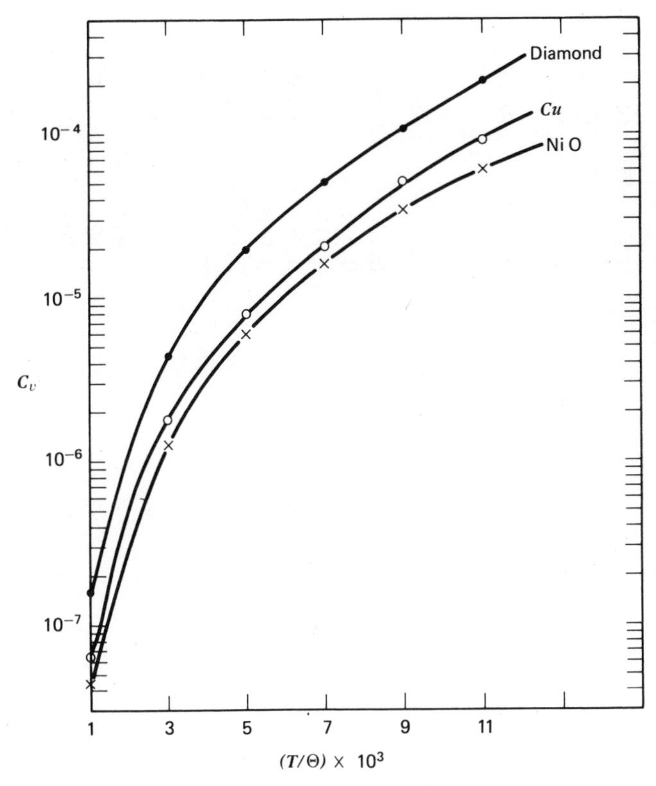

FIGURE 20. Heat capacity according to the Debye T^3 approximation.

TABLE 10 Specific heats (c_p) in J/g K

	10	15	20	40	60	80	100	200	300 K
Diamond	0.000018	0.000053	0.000121				0.021	0.194	0.518
B							0.010	0.015	0.027
Ge			0.013	0.062	0.108	0.153	0.192	0.288	0.322
GaAs							0.047	0.070	0.077

	100	200	298	400	500	800	1000	1500 K
TiO_2			0.706	0.803	0.837	0.910	0.930	0.964
V_2O_3	0.192	0.540	0.694	0.780	0.825	0.900	0.990	1.015
V_2O_4	0.224	0.540	0.715	0.810	0.870	0.940	0.969	1.02
Cr_2O_3	0.161	0.504	0.71	0.746	0.776	0.820	0.838	0.872
MnO			0.630	0.665	0.691	0.740	0.761	0.824
Mn_3O_4	0.264	0.500	0.650	0.690	0.718	0.786	0.830	
Mn_2O_3	0.266	0.500	0.620	0.686	0.730	0.815	0.860	
FeO	0.342	0.692	0.725					
Fe_3O_4	0.244	0.504	0.654	0.739	0.830	1.093		
Fe_2O_3	0.198	0.480	0.658	0.754	0.823	0.990		
NiO	0.187	0.444	0.594	0.698	0.860	0.710	0.730	
CoO	0.255	0.604	0.702	0.704	0.710	0.737	0.760	0.816
Co_3O_4	0.142	0.363	0.513	0.592	0.645	0.758	0.824	
Cu_2O	0.277	0.378	0.445	0.502	0.519	0.569	0.602	
CuO	0.200	0.437	0.562	0.588	0.614	0.690	0.740	

TABLE 11 Specific heats (c_p) in J/g K.

	100	200	298	400	500	800	1000	1500
PbO			0.205	0.217	0.229	0.265	0.289	
ZrO_2	0.155	0.360	0.457	0.520	0.552	0.599	0.617	
CeO_2			0.382	0.389	0.395	0.413	0.425	0.456
ThO_2			0.236	0.253	0.264	0.283	0.294	0.318
$MnFe_2O_4$				0.682	0.728	0.869	0.960	
$NiFe_2O_4$	0.192	0.460	0.620					
$FeCr_2O_4$	0.248	0.452	0.596	0.678	0.720	0.785	0.820	0.870
$FeCo_2O_4$	0.189	0.451	0.630					
$FeTiO_3$	0.265	0.516	0.655	0.732	0.775	0.844	0.876	0.943
$BaTiO_3$	0.185	0.355	0.439	0.483	0.503	0.537	0.550	0.577
$SrTiO_3$	0.247	0.439	0.536	0.592	0.620	0.658	0.672	0.698
ZnO			0.495	0.557	0.598	0.636	0.654	0.655

bution is small compared to the thermal vibrations of the atoms in the lattice. This applies to electronic diffusion. Electronic convection currents might indeed make a contribution in semiconductors.[135]

The mean free path of the energy quanta of the lattice vibration (phonons) before scattering (10–100 Å) determines the heat conductivity of a solid without electronic component of heat conduction. This process defies a simple theoretical treatment even for single crystals—more so for porous sinter bodies. Therefore it is necessary to rely predominantly on empirical data, which, however are still scarce. The thermal conductivity of a polycrystalline solid is in the first approximation the same as for each crystallite, taking into account their packing density. The additional boundary heat resistance increases with decreasing crystallite size. The heat conductivity of sintered Al_2O_3 at high temperatures is about half that of single crystals (sapphire). At low temperatures this factor is only $\sim 10^{-2}$, corresponding to a phonon mean free path of 2.5×10^{-3} cm, identical with the size of the crystallites.

Model experiments to study the influence of particle size have been made with large crystals that could easily be crushed to powder. Compressed $NH_4Al(SO_4)_2$ with 10–100 μm crystallite size had about one-tenth of the heat conductivity of the large crystal.[136,137]

Sintered oxides consist of crystallites of similar size. The packing density is, of course, also important in estimating the ratio of heat conductivity of sinter body to crystal. Experience with MgO has shown that this ratio increases rapidly above 60% of the theoretical density.[138] Sintered oxide thermistors normally cover the range from 50 to 80% of the theoretical density determined and taken from x-ray data. For example, the heat conductivity of MgO increases by a factor 2.7 when the relative density increases from 50 to 65%. The defect structure of the grains or crystals also adds to the heat resistance, a factor certainly important for sintered oxide materials. Eucken and Kuhn[139] had already found that the heat conductivity of mixed crystals (KCl-KBr) had a minimum that was independent on temperature. This was confirmed later with diamond containing impurities or imperfections[140] and solid solutions of $BaTiO_3$ and $SrTiO_3$ between 298 and 403 K where a minimum at 15 at. % was found.[141] At constant Ba/Sr ratio the heat conductivity decreased with temperature, and the anomaly in the Curie-point region, previously found as a reduction of the thermal conductivity, was a function of the ratio.[142] Data related to thermistor materials can be found in thermal conductivity investigations between 300 and 1000 K with spinel phases of the system $M^{2+}\ Fe_2^{3+}\ O_4^{2-}$ and $M^{2+}\ Cr_2^{3+}\ O_4^{2-}$ ($M^{2+} = Mg^{2+}, Ni^{2+}, Cd^{2+}$) or mixed systems with different fractions of M^{2+}. Some experimental data follow in Tables 12 and 13 and Figure 21.

TABLE 12 Heat conductivity of Si and Ge (cal/cm sec K).

	4 K	6 K	8 K	10 K	15 K	20 K	30 K	40 K	60 K	100 K
Si	0.311	0.715	1.27	1.79	2.67	2.98	2.86	2.44	1.90	1.19
Ge	0.597	1.190	1.79	2.22	2.99	2.99	2.15	1.40	0.95	0.48

Si and Ge have both a maximum at about 20 K.

TABLE 13 Heat conductivity of a few compounds (cal/cm sec K).

	300 K	400 K	500 K	700 K	1000 K
NiO compact	0.033	0.030	0.023	0.015	0.012
26% porosity	0.025	0.021	0.018	0.012	0.010
NiO·MgO	0.0096	0.0084	0.0078	0.0065	0.0056
$BaTiO_3$	0.0064	0.0072	0.078		
$SrTiO_3$	0.014	0.012			

Diamond II b	100	200	250	300	400	500
	13.5	5.2	4.0	3.2	2.3	~2.00

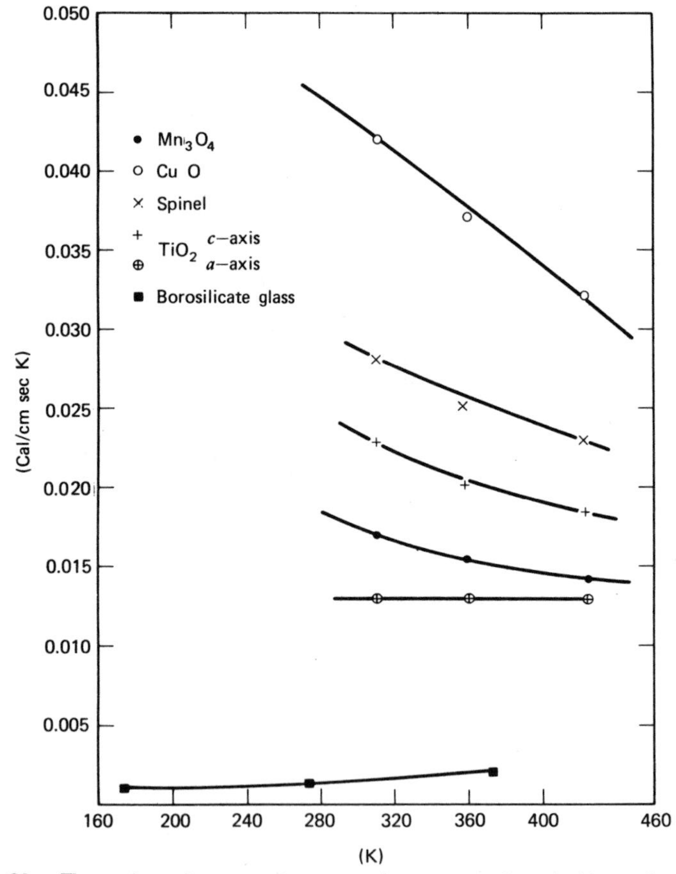

FIGURE 21. Thermal conductivity of some oxidic compounds and of borosilicate glass. After Kittel.[16]

In projecting wider use of III–V semiconductors as temperature sensors, their heat conductivity and its temperature dependence is of interest. For InSb and InAs the heat conductivity goes through a maximum at about 10 K for *n*- and *p*-type crystals. In temperatures below the maximum, the heat conductivity of *n*-type crystals can be 10–12 times higher for *p*-type crystals. For GaAs a similar relationship can be expected.[143]

3.4.3. Thermal Expansion

Thermal expansion can be of interest for various reasons:

1. Cracking might occur during cooling after sintering, especially for shapes more critical for heat stresses (washers or rods)
2. Fatigue cracking could occur in frequent and wide thermal cycles.
3. Mounting in contact jigs and probes could produce mechanical stresses.

Data on thermal expansion of oxides are scarce since it is difficult to obtain suitable rod type specimens of sufficient homogeneity. Therefore theoretical considerations leading to a reasonable approximation are very useful. The Grüneisen relation[144] for the linear thermal expansion coefficient is given by

$$\beta = \frac{K\gamma C_v}{3V}$$

where K is the compressibility, V is the volume, and γ is the Grüneisen constant, dependent on the type of crystal lattice. γ is determined by the valence of the metal atom and the exponents of the potential energy versus interatomic distance relation. Barron has treated the low-temperature range.[145] The Grüneisen equation shows that the linear expansion is proportional to the specific heat at any temperature and therefore also determined in its temperature dependence by the Debye function. For many simple cases, such as metals, the relation α/C_v = constant has been found valid between 100 and 600 K ($\alpha = 3\beta$).

Table 14 gives a few data on linear expansion coefficients, some for materials used in thermistor compositions and related compounds.[146]

3.5. ELECTRICAL PROPERTIES

3.5.1. Semiconductivity of Selected Oxide Systems

It has been previously pointed out that the resistivity of polycrystalline materials can be influenced by grain boundaries or surface states, beyond

TABLE 14 Linear thermal expansion coefficient $\beta \cdot 10^6$.

Oxides	MnO		FeO	NiO		CoO	SnO
	12		12	14		10	3
Spinels	$NiFe_2O_4$	$FeTiO_3$ (Ilmenite)					
	10	10	14	25	130	40	
		300	500	700	800	900	1000 K
Perovskites		$BaTiO_3$		$SrTiO_3$	$Ba_{1-x}Sr_xTiO_3$		$Ba_{(1-x)}Pb_xTiO_3$
100–300 K		6		9.5	11		10
300–700 K		10					
Germanium: increasing between 50 and 390 K from 3.6 to 5.4.							
Diamond		100		400	600		800
		<6		20		40	

the naive effect of porosity. While the influence of doping additions is in many cases well understood (see Section 3.5.2) the effect of grain boundaries is more complex. It has been always useful in solid-state physics to gain knowledge of the properties of single crystals before dealing with polycrystalline specimens of the same systems. Though monocrystals of transition metal oxide system have not yet attained practical importance, their resistivity in dependence on temperature and composition is of considerable interest, especially in systems of unlimited miscibility.

In Figure 22 the resistivity-versus-temperature relations are shown for single crystals of NiO, CoO, and MnO with small, but well-defined stoichiometric deviations, determined with a maximum error of ± 0.1 milliatoms/mole.[147]

In all cases the resistivity at 300 K exceeds $10^7\ \Omega$ cm. This is higher than observed for sintered polycrystalline oxides, even if these have a low impurity profile and no detectable stoichiometric deviations.

It is of great theoretical and practical interest that nearly stoichiometric single crystals representing a binary system of NiO and CoO have much lower resistivities than the pure components.[148] This system has a broad resistivity minimum between 40 and 60 mole %. In this range at temperatures below 300 K resistivities are $2\frac{1}{2}$–4 orders of magnitude smaller than those of the pure components (Figure 23). Figure 24 applies to sintered polycrystalline specimens. Their B values (activation energies) also have a minimum in this range. They drop from 9000 for the pure oxides to 2000–4000 at the minimum, depending on the temperature (Figure 25). It has

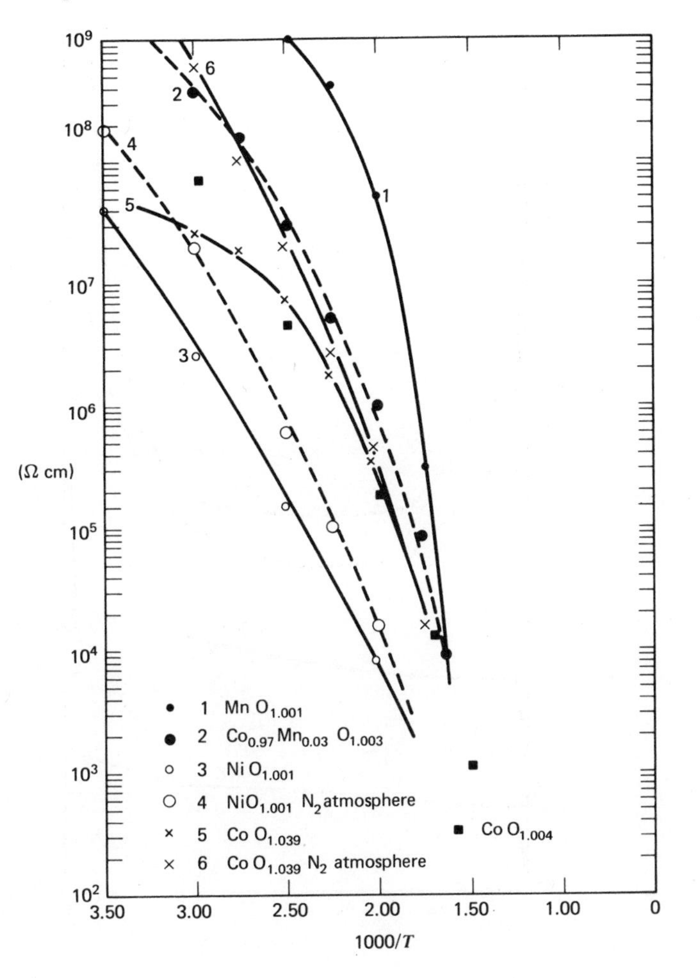

FIGURE 22. Resistivity of oxide single crystals with slight stoichiometric deviations as function of the temperature between 300 and ≈650 K.

been found that the resistivity of single crystals of NiO, CoO, and MnO even with a high degree of stoichiometry (oxygen-to-metal ratio <1.002) can still be increased by lowering the oxygen partial pressure in controlled atmospheres of N_2 or wet H_2, indicating further reduction of the acceptor concentration, which is determined by the O excess above the monoxide composition. In the system CoO-MnO only one composition with 3 at. % Mn has been measured with 3 milliatoms/mole O excess. Only below 330 K a slight resistivity decrease against CoO was observed.

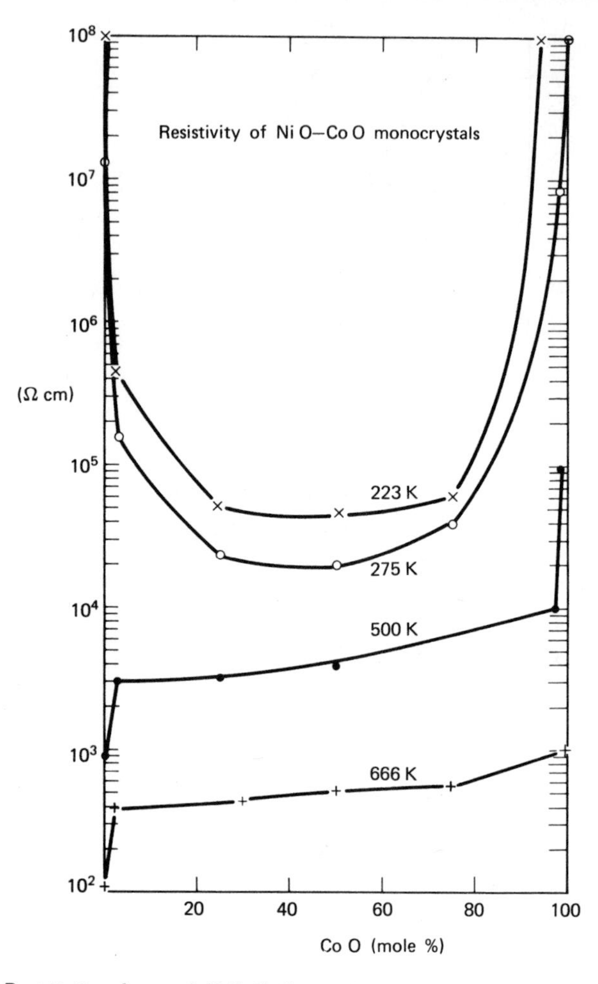

FIGURE 23. Resistivity of mixed NiO-CoO single crystals at temperatures between 223 and 666 K. Results of Smakula.[148]

Many data are available on polycrystalline binary and ternary oxide systems—in most cases not mixed crystals of monoxides—but spinel-type compositions. However, their value in absolute terms is often limited not only by their polycrystallinity with the inevitable effects of grain boundaries, but more so by inhomogeneity caused either by incomplete reaction or limited solid-state miscibility of the components (Section 3.2).

Grain boundaries can have a considerably lower resistivity than the crystallites.[149]

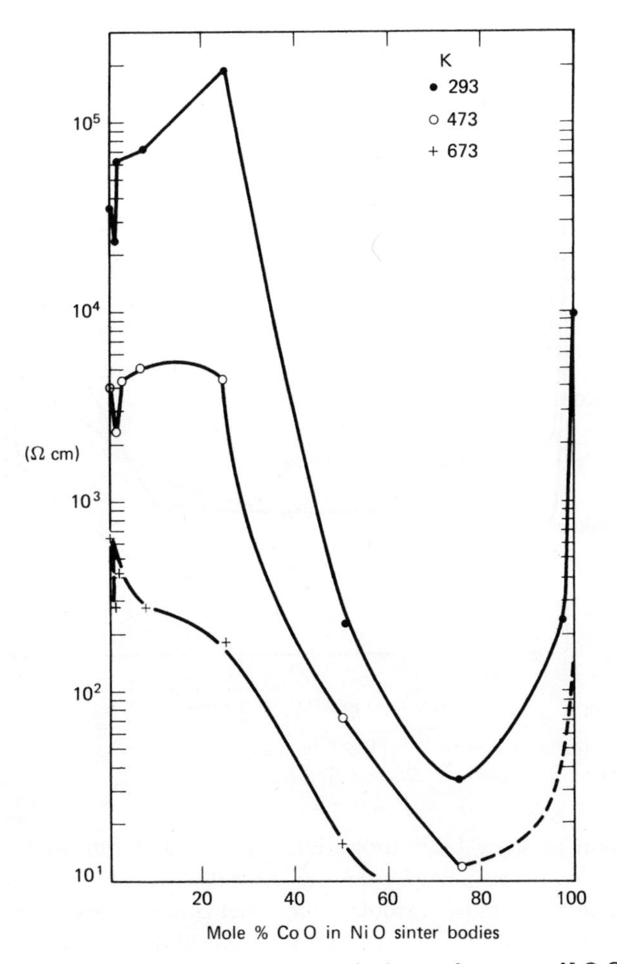

FIGURE 24. Resistivity of polycrystalline sinter bodies in the system NiO-CoO at temperatures between 293 and 673 K.

In model experiments, NiO bicrystals measured across or along the grain boundary were more or less conductive than each of the bulk crystals depending on temperature and O_2 partial pressure. The boundary region with different conductive properties can be rather wide. It is characterized by impurity segregation and increased boundary diffusion of defects. Theoretically, segregation of 50 ppm impurities could result in resistivities differing by 4 orders of magnitude.[150]

Similar effects can be expected in other single-component oxide systems. In multicomponent systems more complex effects must be expected.

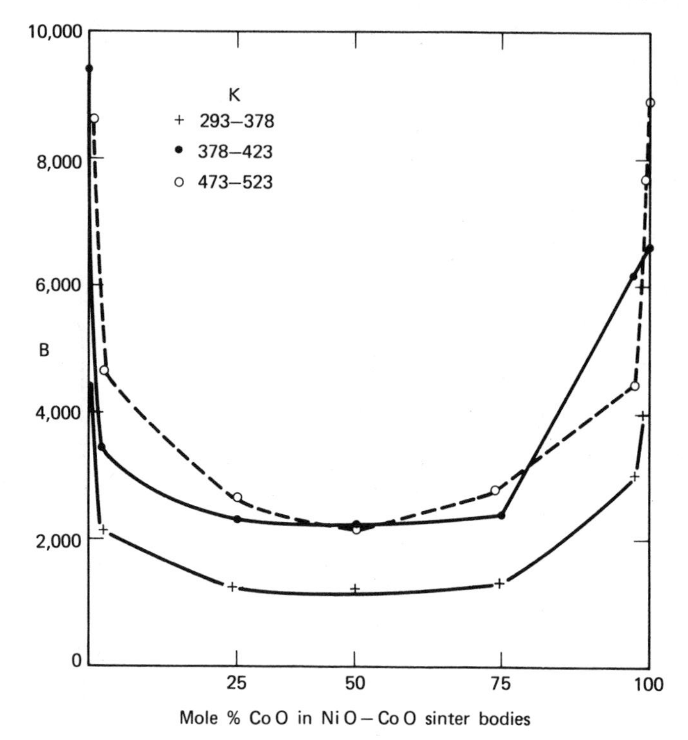

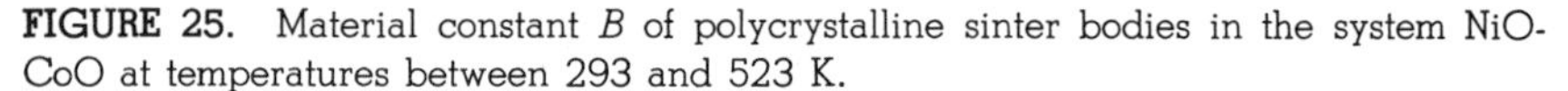

FIGURE 25. Material constant B of polycrystalline sinter bodies in the system NiO-CoO at temperatures between 293 and 523 K.

In addition to these basic limitations, the variable influence of presinter conditions (submicron particle size, molding density, etc.) would introduce other variables. Therefore it would be a rather dubious task to record in this book the resistivity data measured under these ill-defined conditions. However, a modest effort is made to present at least a few data on binary and ternary oxide combinations, where the secondary effects mentioned before are submerged into the larger effects of composition. Additionally a number of references are tabulated for various investigated oxide systems.

Binary Systems. NiO-CoO was coprecipitated as oxalate or hydroxide for maximal homogenization. After calcination at 800 K it was molded and sintered in air at 1270 K, then rapidly cooled at a rate of 100 K/min, to minimize reabsorption of oxygen from the air. Compared to the monocrystalline phase, all resistivities were smaller by orders of magnitude. It is probable that surface states at grain boundaries together with traces of alkali contributed to this drastic decrease. Also, the resistivity minimum was shifted from 50 to 75 mole % CoO. This effect, together with the relatively much lower resistivity for the pure CoO, is probably caused by readsorption of oxygen in the Co region. As shown in Chapter 3.2.1 CoO reabsorbs oxygen from air at ~1180 K (Figures 17 and 18).

Whenever Mn is a partner in a binary or ternary oxide system, its valence is always >2, at least for sinter temperatures <1500 K. In the temperature region for sintering commercial thermistors, spinel formation has to be assumed with the Mn valence ~3. Binary compositions containing less than the theoretical ratio Mn^{3+}/Me^{2+} ($Me^{2+} = Co^{2+}, Ni^{2+}, Zn^{2+}$) tend to form two phases: (Spinel + monoxide) (Section 3.3).

On the other hand, binary systems with CuO as partner tend to lose O_2 at temperatures >1270 K, resulting in the formation of Cu_2O as second phase.

The oxide systems Co-Mn and Cu-Mn in Figures 26 and 27 have been also prepared by coprecipitation of nitrates with alkali and subsequent

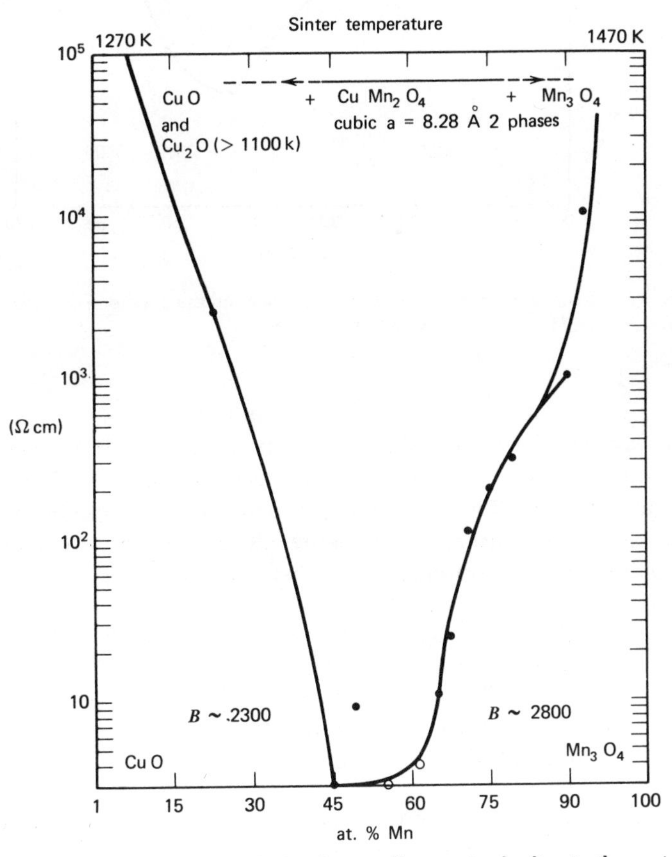

FIGURE 26. Resistivity at ~300 K of polycrystalline sinter bodies in the system CuO-Mn_3O_4. Results of Kolomiets.[151]

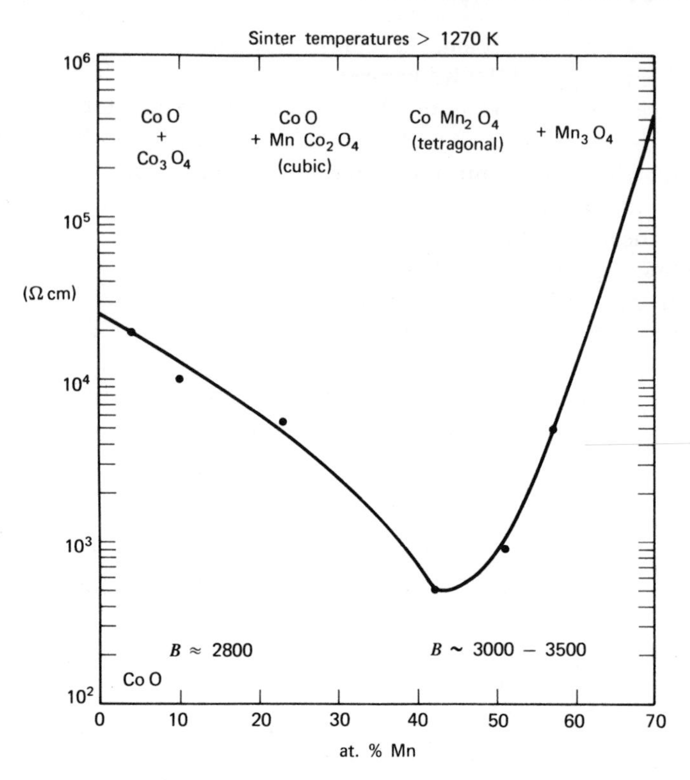

FIGURE 27. Resistivity of polycrystalline samples sintered at temperatures $\gtrsim$1270 K in the system CoO-Mn_3O_4. Results of Kolomiets.[151]

careful washing. They were sintered in air at 1270–1470 K depending on their composition (Cu-Mn systems require lower temperatures) and annealed 100 hr at 530 K.[151] The resistivities in ohm cm in a selected number of ternary systems are displayed in Figures 26–31 for sinter temperatures of 1400–1500 K in air or nitrogen atmosphere. These are

Mn_3O_4-CoO-ZnO	(Figures 28 (Air) and 29 (N_2)
Mn_3O_4-NiO-TiO_2	(Figure 30)
CoO-NiO-TiO_2	(Figure 31)
CoO-NiO-ZnO	(Figure 32)
Cr_2O_3-CdO-TiO_2	(Figure 33)

A careful study of the structural and electrical properties of the system NiO-Mn_2O_3 has been recently reported.[152]

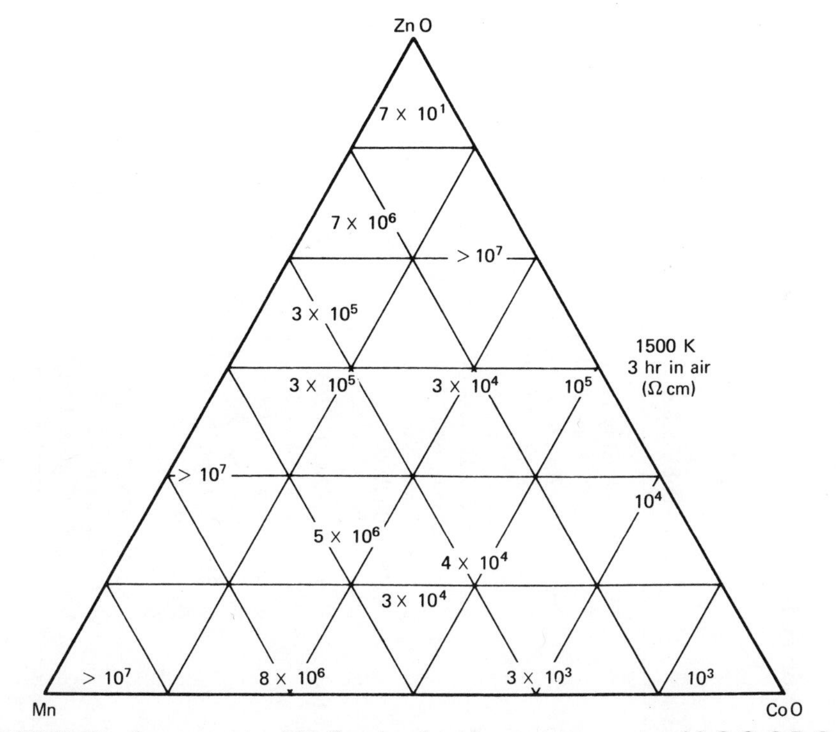

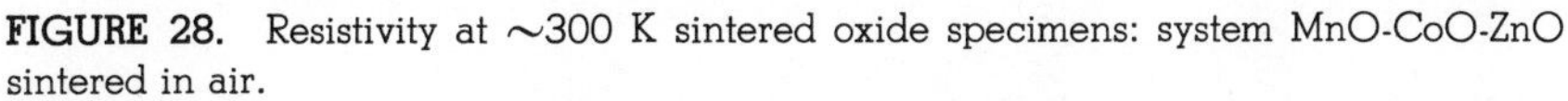
FIGURE 28. Resistivity at ~300 K sintered oxide specimens: system MnO-CoO-ZnO sintered in air.

Tables 15–18 give a survey of investigated systems with references. They deal mainly with transition metal oxides preferentially used in conventional thermistors. Other components are listed in columns A. Data on rare-earth oxides are given in Section 4.2. Not all the data in these references cover a wide scope of compositions of the listed oxide components Patent data are often especially scanty and restricted to optimal values.

3.5.2. Doping Effects on the Resistivity of Oxides

The dramatic effect of alkali doping on the resistivity of Cu and Ni oxides was discovered in the late thirties. During the development of Cu oxide thermistors striking differences in resistivity were found in spite of otherwise equal molding and sinter conditions and the use of "pure" Cu oxides with a minimum of other metal traces. Finally the large resistivity differences could be related to variable contents of alkali, preferentially Na and stimulated the production of a "standard" Cu oxide with a minimal content of sodium.

TABLE 15 Literature on the electrical resistivity of oxide systems.

Oxide System									Reference
Ti	Cr	Mn	Fe	Co	Ni	Cu	Zn	Addition	× = Publication + = Patent
		+			+	+			US 2,282,944, May 12, 1942
		+			+				US 2,419,537, April 29, 1947
	×		×				×	(Mg, Al)	*Philips Tech. Rev.*, **9,** 239–248 (1947)
		+	+					(Mg)	US 2,475,864, July 12, 1949
		×	×				×	(Mg)	*J. Chem. Soc.*, 1729–1741 (1948)
				×	×	×	×	(Mg)	*Compt. Rend.*, **233,** 736–738 (1951)
						×	×		*Z. phys. Chemie*, **198,** 30–40 (1951)
		×	×				×	(Mg)	*J. Rech. Centre Natl. Rech. Sci. Labs. Bellevue (Paris)*, **18,** 118–130 (1952)
+	+						+	(Pr,V,Al,Th)	F. 991,891, Oct. 11, 1951
×	×								*J. Electrochem. Soc. Japan*, **19,** 145–148 (1951)
×	×								*J. Electrochem. Soc. Japan*, **19,** 230–232 (1951)
		×		×	×				*J. Electrochem. Soc. Japan*, **19,** 295–296 (1951)
+			+						G. 815,062, Sept. 21, 1951
		×		×					*Oyo Butsuri (J. Appl. Phys. Japan)*, **21,** 312–314 (1952)
		×			×				*Oyo Butsari (J. Appl. Phys. Japan)*, **21,** 400–402 (1952)
+			+						US 2,616,859, Nov. 4, 1952
		×		×	×				*J. Coll. Arts and Sci. Chiba Univ.*, ***I*****,** 133–138 (1953)
		+	+	+	+	+	+		US 2,633,521, March 31, 1953

TABLE 16

Oxide System									Reference
Ti	Cr	Mn	Fe	Co	Ni	Cu	Zn	Addition	× = Publication + = Patent
		+	+		+				US 2,645,700, July 15, 1953
		+	+		+				US 2,694,050, Nov. 9, 1954
		+	+		+			AlSi	F 1,052,015, Jan. 20, 1954
		+							US 2,703,354, March 1, 1955
	+								US 2,700,720, Jan. 25, 1955
	+								US 2,714,054, July 26, 1955
+	+	+	+	+	+				US 2,720,471, Oct. 11, 1955
				×	×				*Doklady Lvov. Politekh. Inst.*, **1,** 13–18 (1955).
		×		×					*Nagoya Kogyo Gijutsu Shikensho Hokoku*, **5,** 613–619 (1956)
		×			×				*Trudy Odessk Gidrometorol. Inst.*, 45–46 (1956)
		+		+	+		+		Addn. to F. 1,052,015—Nr. 65,137, Jan. 26, 1956
+									Ind. 53,608, Jan. 18, 1956
		+		+	+			Al	F. 1,112,965, March 21, 1956
+									G. 1,029,450, May 8, 1958
				+		+		Li	F. 1,165,582, Oct. 27,1958
		×		×	×	×			*Oyo Butsuri* (*J. Appl. Phys. Japan*), **28,** 117–119 (1959)
					×	×		Li	*Bull. Inst. Politech. Bucuresti*, **21,** 73–86 (1959)

TABLE 17

Oxide System									Reference
Ti	Cr	Mn	Fe	Co	Ni	Cu	Zn	Addition	× = Publication + = Patent
+				+					U.S.S.R. 134,307, Dec. 25, 1960
×			×				×		*Zh. Priklad. Khim.*, **34,** 1880–1883 (1961) (Russ.)
	+	+	+	+	+				US 3,016,506, Jan. 9, 1962
×			×						*Hermsdorfer Techn. Mitt.*, **3,** 98–105 (1962)
		+		+					Pol. 48,183, March 18, 1963
		+		+	+	+			Br. 922,491, April 3, 1963
		×		×	×				*Compt. Rend. Acad. Bulgar. Sci.*, **18**(6), 525–528 (1965)
		×			×		×		*Compt. Rend. Acad. Bulgar. Sci.*, **18**(6), 525–528 (1965)
		×			×	×			*Silikaty*, **10**(1), 48–62 (1966) (Czech.)
		×		×	×	×			*Elektroprom. Priborst*, ***I***(4), 115–117 (1966) (Bulg.)
×						×			*Kogyo Kagaku Zasshi*, **70**(6), 844–861 (1967)
+								Sn/Sb	US 3,341,473, Sept. 12, 1967
+								Sn/Sc/Y/Zr	
								Sb/Ta/Mo	Br. 1,093,073, Nov. 29, 1967
		+	+	+				Mg	Br. 1,138,719, Oct. 2, 1967
		×			×	×			*Izv. Fiz. Inst. ANEB Bulg. Akad. Nauk*, **17,** 33–39 (1968)

TABLE 18

Oxide System									Reference
Ti	Cr	Mn	Fe	Co	Ni	Cu	Zn	Addition	× = Publication + = Patent
		+		+					US 3,408,311, Oct. 29, 1968
		+	+	+	+		+	Mg	Neth. 66,14,015, April 8, 1968
		+			+		+		US 3,430,336, March 3, 1969
+			+					Sr	US 3,511,786, May 12, 1970
				+	+				Pol. 60,173, June 30, 1970
			×		×		×		*Poluprovod. Tekh. Mikroelectron.*, 102–104 (1970) (Russ.)
+			+	+	+	+		Mo/W	Br. 1,224,422, March 10, 1971
		+		+			+		US 3,652,463, March 28, 1972

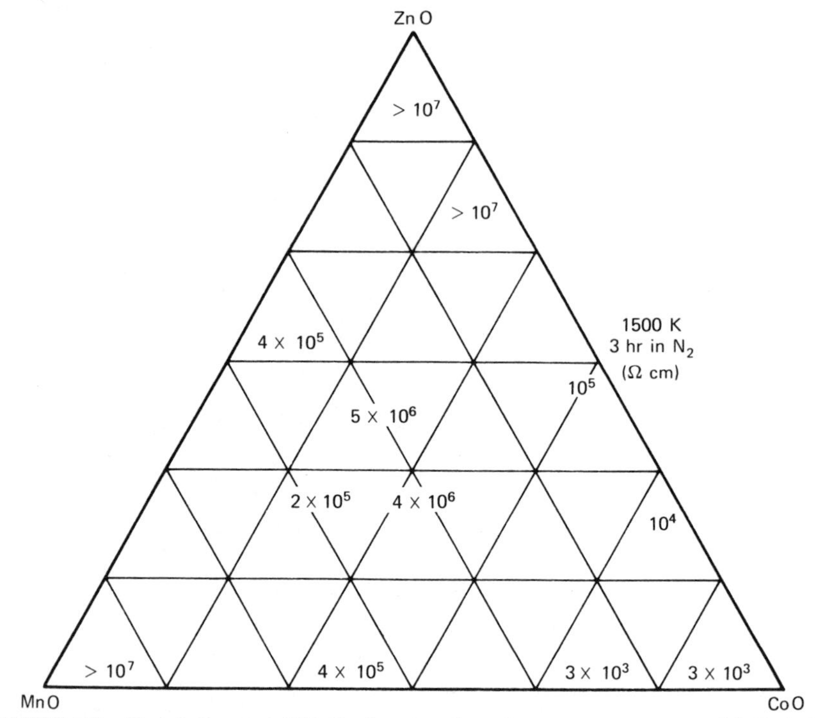

FIGURE 29. Resistivity at ~300 K of sintered oxide specimens: system MnO-CoO-ZnO sintered in N_2. Figures 28 and 29: Figures in triangles represent Ω cm measured at 300 K. MnO present as Mn_2O_3 in spinel phase.

The ubiquitous presence of alkali in commercial otherwise pure Cu oxides was a direct consequence of the usual precipitation process with caustic alkali or soda from Cu^{2+} solutions. Recognition of the relationship between alkali content and resistivity not only led to more uniform CuO raw material, but also offered a possibility to control the resistive properties of the sintered oxide in a desired manner, a method which was practiced on a large scale during the years of war-time production of thermistors, using Li or Na as dopants. It was quite natural to apply the same method to Ni and Co oxides, though only on a smaller scale. The principle of valence-controlled resistivity was born—though never published during the time of war and final holocaust. Some information seeped into the files of CIA and FIAT, the agencies involved in taking inventory of German technical and scientific achievement at the end of WWII.

Verwey et al.[153] reported studies of the resistivity in mixed oxide systems and formulated[154] the principle of induced-valence semiconductivity with the well-known classical example of Li-doped NiO.

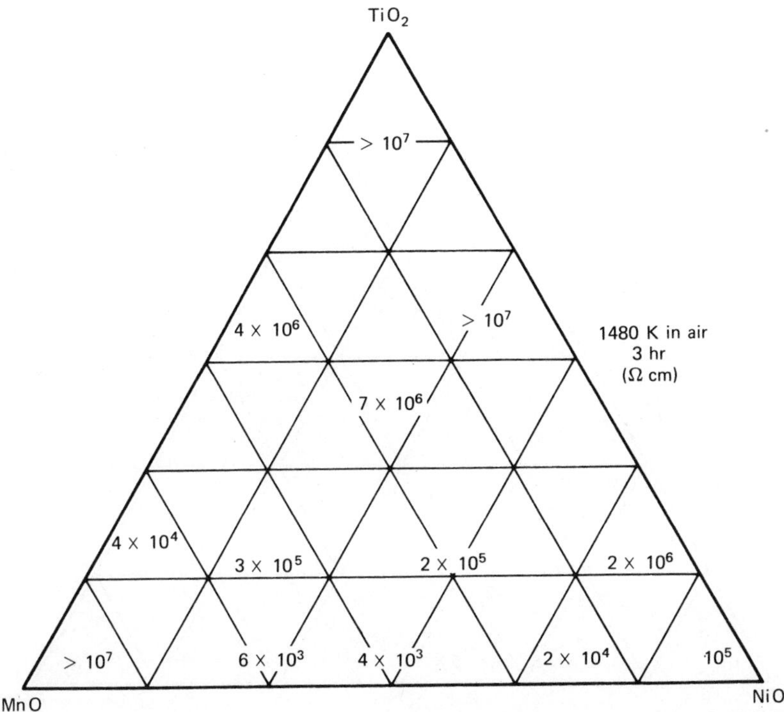

FIGURE 30. Resistivity at ~300 K of sintered oxide specimens: system MnO NiO-TiO_2 sintered in air. Figures in triangles represent Ω cm measured at 300 K. MnO present as Mn_2O_3 in spinel phase.

Verwey's work has stimulated other systematic investigations on this topic by the Westinghouse group of Heikes, Johnston, Miller, and Sestrich.[155–158,112,159,160] According to Verwey the substitution of Li^+ into a divalent transition metal oxide such as NiO leads to the formation of triply charged transition metal ions. Assuming that oxygen in this lattice retains its double negative charge, and Li its monovalent positive charge (because of the high value of the second ionization potential of Li, about 76 eV), the formation of triply charged transition metal ions is necessary to conserve electroneutrality. This concept is shown in the defect models in Figures 34(*a*) and 34(*b*). The opposite effect of trivalent ions is shown in Figure 35.

The Ni^{3+} ions act as acceptors resulting in holes and *p*-conduction and the conduction process would be a transfer of electrons from Ni^{2+} to Ni^{3+} ions or the movement of holes in the opposite direction.

Heikes and Johnston felt that this model was not adequate for explaining the experimental facts. They treated the conduction process as thermally activated diffusion of positive holes that are trapped by the local strain caused

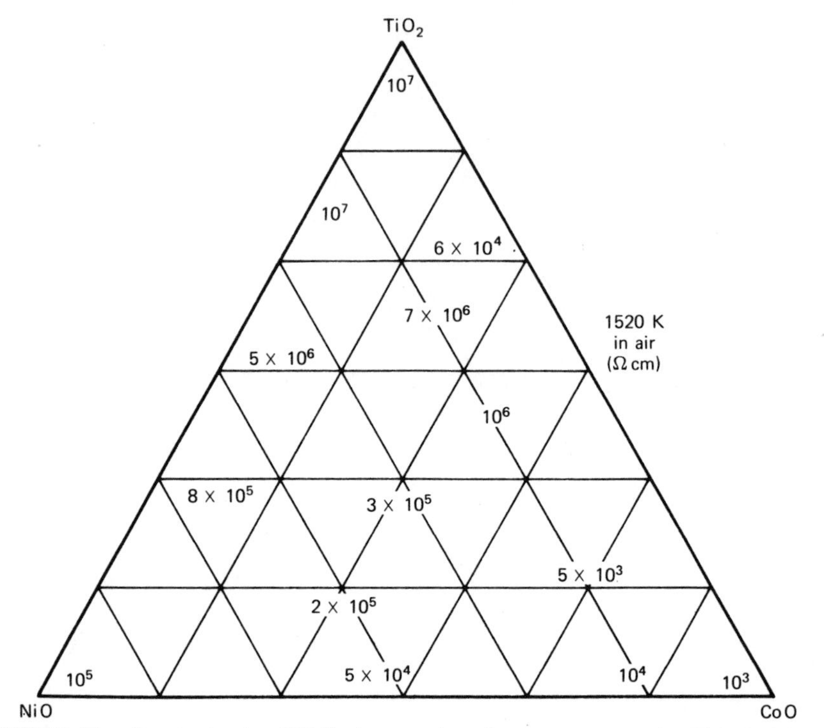

FIGURE 31. Resistivity at ~300 K of sintered oxide specimens: system NiO-CoO-TiO_2 sintered in air.

by their own polarization field (polarons). The temperature dependence of the conductivity is determined by an activation energy of this diffusion, which corresponds to the energy necessary to replace the surrounding ions into unstrained positions, as they exist around the divalent ions. The carrier (hole) concentration does not change with temperature. Their experimental results indicate that the initial decrease in resistivity by Li doping levels off above 2 at. % Li for the oxides MnO, NiO, CoO, and CuO. This applies also to the activation energy, which tends, however, to increase again above 10%. With increasing atomic number of the transition element between Mn and Cu, this activation energy was found to decrease in a linear manner for 2 at. % doping.

New experimental data, such as conductivity and Hall-effect measurements by Zhuze and Shelykh[161] and extended by thermoelectric measurements by Kshehdzov, Ansel'm, Vasil'eva, and Latysheva[162] changed this concept completely and forced investigators to abandon the conduction theory by thermally activated polarons, at least as far as NiO is concerned.

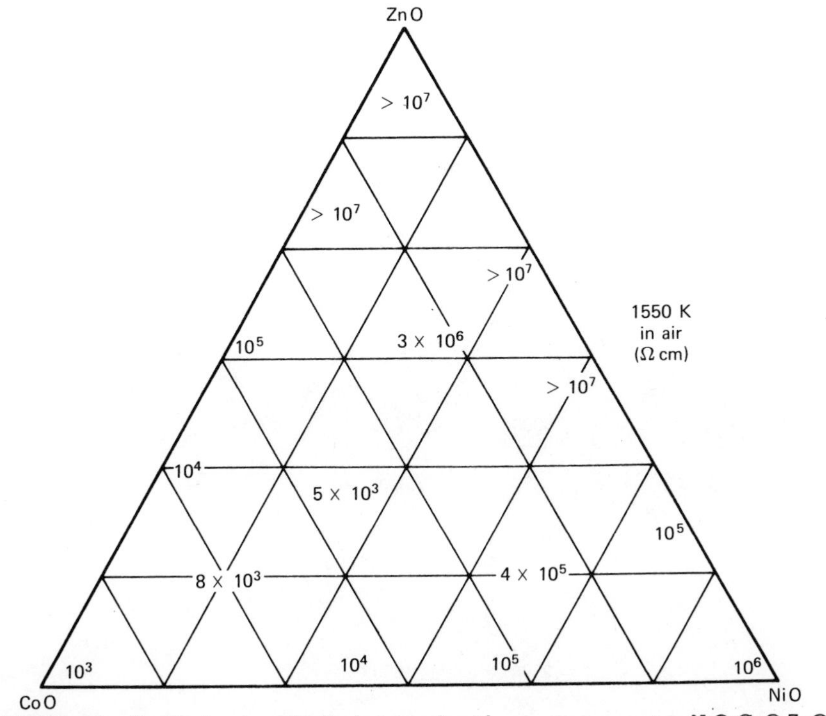

FIGURE 32. Resistivity at ~300 K of sintered oxide specimens: system NiO-CoO-ZnO sintered in air.

For undoped NiO single crystals and polycrystalline NiO doped with 3.4 at. % Li sintered 6 hr in air at 1370–1570 K, they found an increase of the carrier density and decrease of the hole mobility, contrary to the earlier concept of a temperature-activated diffusion-type polaron mobility. The radical difference between these two viewpoints stimulated more investigations on the electronic transport phenomena in NiO and CoO. Austin, Springthorpe, Smith, and Turner[163] flatly conclude that the hopping model is not applicable and prefer the concept of a narrow-band polaron conduction assuming a polaron size that is small comparec to the lattice spacing and warranted by the large coupling constants $Fm^{*-1/2}$ of 1.7 (NiO) and 2.0 (CoO) according to Fröhlich.

$$F = \frac{e^2}{h}\left(\frac{m_0}{2h\omega_L}\right)^{1/2}\frac{\epsilon - n^2}{\epsilon n^2}(m^*)^{1/2}$$

There are certain limitations to be considered for the small-polaron band model, as pointed out by Lang and Firsov.[164] Van Daal and Bosman[165]

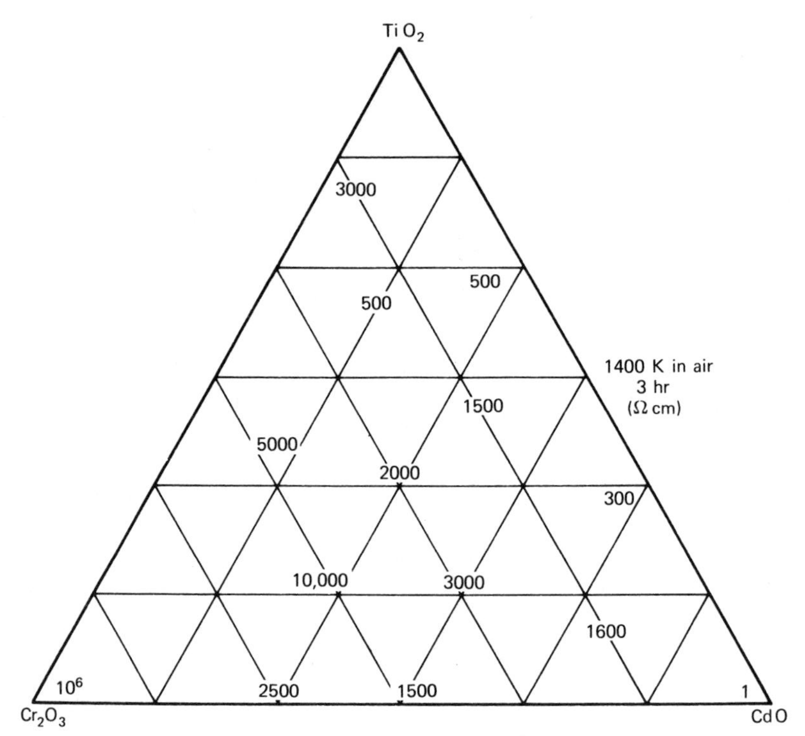

FIGURE 33. Resistivity at ~300 K of sintered oxide specimens· system Cr_2O_3-CdO-TiO_2 sintered in air.

Ni^{2+}	O^{2-}	Ni^{2+}	O^{2-}	Li^+	O^{2-}	Ni^{2+}	O^{2-}
O^{2-}	□	O^{2-}	Ni^{2+}	O^{2-}	*Ni^{3+}*	O^{2-}	Ni^{2+}
Ni^{2+}	O^{2-}	Ni^{2+}	O^{2-}	*Ni^{3+}*	O^{-2}	Ni^{2+}	O^{2-}
O^{2-}	Ni^{2+}	O^{2-}	*Li^+*	O^{-2}	Ni^{2+}	O^{2-}	Ni^{2+}
Ni^{2+}	O^{2-}	□	O^{2-}	Ni^{2+}	O^{2-}	Ni^{2+}	O^{2-}

(*a*)

Ni^{2+}	O^{2-}	Ga^{3+}	O^{2-}	Ni^{2+}	O^{2-}	Ni^{2+}	O^{2-}
O^{2-}	Ni^+	O^{2-}	Ni^{2+}	O^{2-}	Ni^{2+}	O^{2-}	Ni^{2+}
Ni^{2+}	O^{2-}	Ni^{2+}	O^{2-}	Ni^{2+}	O^{2-}	Ni^{2+}	O^{2-}
O^{2-}	Ni^{2+}	O^{2-}	Ni^+	O^{2-}	Ga^{3+}	O^{2-}	Ni^{2+}
Ni^{2+}	O^{2-}	Ni^{2+}	O^{2-}	Ga^{3+}	O^{2-}	Ni^+	O^{2-}

(*b*)

FIGURE 34. NiO with Ni^{2+} vacancies has an equivalent concentration of Ni^{3+} to establish electroneutrality resulting in *p* conduction. (a) Li^+ doping increases Ni^{3+} concentration, thus increasing *p* conductivity. (b) Cr^{3+} (or Ga^{3+}) doping decreases Ni^{3+} concentration with resulting decrease in *p* conductivity. Each Ni^+ eliminates one Ni^{3+} corresponding to one hole.

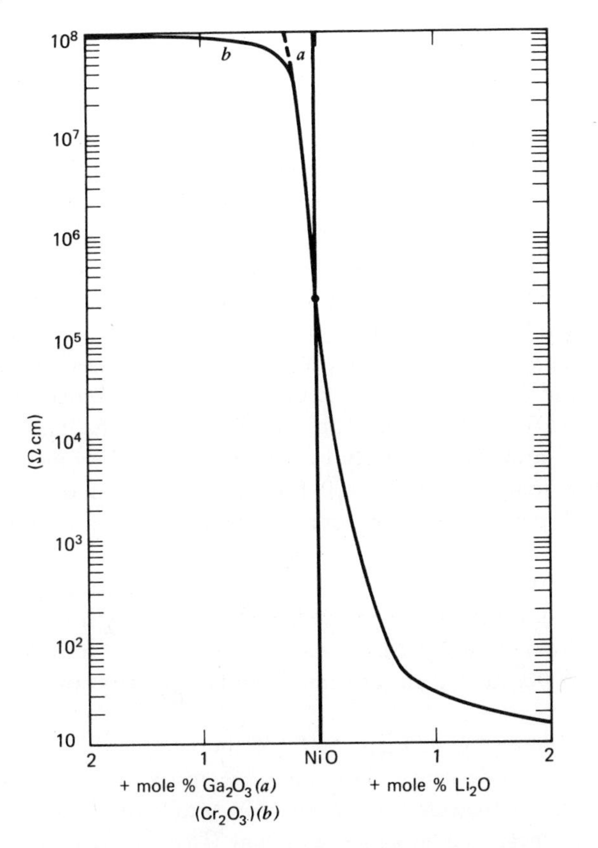

FIGURE 35. Resistivity at ~300 K of NiO doped with Ga^{3+} or Li^{+}. After Hauffe.[105]

follow this line, but also point out that interaction between charge carriers and the applied magnetic field in the Hall mobility measurement can result in anomalies of the Hall effect, especially for NiO and Fe_2O_3. The magnetization induced by the magnetic field could influence the drift current.

Maranzana has presented a theory of such interactions for ferromagnetics and antiferromagnetics that in several points is in agreement with the Hall mobility data for NiO below and above the Neel temperature T_N and for Fe_2O_3 above T_N.[166] The influence of inversion layers at grain boundaries is discounted by the fact that single crystals and ceramic samples give identical results for NiO.[167]

Resistivity at Temperatures below 200 K. All the preceding discussions are based on measurements at temperatures below and above T_N, but never extending below 150 K. An intensive theoretical and experimental study[168] of undoped CoO_{1+x} with stoichiometric deviations x ranging from 0 to 0.013, though mainly dealing with the resistivity as function of temperature between 1173 and 1723 K at O_2 partial pressure from 10^{-12} to 1 atm, was also extended to temperatures between 200 and 250 K. It was found that the activation energy decreased from 0.51 eV for $x = 10^{-3}$ to 0.22 eV for $x = 0.013$.

For temperatures below 150 K a decrease of the slope of log ρ versus $1/T$ for NiO was first observed by Morin[169] and confirmed by others.

At first, grain boundary conduction was considered as responsible for this effect, but later was rejected by Springthorpe et al.[170] based on the resistivity if Li-doped and NiO crystals down to 20 K. The slope decreases from 0.2 above 200 K to 0.007 eV at 20 K and still further at 4 K. The influence of x between 0.002 and 0.032, clearly evident at 200 K, becomes negligible at 25 K. At low temperatures (<100 K) impurity conduction is apparently prevalent with possible compensation effects by impurities acting as donors. This has been confirmed over the years by experience in making oxide thermistors for the liquid He and H_2 range on transition metal basis. Compensation effects at high temperature have been directly measured by chemical determination of the Ni^{3+}/Li^{+} ratio, which decreased from 1 for 0.01 at.% Li to 0.82 for 0.3 at. %, in agreement with resistivity and Seebeck effect data.[159]

Heikes and Johnston in their study of the conduction in Li-substituted transition metal oxides have shown a large resistivity drop together with reduction of the activation energy also for MnO. More recent Hall-effect and resistivity measurements of the system ($Li_{0.001}$, $Mn_{0.999}$) O by Nagels and Denayer[171] at temperatures up to 1073 K are not in agreement with a conduction model by hopping of small polarons,[162] and they assume that the conductivity increase with temperature is caused by increasing hole concentration. The activation energy to free a hole from the Ni^{3+}-Li^{+} acceptor center is 0.34 eV. The hole mobility of $5.5.10^{-3}$ cm^2/V sec. is smaller than for the (Li_x $Co_{(1-x)}$) O system with equal x. Only 25% of the acceptor centers are ionized at 800 K, not 50% as expected from $n_h = Na \exp(-0.37/RT)$. This small discrepancy might be caused by partial compensation.

In all previously mentioned investigations, the doping concentration was the dominant factor—the partial pressure of oxygen being considered equal. A fuller understanding of these oxide systems requires the thermodynamical treatment of the defect equilibria caused either by controlled valence doping or by stoichiometric oxygen deviations. A basic theoretic and experimental study of this type was already mentioned dealing with polycrystalline CoO_{1+x} at pressures between 1 and 10^{-12} and in the temperature range

1175–1475 K which would include the usual sinter temperatures for thermistors. The resistivity at temperatures between 1259 and 1620 K decreases linearly with the excess of oxygen in CoO produced by increasing O partial pressure. This work was extended by Fisher and Wagner[172] to Li-doped CoO single crystals and led to the conclusion, that for low doped crystals (0.04 milli at./mole) at high O_2 pressure the excess oxygen is nearly a linear function of $p^{1/4}$ for most of the investigated temperatures from 1160 to 1460 K, confirming the concept that singly ionized Co vacancies dominate the defect equilibrium.

$$[V_M'] = -\tfrac{1}{2}[F_M'] + \{\tfrac{1}{4}[F_M']^2 + K_x K_7 P(O_2)^{1/2}\}^{1/2},$$

(V_M') = mole fraction of singly ionized Co vacancies
(F_M') = mole fraction of ionized foreign atoms such as Li
p = number of holes per molecule: $p = [V_M'] + [F_M']$.
K_7 and K_x = equilibrium constants for the equilibria

$$p[V_M']/[V_M] = K_7 \qquad [V_M] = K_x P(O_2)^{1/2},$$

$[F_M]$ = total concentration of foreign (Li) atoms = $[F_M']$ at high temperatures.

At low oxygen pressure the x *vs* $P^{1/4}$ curves for heavily doped CoO (1.44 milli at./mole) seems to indicate oxygen deficiencies and the coexistence of ionized Co interstitials and free electrons. This is also supported by the shallow maximum of the Seebeck coefficient with a negative slope starting at $\sim 10^{-6}$ atm.

Fisher and Wagner[173] also discussed the problem of doping semiconducting oxides and the resulting electrical properties at low ($<10^{-2}$ atm) and high ($>10^{-2}$ atm O_2) pressures for Li-doped NiO at Li concentrations <1 at. %. Their considerations shed new light on previous work by Bosman, van Daal, and Knuvers[174]; however, it would have only minor importance for doped thermistors of this type, which are normally sintered in the O_2 high pressure range. The influence of Al_2O_3 or SiO_2 additions on the nonstoichiometry of CoO at 1425 K and 1 atom O_2 can be estimated from tensivolumetric measurements. The fraction of doubly ionized Co vacancies and the equivalent number of holes decreases from 10^{-2} for CoO to 10^{-3} for $CoAl_2O_4$ and 10^{-5} for Co_2SiO_2.[175] This effect, resulting in a resistivity increase, could be important for thermistors with a high fraction of CoO.

Doping of Iron Oxide. Iron oxides have always been attractive as raw material for thermistors because of their low price. However, the pure commercially available oxides Fe_2O_3 and Fe_3O_4 have some shortcomings: (1) The resistivity of pure Fe_2O_3 at room temperature exceeds 10^{12} Ω cm and has an activation energy of about 1 eV. (2) The resistivity of Fe_3O_4, being much smaller, has also a very small temperature dependence. It is semiconductive up to 250 K with a resistivity drop by a factor 10 at 119 K. Within the practical application range above room temperature its resistivity is practically constant, about 1.10^{-3} Ω cm. Furthermore, Fe_3O_4 tends to oxidize in air to Fe_2O_3 with corresponding change of its electrical properties.

There are two ways to reduce the resistivity of Fe_2O_3: (1) partial reduction, and (2) doping to induce controlled valence.

X-ray studies by Barth and Posnjak[176] have shown that the well-known mineral Ilmenite represents an isomorphous compound $FeTiO_3$ with Ti^{4+} and Fe^{2+} cations at room temperature, the latter acting as donor centers. The resistivity of Fe_2O_3 as function of the Ti content drops drastically by more than 5 orders of magnitude from "pure" Fe_2O_3 by doping with 1 at. % Ti. Ti-doped Fe_2O_3 can be represented by the general formula

$$(Ti_x^{4+}\ Fe_x^{2+}\ Fe_{1-2x}^{3+})\ O_3^{2-}$$

The presence of Fe in two valence states at crystallographically identical lattice points is causing the electronic conductivity in the *d*-levels of iron. The ρ versus $1000/T$ slope and therefore the activation energy decrease with increased Ti doping (Figure 36).[177] This effect has also been used by Philips to make patent claims for the production of valence-controlled Fe_2O_3 thermistors (U.S. 2,735,824, Feb. 21, 1956). Mg doping results in *p*-conduction by formation of acceptor centers. The variable sign of the Seebeck effects in undoped Fe_2O_3 emphasizes the amphoteric character of its conduction mechanism, which depends on the relative contributions of donors and acceptors.

For thermistor applications only heavily doped Fe_2O_3 is of interest, which simplifies the picture.[153] Wagner and Koch[178] found that the conductivity of Fe_2O_3 at 1275 K was independent of the O_2 pressure between $3.6.10^{-3}$ and $2.1.10^{-1}$ atm., suggesting intrinsic semiconduction. This implies, indeed, amphoteric behavior with respect to doping with cations of higher (Ti^{4+}) or lower (Ni^{2+}, Mg^{2+}) valence. (Doping of earthalkali titanate to make PTC thermistors is treated in Section 2.2.)

Valence-controlled resistivity has been investigated[179,180] with other oxides, using dopants that either decreased or increased resistivity. For a clear interpretation of the doping effects, the solubility limits of the dopant in the host lattice had to be known.

System TiO_2 + WO_3 (≦0.7 mole %). W on Ti sites acts as electron donor according to the equation

$$WO_3 = W^{2+}\ (Ti) + 2e + TiO_2 + \tfrac{1}{2}\ O_2{}^{g}$$

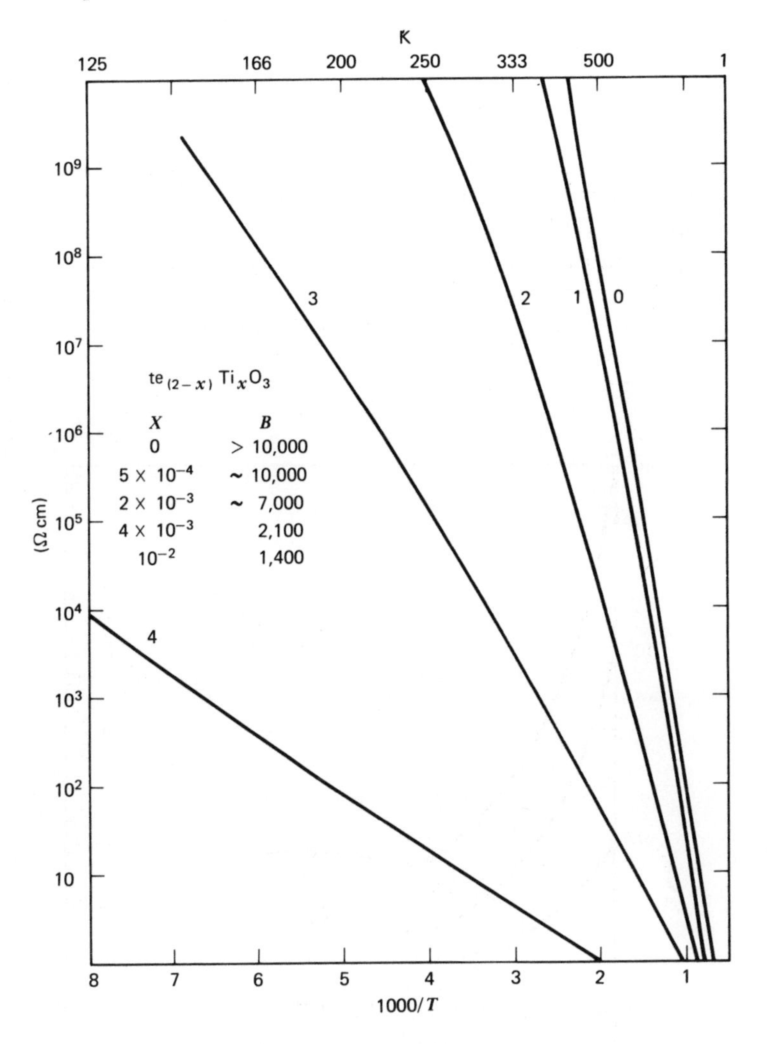

FIGURE 36. Resistivity versus temperature between 140 and 1000 K of sintered Fe_2O_3 doped with Ti^{4+}. Results of Morin.[177]

Figure 36 shows the resistivity ratio of doped and undoped TiO_2 for different temperatures over the usable doping range.

System $TiO_2 + Ga_2O_3$ *(*Al_2O_3*) (*≤ 2 *mole %).* According to the following equation a reduction of the electron concentration would result:

$$\tfrac{1}{2}O_2^{g} + 2e + Ga_2O_3 = 2\,Ga^{1-}\,(Ti) + 2TiO_2$$

The observed resistivity increase is less striking than the decrease with W doping. However, other factors, for instance ionic contributions from anion mobility, could obscure the controlled-valency effect.

System TiO_2-Cr_2O_3 ($\leqq$2 mole %). The substitution of Cr on Ti sites should decrease resistivity by consumption of electrons. This happens indeed for temperatures >1100 K (Fig. 37).

For lower temperatures, however, the resistivity is drastically increased by doping with Cr, which obviously acts in this case as ion with higher valence (Figure 37). This effect is dependent on the O_2 pressure.

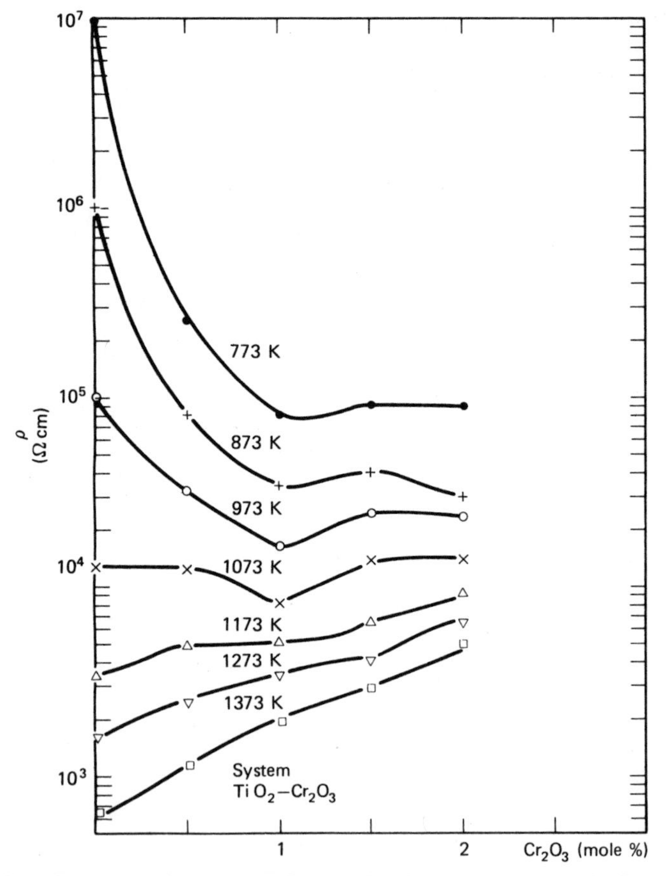

FIGURE 37. Resistivity of sintered TiO_2 doped with up to 2 mole % Cr_2O_3 at temperatures between 773 and 1373 K. Results of Grunewald.[180]

At $\sim\frac{1}{5}$ atm O_2, the Cr ions have a higher valence than 3. Quantitative relations for the equilibrium of Cr^{3+} and Cr^{3+x} ions in the TiO_2 host lattice have been found by measuring the electronic conductivity of TiO_2 specimens doped with 0.5 and 1 mole % Cr_2O_3 for O_2 partial pressures ranging from 10^{-6} to more than 100 torr. Plotting these data together with the conductivity of pure TiO_2 for the same temperature and pressure ranges permits one to read out equilibria conditions for which the doping with Cr_2O_3 either increases or decreases the conductivity. The line connecting the intersections of the curves for pure and doped TiO_2 clearly indicates that with increasing temperature, higher O_2 pressures are necessary to reduce the resistivity by Cr doping.[180]

3.5.3. Resistance and Capacitance under AC Conditions (10^2–10^{10} Hz) (Dielectric Properties)

The resistivity of nearly perfect single crystals exhibits a certain intrinsic dispersion with increasing frequency, more so for imperfect crystals and polycrystalline materials. In sintered oxides, conditions are still worse. Due to incomplete reaction or multiple phases even in equilibrium, zonal resistivity differences could exist owing to a system of grains in a matrix with different resistivity or surrounded by boundary layers. The case of well-conducting grains separated by layers with lower conductivity has been treated by Koops[181] with $Ni_{0.4}Zn_{0.6}Fe_2O_3$ as an example. He found a strong dispersion starting at 100 Hz, in good agreement with the classical Wagner[182] model of poorly conducting layers separating the conductive grains. Whenever such a dispersion curve is found in sintered semiconductors, the same model must be adopted (and has been done successfully for PTC materials). Its equivalent circuit is a resistance and a capacitance in series. The resistivity drop between 10^2 and 10^5 Hz can be very large, depending on the thickness of the interlayer and their dielectric constant.

The observed dispersion in this frequency range is "nonintrinsic," but is caused by the effect of frequency on the impedance of the resistance–capacitance network present in sintered materials. The intrinsic dispersion is completely submerged into these effects and would surface only at frequencies greater than 1 MHz at which the reactance of the interface layer becomes negligible compared to the bulk resistivity of the grains. The dielectric constant of the bulk material also contributes to the impedance of a semiconducting monocrystal or sinter body and determines its capacitance. Application of thermistors in hf applications has stimulated interest in their capacitance, and its dependence on temperature and frequency.

Contrary to Koop's results, recent measurements of the hf resistivity of the ferrite $Co_{3-x}Fe_xO_4$ indicated a smaller dispersion between dc and 10^2 Hz.

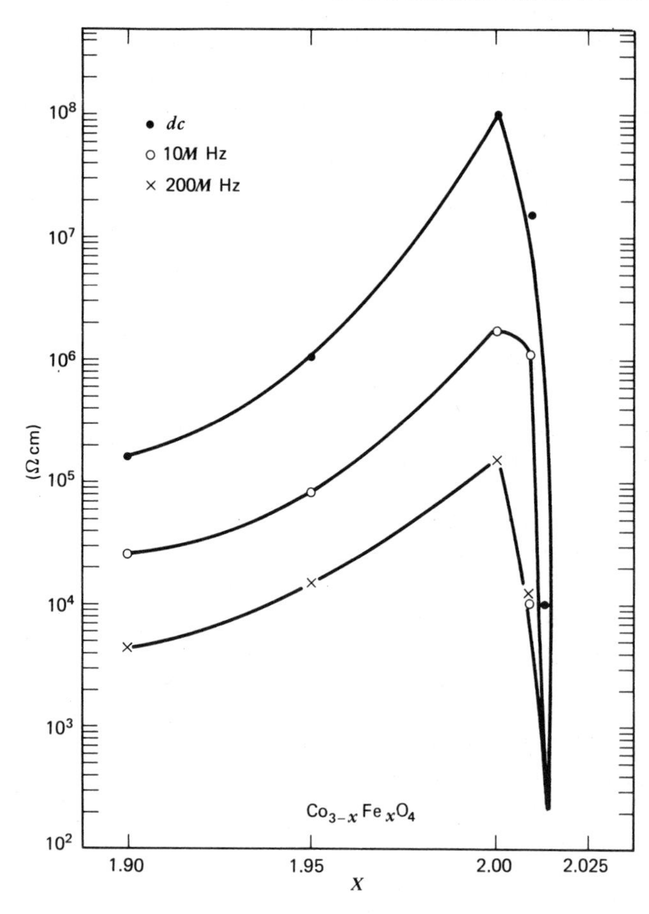

FIGURE 38. Dispersion of the resistivity between dc and 200 MHz for $CoFe_2O_4$ with small variations of the Co/Fe ratio. Results of Yamazaki.[183]

For $x = 2.03$–2.08 it has a minimum (Figure 38). Small deviations from the stoichiometric composition less than 0.01 result in a much larger dispersion between 10^3 and 10^8 Hz. Above 200 K the temperature dependence of resistivity is the same from dc to 2.10^8 Hz. Below this temperature it decreases rapidly between dc and 7.10^7 Hz.[183]

Data on dielectric constants of transition metal oxides and their mixed systems (spinels) are scarce. The dielectric constants of CoO, NiO, and 50 CoO · 50 NiO single crystals at 300 K are given in Table 19.

With increasing temperature all ϵ values increase exponentially with temperature with activation energies of 0.33, 0.20, and 0.16 eV for CoO,

TABLE 19

Hz	100 CoO	75 CoO · 25 NiO	50 CoO · 50 NiO	25 CoO · 75 NiO	100 NiO
10^2	12.9		200		18
10^6	12.9	40.2	45.6	38.1	11.9
10^8	12.9	15.3	16.5	15.1	11.9

CoO–NiO, and NiO at >313 K. This increase is also strongly frequency dependent, being steeper at lower frequencies. Below 313 K a frequency-independent region exists with activation energies of 42, 27, and 16 meV.[184]

Similar relations exist for the systems MnO-CoO and MnO-NiO as single crystals. At ~300 K, MnO has a dielectric constant of 18.1, practically frequency independent between 10^2 and 10^7 Hz. In the same temperature region, solid solutions have ϵ values between those of the components.[185] However, at 350 K a strong dispersion starts, resulting in ϵ values of >100 for equimolar ratios at frequencies $>10^3$ Hz and increasing to over 300 at 430 K.

Even for frequencies up to 10^6 Hz the dispersion at temperatures over 350 K remains rather large. Since this is the temperature range of practical thermistor applications, the capacitance effect is not negligible.

Iron-containing semiconductor compositions such as Mn, Ni, or Zn ferrites and TiO_2-doped Fe_2O_3 are typical representatives of the Wagner–Koops model quoted above. In these cases dielectric constants up to 10^5 at 5 KHz have been observed. This also applies to NiO doped with 10^{-2} at. % Li.[186] Most interesting is the fact that even in Ni and Mn ferrite single crystals, supposed to be without boundary layers, ϵ values up to 10^6 were found.[187]

Small dispersion has been found in single-component hot sintered oxides such as ZrO_2. At 1270 K the dispersion between 2 and 20 KHz was only ~5% for a resistivity varying between 10^4 and 10^5 Ω cm with the O_2 partial pressure. Sintered pure Ni and Co oxides have also small dispersion over a wide frequency range.

The dispersion of ZrO_2 doped with trivalent rare-earth ions would be of interest, since it has a potential as thermistor material for temperatures over 1200 K (see Section 4.2.1) and it was found that its ionic contribution to the total conduction can be less than 1% at 1270 K. (For lower temperatures and higher O_2 pressures the conduction mechanism is more complex).[188–190]

If thermistors are to be used for continuously variable amplitude regulation of rf circuits, their dispersion in this frequency range is important especially if the application is more demanding than amplitude control "by ear" in radio receivers. For precision rf test equipment such as impedance bridges Co oxide thermistors containing up to 10 at. % NiO have shown great promise not only in sufficiently small dispersion, but also in their excellent stability after sealing in glass envelopes. They have been used in rf bridges permitting a resistance ratio of ~600 for a regulating power input of ~2 W applied by self-heating with an af current.

3.5.4. Electric-Field Dependence

When a voltage is applied to a thermistor, the resistance can decrease not only by self-heating, but also by true field dependence. Such effects can play a role for high-resistance units or in application with voltage pulses such as in surge protection and time delay.

If the composition or the sinter process favors the formation of blocking layers between the conductive grains, in the first case by addition of glass or insulating components and in the second case by insufficient sintering, then voltage dependence can be expected already at relatively low fields. For pure and well-sintered Ni-Mn-spinel $Ni_{0.6}Mn_{2.4}O_4$, the very small frequency dispersion of resistance and capacitance up to 50 KHz indicated the absence of grain-boundary blocking layers. Therefore a true electric-field dependence of the resistivity could be determined by using 25-nsec-rise time pulses. It

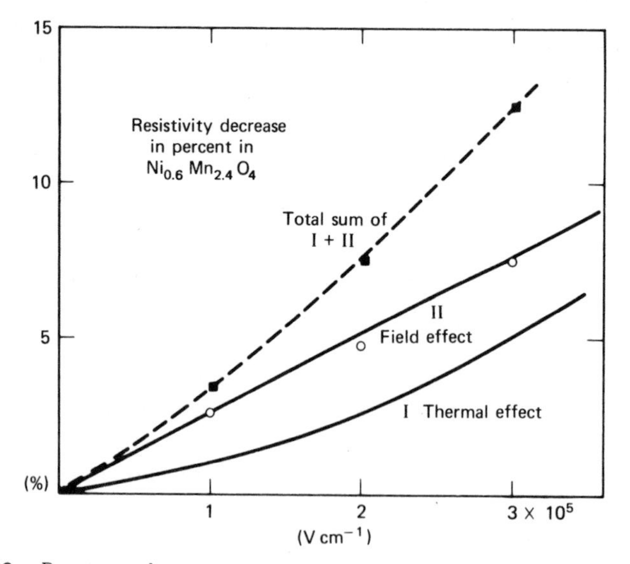

FIGURE 39. Resistivity decrease in percent of $Ni_{0.6}Mn_{2.4}O_4$. After Freud.[191]

decreased linearly with the field up to 3.10^5 V cm^{-1} with a slope of 2.4×10^{-5} % V^{-1} (Figure 39).[191]

3.5.5. Mechanical Properties

It is taken for granted that semiconducting monocrystalline or polycrystalline sensor materials have sufficient mechanical strength to undergo the necessary production steps, such as contacting, measuring, and encapsulation. More critical is the pulling or shearing strength of attached wire contacts, which often are incorporated into specifications (usually 1.3–2.7 kg = 3–6 lb). Furthermore, stability under vibration and acceleration conditions existing in air and space craft have been made mandatory for certain types. The bulk strength and cohesion of sintered oxides is normally very good. Failures can be caused by microcracks, either originating from molding or resulting from heat stresses after sintering. Data on the bulk elastic properties of oxides are scarce. Some recent measurements with pressure-sintered pure and Li-doped NiO indicate a striking increase of Young's modulus between 300 and 540 K, where a sharp maximum is reached at the Néel temperature. This increase corresponds to a factor of about 5 in both cases. Up to 8 at. % Li, the modulus increases below 350 K by a factor about 2, but decreases at higher temperatures (over 400 K).[192] The built-in stresses acting across contact interfaces are related to the bonding strength in metal-semiconductor contacts. They are of the order of 10^7–10^9 dyn cm^{-2} and could influence the contact resistance. A critical evaluation of such effects was made for MOS structures.[193] The creep behavior of oxides at high temperatures commonly used for sintering or in application of high temperature thermistors is very sensitive to their stoichiometry. This was investigated for CoO, FeO, CeO_2, TiO_2, and UO_2.[696]

4

Sensor types

4.1. CRYOGENIC SENSORS—PRINCIPLES OF LOW TEMPERATURE THERMOMETRY

4.1.1. Thermistors

Since the resistivity of semiconductors follows in the first approximation the equation

$$\rho = A \cdot e^{B/T}$$

A and B being material constants, it is evident that, with decreasing temperature, materials with smaller B values have to be chosen to obtain the same resistance value at a certain temperature. Figure 40 shows clearly that extreme resistivities would result, approaching insolation, if semiconductors with B values suitable for room temperature would be applied to lower temperatures. The sensitivity would also increase beyond any necessary range. A temperature coefficient of 10% K^{-1} would require the B values given in Table 20.

TABLE 20

At T (K)	4	10	20	77	90	200
B (K)	1.6	10	70	570	800	4000

The constant A has a smaller influence on the temperature dependence of resistivity, resulting from the temperature dependence of the carrier mobility (see Section 2.1). The temperature coefficient per degree, given by $\alpha = -100B/T^2$, is shown in Figure 41 in dependence on B and T for a wide

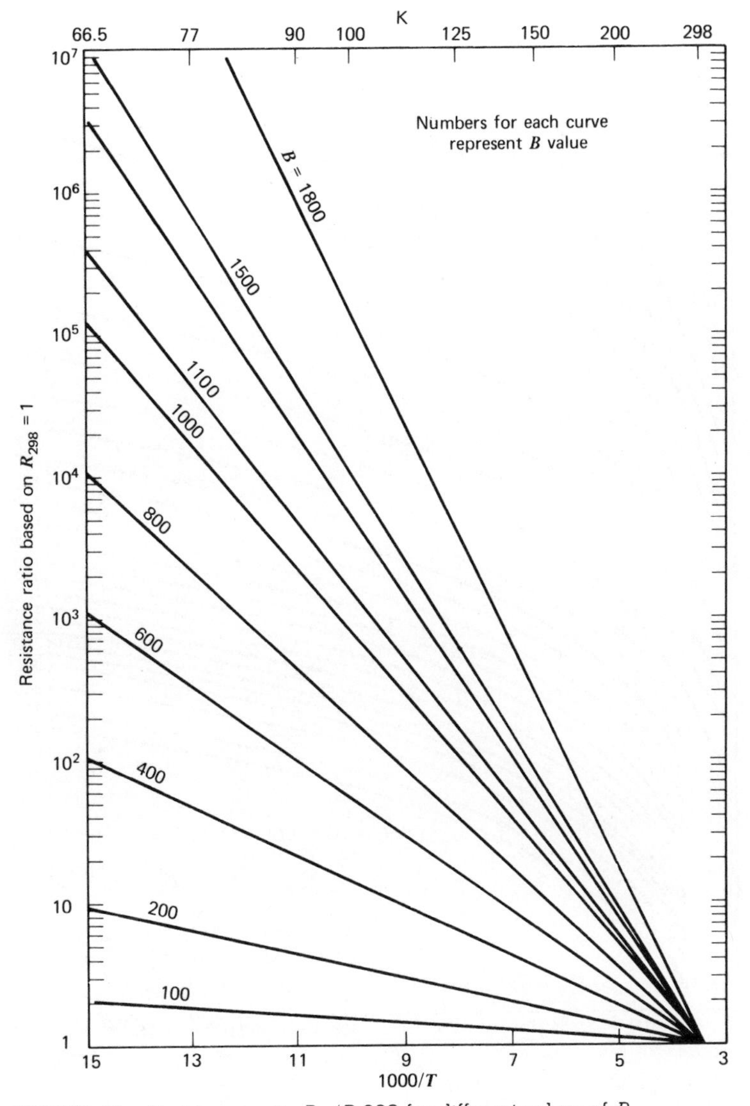

FIGURE 40. Resistance ratio R_T/R 298 for different values of B.

temperature range. The intrinsic linearity between α and B is distorted by the logarithmic α scale.

It is possible to make semiconductor materials with different B to meet the condition of Table 20 and Figure 40. However, developed materials have not always found commercial applications lucrative enough to stimulate large-scale production. At the present time certain types with B values of

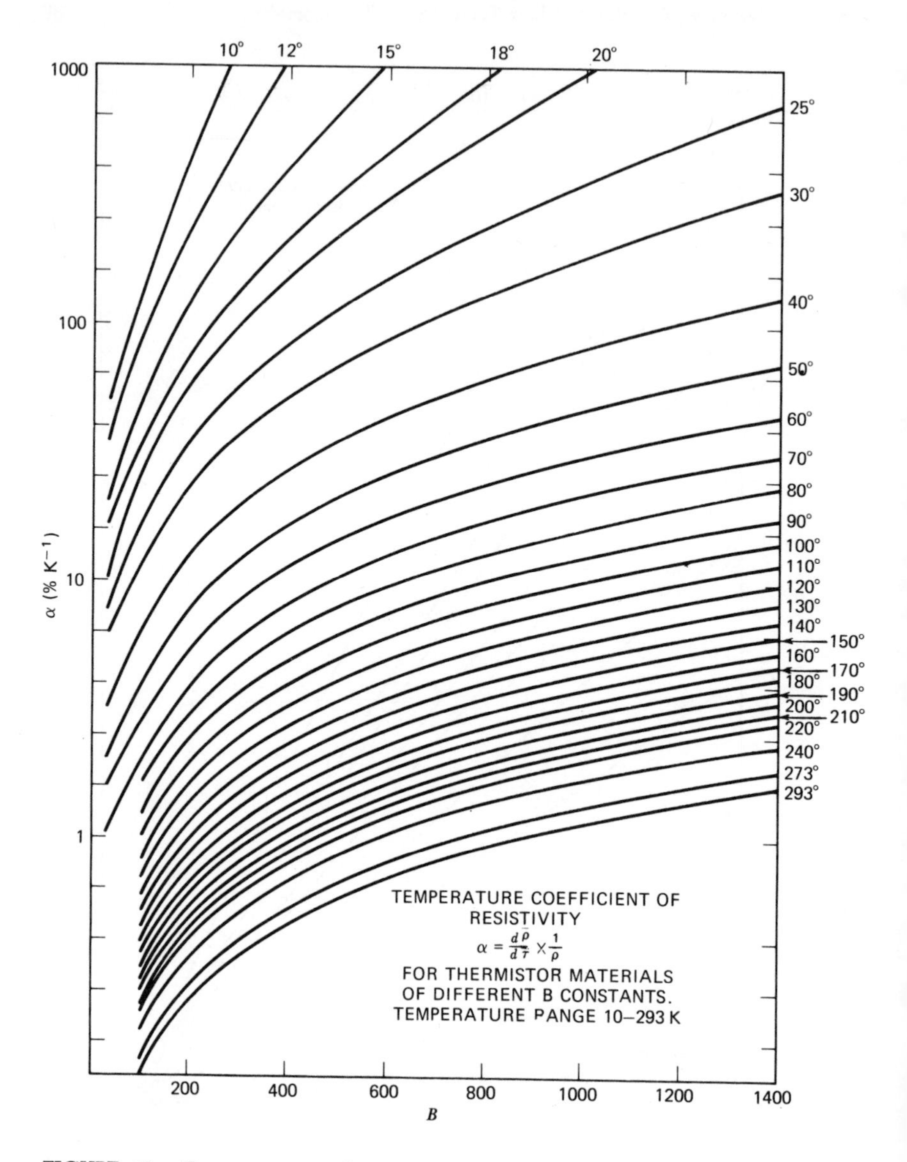

FIGURE 41. Temperature coefficient of the resistivity between 10 and 293 K for thermistor materials with $B = 100–1400$ K.

2000, 900–1200, 200–300, and 60–80 K have appeared on the thermistor market.

The chemical principles for making and modifying materials with such low B are somewhat different from those applied for B over 2000. In this range the systematic variation of the metal oxide components and to a lesser degree the sinter atmosphere determine the resistivity parameters.

Between $B = 2000$ and 900, iron oxides with variable but well-controlled nonstoichiometry have been used successfully. Cryogenic thermistors with B values small enough to limit their resistance at 4 K to values less than 1 MΩ with a wide variety of desirable values for the range 4–150 K have been developed and manufactured. In all cases the p-conduction in iron, cobalt, or nickel oxides was used and the acceptor and hole concentration adjusted by sinter conditions, controlled-valence doping, and a well-defined impurity profile.[194–199]

Table 21 gives up-to-date information on the commercially available types, which represent only a small part of the possible modifications.

Other types with temperature coefficients between 17 and 23% K^{-1} at 90 K can easily be produced with resistance values between 1 and 20 kΩ at 90 K, and have been mainly used in ordnance equipment for Jupiter missiles during the early 1960s.[200]

Schlosser and Munnings[201] have concluded, as the result of their repetitive dippings of L0904-HE-T_2 thermistors with 100 kΩ nominal resistance at 4.15 K in liquid He, that reproducibility of 0.5 mK can be safely assumed. This is of the same order of magnitude as that found with Ge resistance thermometers and better than that reported by the author, who specified a more conservative value of ±15 mK in view of the difficulties of eliminating temperature inhomogeneities in the He bath.

After 7 days no change of thermistor resistance in He was detectable within the accuracy of the British Physics Laboratory comparator CZ 457 M5 with a resistance ratio sensitivity of 2.10^{-5} that was used.

The authors[202] corrected their bath temperature by determining the heat effect for the immersion depth of the sensor, which was 24.4 Ω cm^{-1} (~0.024% of the nominal resistance at 4.2 K) or 0.2 mK cm^{-1}). The barometric pressure sensitivity was 265 Ω mm^{-1}. The magnetoresistive effect up to 100 kG is negative, was approximately proportional to H^2, and accounted for temperature errors of ~10 mK at 50 kG and ~50 mK at 100 kG. They used a figure of merit defined by the ratio of the magnetoresistive term $R_R = R(H,T - R(0,T)/R0,T$ and the temperature coefficient $(\delta R/\delta T)(1/R)$. It has the dimension of K, spells out the temperature error caused by the magnetic field, and is of course field dependent. It can change sign, depending on the signs of magnetoresistance and temperature coefficient. For all tested thermistors with 20, 100, and 3000 KΩ in liquid He

TABLE 21 Commercially available cryogenic thermistors.

Type	Nominal Resistance (kΩ)	Temperature Coefficient α(% K^{-1})	at K	Nominal B (K)	Range (K)	Behavior in a Magnetic Field
Keystone						
RL10X04-10K-315-S5	10	−8.4	90	650	77–90	Not reliable
	31.5	−10.4	77			
RL06X0628-31x7K-315-S5	31.7	−8.4	90	650	77–90	Not reliable
	100	−10.4	77			
L0904-6K-H-T2	6.6	−13	20.25	54 + 7	18–22	Usable up to 100 kG
L0904-125K-H-T2	125	−17	20.25	61 ± 7	18–22	Usable up to 100 kG
L0904-180K-H-T2	180	−18	20.25	64 ± 7	18–22	Usable up to 100 kG
L0904-100K-HE-T2	100	−275	4.15	58 ± 10	3–5	Usable up to 100 kG
L0904-3meg-HE-T2	3,000	−300	4.15	53 ± 10	3–5	Usable up to 100 kG

and 6.6 KΩ in liquid H_2, the F values were less than $\frac{1}{2}$ of those for Pt-Ge and carbon thermometers compiled by Neuringer and co-workers,[203] who have made extensive studies on the influence of high static magnetic fields up to 150 kG on cryogenic sensors[204,205] (see Table 22). F is the figure of merit.

Cryogenic thermistors are available with a mass of ~20 mg. Therefore their heat capacity is very small, due to the $(T/\Theta)^3$ dependence of their

TABLE 22 Field-dependent temperature errors for various low-temperature thermometers.

		Magnetic Field		
Type of Sensor	Temperature K	0–2 kG ΔT/T (%)	2–6.4 kG ΔT/T (%)	>6.4 kG ΔT/T (%)
GaAs Diode	4.2	2–3	30–50	100–250
	10	1.5–2	25–40	75–200
	20	0.5–1	20–30	60–150
	40	0.2–0.3	4–6	15–30
	80	0.1–0.2	0.5–1	2–5
Carbon Resistors				
Allen Bradley Co.	4.2	<1	5	10
	10	<1	3	5
	20	<1	1	2
Platinum	20	20	100	250
	40	<1	5	10
	80	<0.5	1	2
Germanium	4.2–20	5–30	not recommended	
$SrTiO_3$ Capacitors				
Corning Glass				
Lake Shore Cryotronics	1.5–20	≪0.05	≪0.05	≤0.05
Thermistors				
Keystone Carbon Company	4.2	<0.05	1	3
	10	<0.05	0.3	1
	20	<0.05	0.1	0.5
	40	<0.05	0.1	0.5
	60	<0.05	0.1	0.3

specific heat. For a L 0904 He-type, the heat capacity drops in the following manner[206]:

300	90	20	4.2 K
11.	2.7	0.4	0.08 mJoule K^{-1}

Since the dissipation constant is limited by heat transfer phenomena and ~1 mW/K, self-heating must be avoided by small power input. This also applies to other miniaturized cryogenic sensors. High-resolution bridges have been developed for this purpose (see Section 6.3).

4.1.2. Germanium Resistance Thermometers

Diode and transistor technology stimulated a huge pool of knowledge in producing, processing, and mechanical handling of the semiconducting elements Ge and Si. The pure elements have intrinsic conductivity with energy gaps of 0.66 and 1.09 eV, respectively, at 300 K. Since *n* or *p*-doping of single crystals results in extrinsic conductivity with activation energies in the meV range, their application to cryogenic thermometry became attractive. After an earlier suggestion by Estermann,[207] Kunzler and coworkers[208] at the Bell Telephone Laboratories built quite reproducible Ge thermometers using bridge-shaped samples of As-doped monocrystalline Ge. This design was chosen to minimize strains, since Fritzsche[209] had found an anisotropic piezoresistance that reveals itself when irreproducible distortions by thermal cycling occur. Commercial Ge thermometers usually attain strain-free mounting by other means.[210] A typical design is shown in Figure 42.

Dopants with high segregation coefficients such as P, As, Sb, and Ga are used to introduce high acceptor or donor concentrations of about 10^{19} cm^{-3} (~200 ppm). The BTL thermometers were As doped and those developed

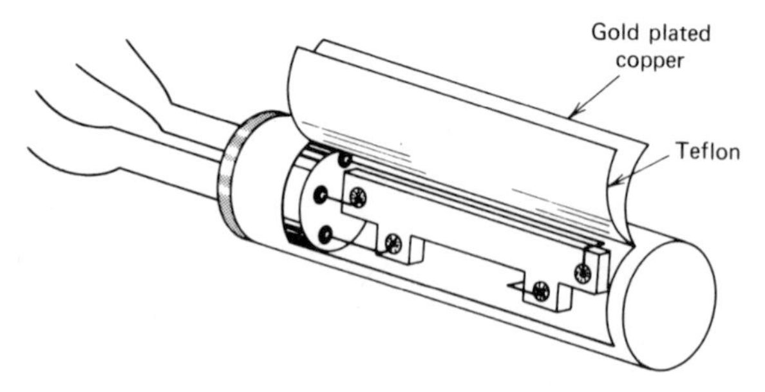

FIGURE 42. Schematic view and design of a Ge resistance thermometer. After Blakemore et al.[210]

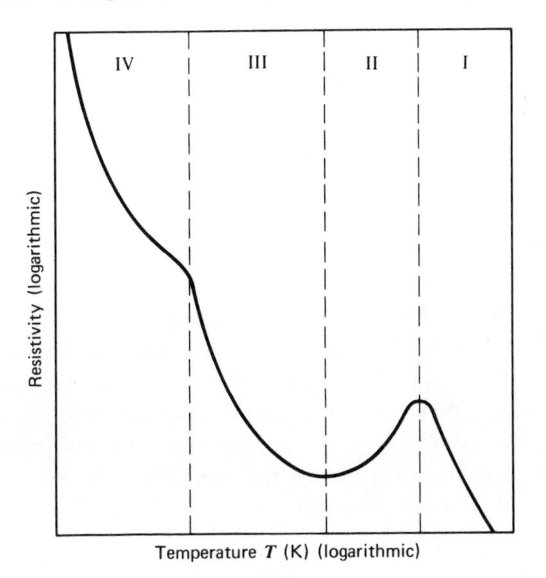

FIGURE 43. Electrical resistivity as a function of the absolute temperature, plotted in a log–log scale, for a semiconductor similar to Ge. Results of Blakemore.[211]

by Honeywell were Ga doped and partially compensated with Sb (3.15×10^{16} cm^{-3} Ga, 3.10^{15} cm^{-3} Sb) to obtain maximal sensitivity in the He–H_2 range and suppress it at temperatures above 40 K. Blakemore[211] has made a comprehensive study on Ge for thermometry. There are four distinct temperature ranges with different types of conductivity between 1 and 400 K, corresponding to Figure 43.

Since cryogenic applications require doped Ge, the intrinsic conduction range I can only be observed at higher temperatures (>400 K). At ~420 K, Ge with 10^{16} cm^{-3} impurity atoms has a transition to a conductivity Type II characterized by free carrier exhaustion and positive temperature coefficient of resistance. The coexistence of donor and acceptor impurities (sometimes unintentional but in most cases deliberately planned and controlled), results in the following processes in Ge containing donors such as Ga or In. The acceptors seize electrons from donors, thus eliminating holes to a negligible concentration. For a compensation of donors by acceptors larger than a few percent, the electron concentration is given by

$$n \approx \frac{Nd - Na}{1 + ANa\, e^{B/T}}$$

A and B depend on the nature of donors.

$ANa\ e^{B/T}$ at temperatures above 100 K is much smaller than unity; therefore $Nd - Na$ remains independent of temperature. Since the mobility decreases with temperature, the resistivity increases in range II. On further cooling below 100 K, n becomes small compared with $Nd - Na$ and the resistivity in the extrinsic range III is determined by

$$\rho_3 = \frac{1}{C} \frac{Na}{Nd - Nc} e^{B^{\mathrm{III}}/T}$$

Most of the temperature dependence originates not from C but from the exponential term. It is in this range where log ρ is a linear function of $1/T$. It extends to ~10 K, where apparently transition to a new range IV occurs with a lesser slope, known as the impurity conduction region. Most of the intrinsic electrons are bound to the donors, but they still possess the ability to hop from one impurity atom to the next. For this temperature range below 10 K another conductivity formula can be applied:

$$\rho_4 = \frac{1}{D} e^{B^{\mathrm{IV}}/T}$$

in which B^{IV} obviously is the activation energy for the hopping process and D and B^{IV} depend not only on Nd and Na, but also on the type of impurities.

Experimental values of B^{III} and B^{IV} for Ge doped with 3.75×10^{16} atoms/cm^{-3} Ga, of which 8% are compensated by Sb, are ~ = 6.4 meV and 0.59 meV, respectively. The transition from the intrinsic to the impurity region can be hard or soft. In the first case the R versus $1/T$ characteristic has a bump (Figure 44); this can be avoided in the second case. Here Ge is doped with 1.8×10^{17} atoms/cm^{-3} As, of which 8% are compensated with Ga. This obviously results in a smaller disparity between the activation energies of region III and IV (region III has 7.0 meV; region IV, ~1.5 meV) (Figure 45). For wide-range thermometry As-doped Ge, and for narrow-range thermometry Ga-doped Ge, have been suggested. A matching of thermometer characteristics depends on a strict control of Nd and Na and avoidance of accidental contamination of Ge. Not only the activation energies for the extrinsic and impurity range are important, but also the constant D in range IV, which increases rapidly with the number of impurities. The equations for the extrinsic and impurity region are only approximations. They will not do justice to units that have been found to be reproducible in the mK range. Therefore strong efforts have been made to obtain precise resistance temperature characteristics for the temperature range 1–100 K. Since there is no simple analytic formula fitting the experimental data over an extended temperature region, polynomial types of

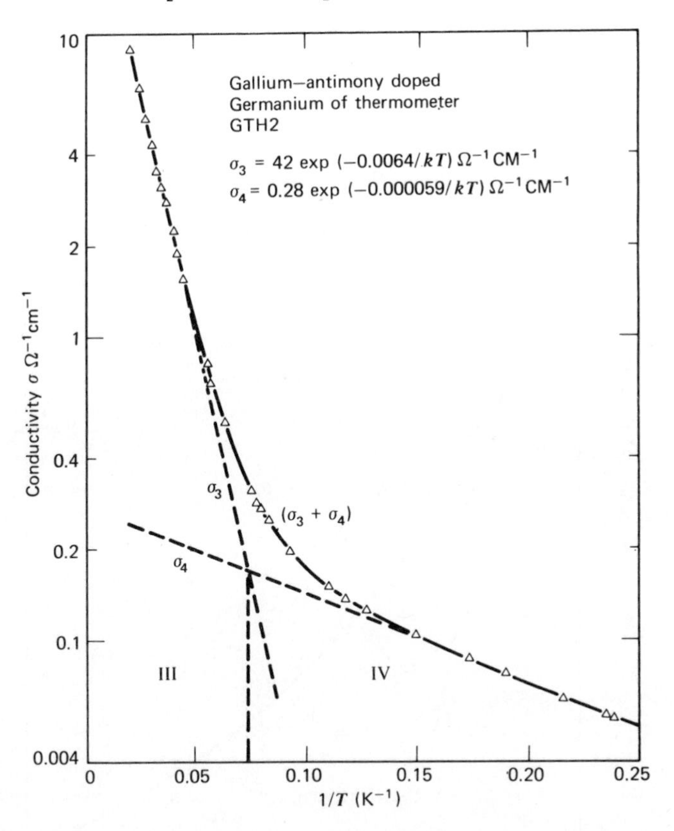

FIGURE 44. Hard transition between conductivity regions for Ge doped with 3.75×10^{18} cm^{-3} Ga compensated with 3×10^{17} cm^{-3} Sb. Results of Blakemore.[211]

computer fit have been suggested and applied, with main emphasis on temperatures below 20 K.[212,213]

A critical survey of these earlier investigations was given at the 5th "Symposium on Low Temperature" in Washington (1971). It confirmed the conclusions of Cataland and Plumb, that for 1–20 K a least-squares analysis based on the polynomial

$$\log R = \sum_{j=0}^{n} Aj(\log T)^j$$

provides a precise fit for As-doped Ge thermometers and a satisfactory fit between 4 and 40 K with an eleventh-degree polynomial resulting in a temperature error of 0.01%. Up to 110 K the best fit can be accomplished

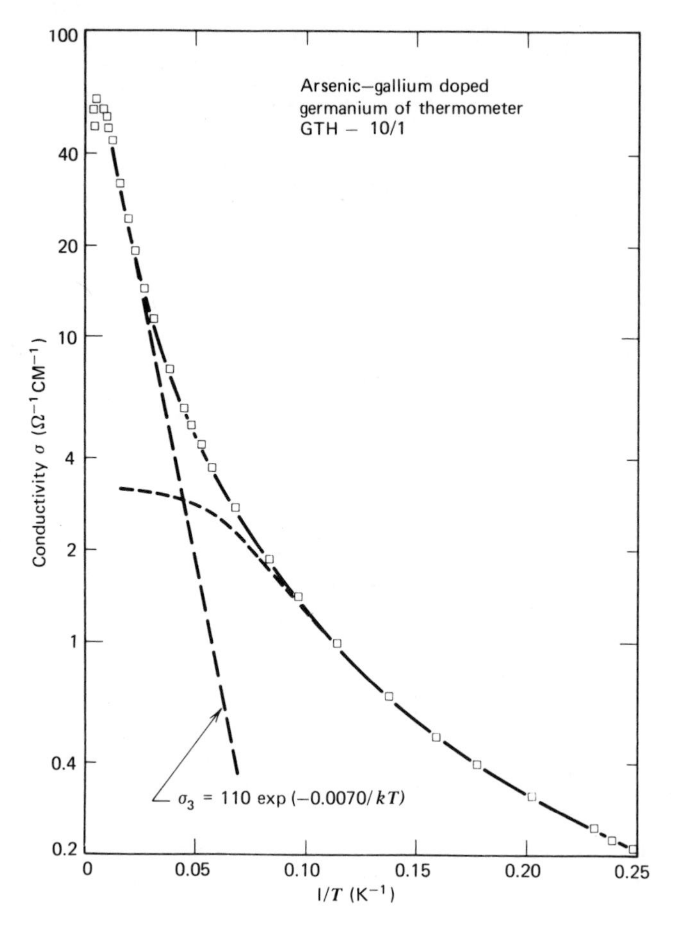

FIGURE 45. Soft transition between conductivity regions for Ge doped with 1.8×10^{17} atoms cm^{-3} and compensated with 1.4×10^{16} atoms cm^{-3} Ga. Results of Blakemore.[211]

if data below 15 K are omitted and ninth-degree polynomials are used. In this case the temperature error is less than 0.01%. The details on the computation procedure are given in the papers. Similar investigation have been made by Collins and Kemp of National Standard Laboratory in Australia.[214]

For Ge-thermometers poorly suitable to least-square fitting a cubic spline method seems to be promising with each spline curve represented by a cubic polynominal for each segment.[697]

Stability investigations with Ge sensors are reported in Section 6.4.

In the Soviet Union, Orlova and co-workers[215] have designed Ge thermometers from *n*-type Ge with $\sim 10^{17}$ atoms cm^{-3} Sb for the range 1.5–20 K.

A bar of this Ge with 0.2-cm diameter is cut into 2-mm-thick disks, which are then polished to a thickness of ~0.1 cm. Current and potential leads are cut out from the disks by an ultrasonic tool. Au-Sb alloy contacts of 0.03–0.04-mm diameter soldered onto the Ge leads in H_2 atmosphere had ohmic performance with no sign of rectification. The encapsulation of these sensors in copper tubing with glass platinum seals and filling capillary stem results in units of 3-cm length and 0.35-cm diameter. It is of interest to note that in ac bridges that have been developed for greater sensitivity, Ge resistance thermometers show slightly smaller resistance values. Fortunately the apparent temperature error decreases considerably with temperature (from 100 mK at 77 K to 0.2 mK below 17 K).[216]

4.1.3. Silicon as Temperature Sensor

In spîte of the advanced metallurgical technology for producing Si of high purity and the extensive expertise for modifying its electric properties by doping, silicon thermometers and bolometers have appeared later on the market than Ge thermometers.

In an early study[217] the resistance of *B*-doped Si crystals compensated to *n*-type with P or Sb has been investigated between 2 and 50 K using a four-terminal method. At 20 K their resistance varied between 7 and 400 Ω and their resistance change dR/dT at 4.2 K between 10^2 and 5.10^5 Ω K^{-1}. For crystals with resistivities of 0.030–0.042 Ω cm, a four-constant semiempirical equation fitted the resistance temperature curve between 2 and 20 K within 0.2% of the temperature. The magnetoresistance was lower than for Ge and as low as for carbon resistors. The apparent temperature shift at 10 kG to the current and to the (111) plane was only 5 mK. Cycling between 300 and 4.2 K over a period of 8 months did not change the resistance characteristics beyond the reproducibility of the temperature scale (0.05%). These results have been found with bulk doped crystals.

Herder, Olson, and Blakemore[218] have tried to make Si and Ge thermometers by diffusion of an electrically active dopant into one face of crystal with the goal of producing small regions of high thermometric sensitivity. *N*-type Si was doped with *B* from a trimethylborate source for 30 min at 1175 K and then annealed 100 min at 1475 K to produce a deeper diffusion and a high resistive junction at a depth of approximately 5 μm. Unfortunately *B*-diffused silicon thermometers lack cycling stability. They increase their resistance up to 5% at 4.2 K, corresponding to temperature errors of 100 mK. This is apparently the result of internal strains and the appearance of minute cracks. A number of remedial steps such as slow cooling (10 K min^{-1}) from the diffusion furnace temperature or from the Al contacting temperature (845 K) were tried, but without success. As-diffused

Ge thermometers also were less stable than those made from bulk doped Ge. However, in this case the temperature errors in cycling did not exceed 0.01%. In spite of this failure new efforts were made to use doped silicon as a low-temperature sensor. Bachmann, Kirsch, and Geballe[219] have built a bolometer by producing a doped surface layer with a low concentration gradient of the dopant. Moving along this flat profile by etching, the desired low-temperature resistance characteristics can be obtained.

For n-type doped silicon below the critical concentration of about 2.10^{18} cm^{-3} excess donors, the low-temperature resistivity is practically infinite. A 5-μm-deep degenerate layer is produced one one surface by diffusing P. After protecting two corners of the surface with wax, the remainder of it is etched until the degenerate layer has become very thin, while the wax-covered parts remain thick and can serve as ohmic contacts. Further heating to 1473 K permits the thin layer to diffuse into the silicon substrate making the doping profile flat. The etching time and the resulting residual degenerate layer permit tailoring the resistance-versus-temperature characteristics. For optimal sensitivity bolometers should have a large fractional temperature coefficient of resistivity,

$$\alpha = \frac{1}{R}\frac{dR}{dT}$$

and a small heat capacity. For a given temperature the latter, according to Debye's theory, is inversely proportional to cube of the characteristic temperature of the material. The Debye temperature of Ge is 370 K; for Si, 658 K. Therefore the heat capacity of the same volume of doped electrically active Si would be $5.9 \times (5.35/2.30) = 13$ times smaller, considering also the lower density of Si of 2.30 against 5.35 for Ge. A bolometer according to reference 219 with a resistance of ≈ 10 kΩ at 4 K was tested in an evacuated chamber immersed in a He^4 bath of 0.98 K. Two gold wires served as support, current leads, and thermal link. For calibration a load curve (μA versus V applied) was measured and the responsivity dE/dQ was determined as $7.9.10^4$ VW^{-1}, where Q is the absorbed radiation power and E is the voltage across the bolometer with a time constant of 155 μsec. Further applications of this principle have been reported by the same authors at the 5th Symposium on Temperature, Washington (1971).[220]

After forming a diffused degenerate layer of 20 μm on the polished and etched face of a Si wafer, it is cut into chips of 0.5×0.5 cm. The next steps are the same as described above, namely, reducing the main portion of the degenerate layer by etching and second shorter P diffusion [see Figures 46(*a*) and 46(*b*)]. An intermediate heating in air before the second diffusion is recommended to prevent evaporation of impurity atoms from the surface. The degenerate layer on the back and sides of the chip is etched off while the

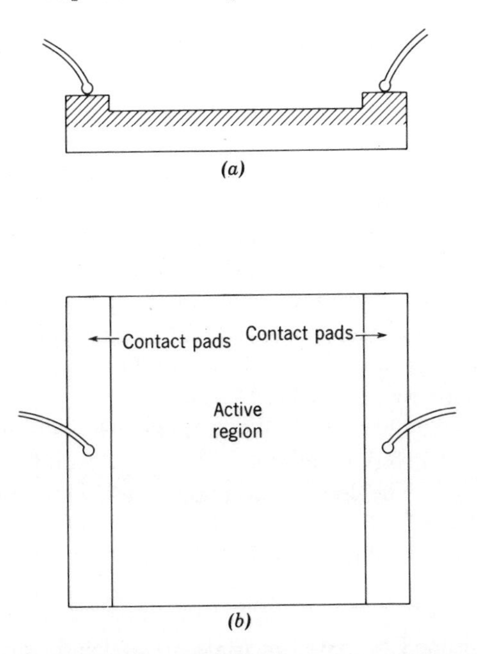

FIGURE 46. (a) and (b) Diffused silicon thermal element. The shaded region represents the diffused surface layer. After Kirsch et al.[220]

contact pads and the active layer on the front face are protected by wax. During the final etch to adjust the resistance of the active layer the contact pads are protected by In solder, which also serves as contact to monitor the resistance at 4.2 K. A resistance of 3 kΩ at 4.2 K for an active region of 0.3 cm length and 0.5 cm width would have a temperature coefficient $\alpha = 100\%\ \mathrm{K}^{-1}$ at 2 K. These elements offer a number of possibilities:

1. Two active elements of different characteristics on the same chip would permit application in two temperature ranges. The separation of these two elements requires a groove cutting through the contact pads and the active region.
2. One of the active elements could be used as heater, the other as temperature sensor in calorimetric applications.
3. In bolometric work in the visible and near IR, detection of 10^{-9} W is possible.

The resistance change up to 8 kG is less than 3%. Cycling stability between 300 and the He temperature is within 10 mK. No visible evidence of surface faults in the active region produced by thermal cycling has been found. Therefore chemisorption of atmospheric contaminants must be responsible for this small instability.

He-cooled Si bolometers seem to be the most sensitive detectors for infrared radiation with a large signal-to-noise figure. Since their sensitivity (responsitivity) is not only proportional to their *TC* of resistance (and therefore to the material constant *B*), but also inversely proportional to the heat capacity of the sensor material, silicon offers a definite advantage. As already pointed out earlier in this paragraph, the application of Debye's theory of specific heats (see Section 3.4.1) results in a ratio of $C_{Si}/C_{Ge} = 1/6$ for equal masses at equal temperatures. Since in a bolometer the volume of the sensors counts, the smaller density of Si reduces the necessary power for a certain temperature increase further to 1/13.

Its optical properties also favor Si as a bolometer material. Bulk doped Si with 10^{18} cm^{-3} carriers has a higher absorption constant for infrared than Ge with 10^{16}–10^{17} cm^{-3}. The lower refraction index (3.3 against 4.0 for Ge) reduces the reflection in the far infrared.

Si bolometers can be obtained with the specifications given in Table 23.

TABLE 23 Standard bolometers (data valid for 1-cm leads)

Model	Dissipation Constant (μW K^{-1})	Thermal Time Constant (μsec)	Responsivity (V W^{-1})	Resistance at (1.4 or 4.2 K) (MΩ)
Si 1.4R	5	300	1.2×10^5	2 to 0.2
Si 1.4R	55	27	1.1×10^4	2 to 0.2
Si 4.2R	16	2500	3.7×10^4	2 to 0.2
Si 4.2T	180	220	3.3×10^3	2 to 0.2

New Models DB1, DB2-1, DB2-10, and DB2-100 have different specifications. For ultrafast response GaAs, a photoconductor with 10 nsec response time (Type SP 40) with a smaller range from 100–370 μm is offered. For a comparison of Si– and GaAs bolometers and pyroelectric sensors inquire: Molectron Corporation, Sunnyvale, Cal. 94086.

The spectral response of the detector covers the range 1–10^3 μm. The noise equivalent power (NEP) of these bolometers is given as $\sim 3 \times 10^{-13}$ W Hz$^{-1/2}$. The contact noise is smaller than for Ge bolometers by bonding the leads to a highly doped degenerate surface layer.

The bolometers are He cooled, but not immersed in liquid He, and vacuum mounted to obtain maximal sensitivity. Figure 47 shows their resistance-versus-temperature characteristic. According to US patent 3,312,572,

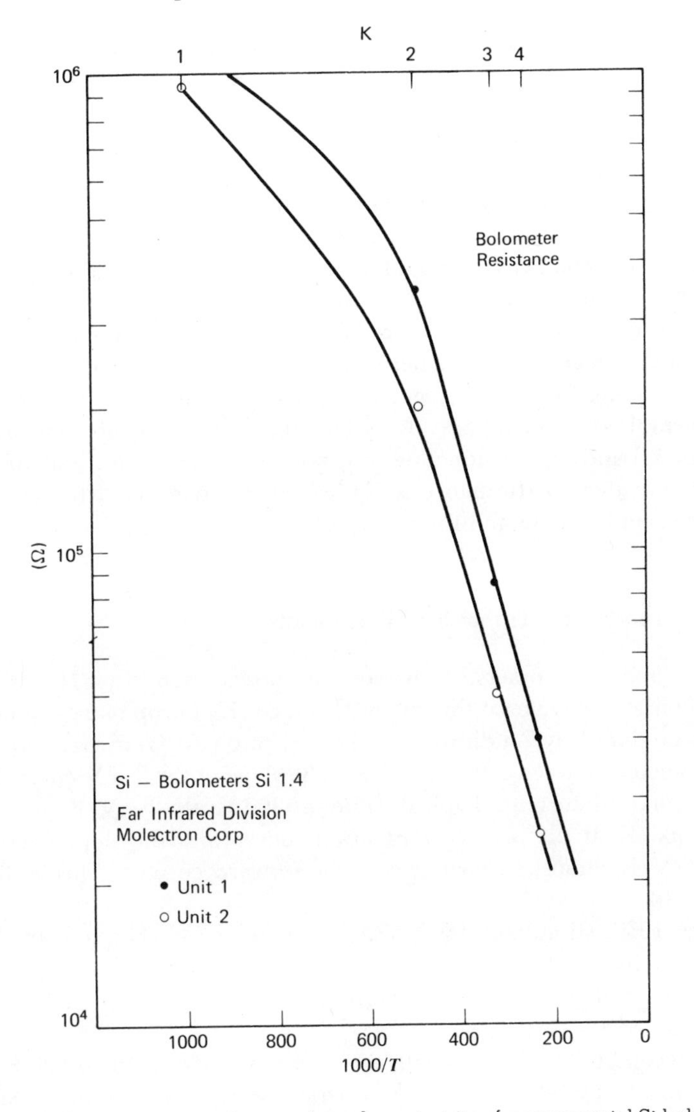

FIGURE 47. Resistance versus temperature characteristic of a commercial Si-bolometer. After Molectron, Sunnyvale, Cal., 94086.

April 4, 1967, 0.4-μm thin-film Ge or Si bolometers can be produced reliably and reproducibly by sequential evaporation of at first Ge or Si onto sapphire substrates preheated to ~400 K and then antimony or bismuth contact zones selected for deposition of 0.12-μm layers by masking. For the deposition of the ohmic contact zones, the substrates are preheated to 420 K. This type of bolometer can be used up to 500 V bias. This is more than for oxide bolometers. Its temperature coefficient at 300 K is -3% K^{-1}. An apparatus for large-scale production is mentioned. Silicon can also be used as temperature sensor for temperatures considerably above ambient where standard oxide thermistors lose sensitivity due to their B of only approximately 4000 K ($\sim$0.33 eV).

If N-type single crystals of Si with an impurity level in the ppb range are doped with 1–100 ppb gold, the effect of the background impurities is quenched by compensation and a semiconductor with a B = 6200 K ($\sim$0.54 eV) is obtained (US Patent 3,292,129, December 13, 1966). Its temperature coefficient is therefore larger by a factor of about 1.5; however, its resistivity is also much higher (10^5 Ω cm) compared to 10^2–10^3 Ω cm for oxide materials. This fact, together with the much higher price, could be detrimental to practical use. A systematic comparison of semiconducting (Ge and C) and superconducting (Sn) bolometers between 1 and 10^5 μm has shown that their performance is equivalent. Carbon and NbN are inferior due to granular noise and fluctuations.[673]

4.1.4. Diode and Transistor Thermometry

Cohen and co-workers,[221] based on observation by Harris[222] and McNamara,[223] have used the forward voltage (FV) drop of a *p-n* junction in diodes produced by Zn diffusion at 1123 K into 0.01 Ω cm GaAs as a linear thermometer between 300 and 1.4 K. The slope of the T-FV curve depends on the forward current. Typical values at 300 K are $2 \cdot mV/K$ at 1 μA and $-3.5 \cdot mV/K$ at 0.1 μA. Toward low temperatures the slope decreases to $-1.5 \cdot mV/K$ almost independent of the forward current. This is shown in Figure 48.

Since 1966, stimulated by NASA,[224] this junction thermometer became commercially available.* Up to date reports on further improvements of GaAs diode thermometers were given at the 5th Symposium on Temperature.[225,226]

The second report[226] deals with the accuracy between 4 and 300 K, which was found to be 30 mK at 20 K where a sensitivity minimum exists and 10 mK above 77 K and below 7 K. This was found with 400-μA forward current. For currents below 100 μA, reproducibility is poor and noise so high that it affects the voltage readings. For 400 μA it is less than 1 μV. Above 150 K the forward voltages drifted in irregular manner. The other paper has two topics: (1) theoretical treatment of the temperature influence on FV above and below 100 K.

* Lake Shore Cryogenics Inc., Eden, N.Y. 14057.

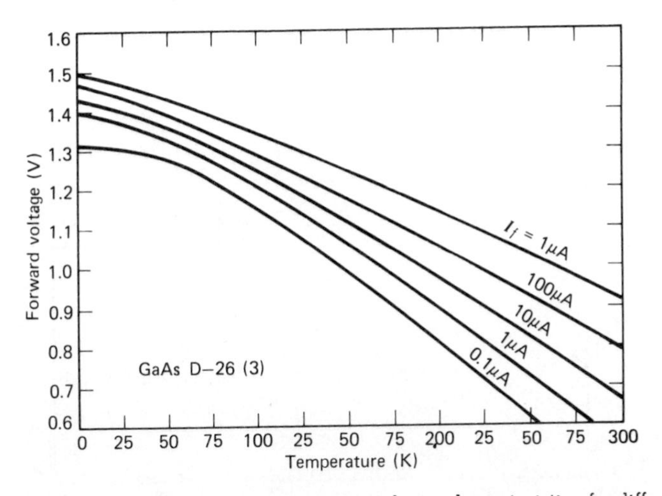

FIGURE 48. Experimental temperature versus voltage characteristic of a diffused GaAs *pn*-junction. After Cohen et al.[221]

The FV-T characteristics of GaAs diodes can be explained from 1 to 400 K. A practical interpolation formula valid between 100 and 350 K gives an error of less than 3 mV corresponding to 2 K:

$$FV_T = 1.522 - \frac{5.8 \times 10^{-4}\,T^2}{T + 300} - CT \ln DT$$

C and *D* are empirical constants to be determined experimentally for a given device and current.

At 300 K the sensitivity is 3.1 mV/K; it changes less than 20% down to <60 K for 10 μA.

For lower temperatures FV is approximately a linear function of ln T; for instance at 10 μA,

$$FV_{(T)} = 1.4785 - 0.0125 \ln T$$

which corresponds to a sensitivity of $-(12.5/T)$ (mV/K) below 20 K.

The influence of magnetic fields up to 10.5 T parallel and perpendicular to the junction: 1 T (tesla) = $10^4/4\pi$ G ≈ 0.8 kG. The apparent temperature error as function of temperature is given for the magnetic fields of 2, 4, 6, 8, 10 T and ranges from ~90 mK at 2 T to ~5000 mK at 10 T with only a slight variation between 4 and 40 K in Figure 49.

Up to 5 T, FV = $FV_{(H)} - FV_{(0)}$ is a quadratic function of *H*. At higher fields the increase is flatter. These slopes are proportional to T^{-2}. Therefore ΔFV at low fields (<5 T) and temperatures below 20 K is proportional to $(H/T)^2$.

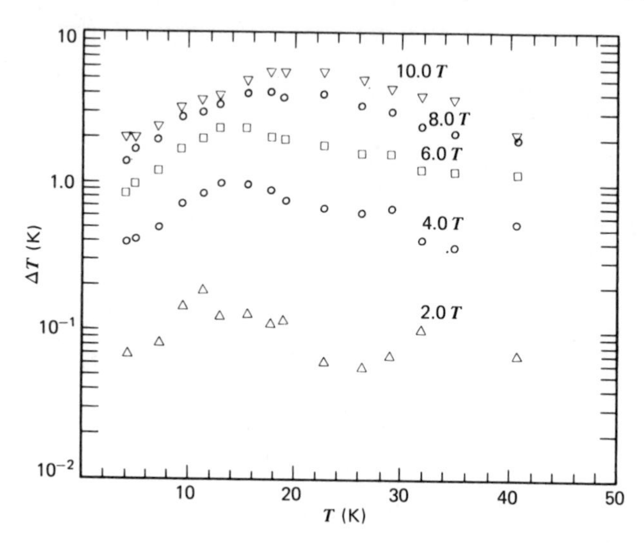

FIGURE 49. Apparent temperature error of GaAs diode thermometer in magnetic fields (2 – 10T or 1.582 – 7.9 μG). Results of Swartz and Gaines.[225]

The increase in forward voltage ΔFV is essentially independent of the current and is not caused by the magnetoresistance of the bulk, but by junction effects.

The resistivity of *n*-type GaAs crystals or films prepared by vapor or liquid epitaxy would lend itself for temperature measurements below 2 K. *B* values below 7 have been measured in this range.[227]

TRANSISTOR THERMOMETER. The forward current across any *p-n* junction consists of three major components: diffusion current or drift current, generation recombination current, and surface leakage current. Each of them is determined by a different mechanism and temperature dependence results in a complex compound behavior. Verster[228] has shown that any semiconductor material operated as transistor simplifies these relationships since only the diffusion component of the emitter is under consideration, while the generation recombination and surface leakage currents are drained away as base current. The Ic-Vbe characteristic of transistors 2 N 1893, 2 N 3904, BFY77, and SX 3704 were measured between 70 and 420 K using four different circuit techniques. For 2 2 N 1893 transistors operated at Ic = 10 μA in a differential thermometer application (Figure 50) the tracking error (Figure 51) was 0.7 mK K^{-1} with an instability of ~1 mK day^{-1}.

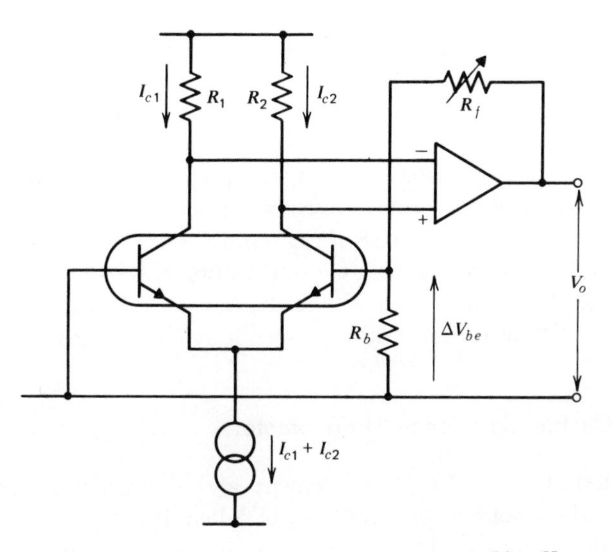

FIGURE 50. Circuit diagram for dual transistor thermometer. After Verster.[228]

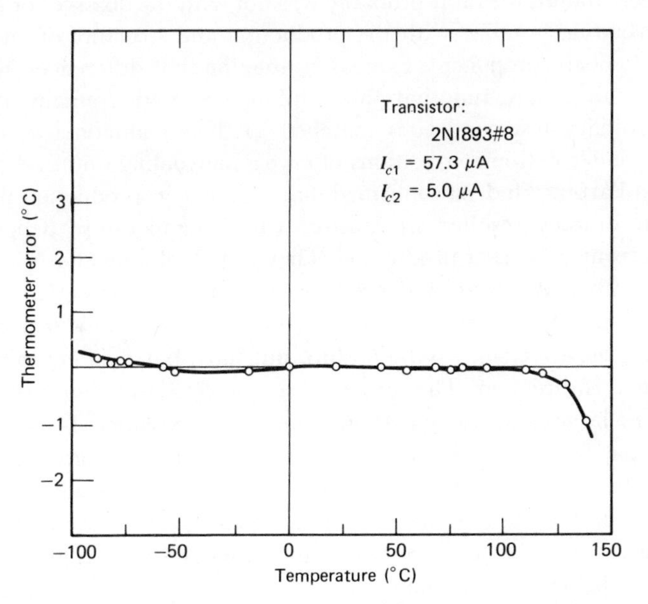

FIGURE 51. Measured error curve for a practical thermometer operating on the collector current step principle. Results of Verster.[228]

An interesting application of Si transistors as temperature sensors at high hydrostatic pressures up to 6 kbar is based on the fact the collector current increases for constant emitter-base voltage, but decreases for constant base current. For mixed-voltage current feed by experimentally matching the resistance in the emitter-base circuit, the pressure effects are compensated and the collector voltage can be used as direct temperature criterion with sensitivity of 1–2.5 V K^{-1}. The upper temperature limit is ~420 K.[229]

The interbase resistance of a Ge unijunction transistor has been used as thermometer between 20 and 300 K. Its sensitivity is rather constant, but much smaller than for most other semiconducting cryogenic sensors.[698]

4.1.5. Carbon Resistance Thermometers

COMMERCIAL UNITS. In 1952 Clement and Quinell[230] introduced commercial carbon composition resistors of Allen Bradley into cryogenic resistance thermometry. They have been in use since that date, despite different opinions on their reliability, reproducibility, and stability. Plumb and Edlow[231] and later Weinstock and coworkers reported[232] favorably on the reproducibility of carbon composition sensors. Whenever less favorable results were found, the fault probably was not with the observer or his equipment. Everyone familiar with the production and structure of these mass-produced cheap components can easily imagine that differences from batch to batch can occur, and that these differences might remain undetected under ordinary test conditions matched to their common use, but could show up under extreme conditions of cryogenic cooling and cycling. Weinstock and Parpia[233] indeed confirmed that a change in production techniques by Allen Bradley resulted in resistors whose temperature dependence is different from the earlier production. They tested Ohmite and Allen Bradley resistors with $\frac{1}{8}$ W rating and nominal resistance values of 12–5600 Ω between 3 and 300 K with the goal of establishing an analytical expression describing the R versus T relationship, but have failed so far as also other efforts in this direction. The lack of a simple R/T function has until now limited the large-scale use of carbon composition resistors. Their initial low price is partially upset by the need for individual calibration. Data on reproducibility presented on a logarithmic resistance scale can be very deluding, considering the small temperature coefficient of these units.

One great advantage of carbon composition resistors is their extension into range below 1 K, which they share with Ge resistance thermometers and ceramic Corning sensors.

Speer resistors have been measured down to 0.02 K[234] and found to be reproducible when cycled between 0.075 and 300 K. In a magnetic field of

6 kG the resistance decreased ~5% at 0.115 K, corresponding to a temperature error ΔT of ~10 mK.[235]

For the magnetic field dependence of Allen-Bradley resistors expressed in percent temperature error for 4.2, 10, and 20 K, see data by Neuringer and Rubin (Table 22).[205]

A comparison of the magnetoresistance of Allen-Bradley and Speer carbon resistors between 0.5 and 4.2 K[688] in fields up to 140 kG had the following results:

1. $\frac{1}{8}$ W Allen-Bradley resistors with 10 Ω nominal resistance had a more reproducible nearly linear magnetic field dependence $\Delta R/R_0$ increasing from 10% at 4.2 K to 36% at 0.49 K for 140 kG.

   The reproducible transverse magnetoresistance of resistors with 2.7 Ω nominal resistance reduces their field induced uncertainty to ±4 mK.[699]

2. At the same field strength for $\frac{1}{2}$ W Speer resistors grade 1002 with a nominal resistance of 220 Ω the magnetoresistance at first increased down to ~1.5 K and then decreased again. Also its sign is negative for fields less than 60 kG and then reserves itself to positive values.

3. The percentage temperature error in the magnetic field for equal temperatures or fields is much smaller for Speer resistors. The negative magnetoresistance of type 1002 Speer resistors between 0.35 and 1 K is 10% less in transverse than in longitudinal direction.[236]

For Allen-Bradley units in this temperature range a positive and isotropic magnetic resistance was found in agreement with Neuringer, Rubin, and other investigators. These variations in magnetic resistance can be caused by inhomogenetics. Contrary to the positive isotropic magnetic resistance, which is a linear function of H^2, the negative effect follows the equation:

$$\frac{\Delta R}{R} = -f\left(\frac{B}{T}\right)$$

and reaches saturation with $B = 7 \times 10^{-3}$ K at ~9 kG. These errors are much smaller than for GaAs diodes or the GRT, but larger than for thermistors L 0904.

It has been suggested to reduce the effective magnetoresistance of carbon resistors (Allen-Bradley) by using them in pairs and measuring the different temperature dependence of resistors of different nominal value or wattage rating. The temperature sensitivity of a single resistor is partially sacrificed to reduce magnetoresistance[237]; compare Table 22 in Section 3.1.1.

Long-range stability of carbon composition resistors in cryogenic measurements has been studied by a number of investigators.[238–241]

Speer resistors 1002 have been widely used as cryogenic sensors. In comparing sensibility and stability of carbon resistors it was found as a general rule that drifting increases with increased sensitivity.[458] The pressure dependence of Allen-Bradley units is given by

$$\frac{\Delta R}{R} = 2 \times 10^{-4}\ \text{atm}^{-1}$$

independent upon temperature,[695] while for Speer units below 1 K no pressure effect was detected up to 30 atm. Irradiation with 10^6 rad at 1 K had no influence in resistance of Ohmite resistors and on only a small one at 20 K. However, a definite effect of a hf electric field was observed by Pearce, Marham, and Dillinger.[242]

Carbon slab thermometers. Standard carbon composition resistors, even their smallest types, are often to large for certain applications, especially if the temperature of very small objects is to be measured. Several investigators have used small slabs or disks cut from commercial carbon resistors.[243–245]

An important problem in this case is proper recontacting; otherwise the temperature characteristic is to a large degree determined by the contact. This problem was solved by cutting $\frac{1}{2}$ W Speer resistors grade 1002 (or 27 or 220 Ω) parallel to their axis into slabs at least 0.018-cm thick including the original Cu contact.[243] The temperature dependence of the slab is a function of the resistance values of the whole initial resistor R_0 and the cut slab R_T and is given by

$$R_{(T)} = \beta(R_{0T})^n \quad \text{for} \quad 0.05\ \text{K} < T < 300\ \text{K}, \quad n = 1.01 \pm 0.02$$

β is approximately 5 and depends on the ratio of the cross sections of the slab and whole resistor. Another solution is possible by starting with $\frac{1}{2}$ W 10 Ω carbon composition resistors Ohmite "Little Devil" from Allen-Bradley; plane parallel slabs are cut leaving one contact lead intact and cutting the other end to a trapezoidal shape.[246] After gold plating the entire carbon pieces, the gold is removed excepting the sloping sides of the trapezoidal shape end of the slab, which is cut away and thinned by abrasion of its base. This leaves a trapezoidal temperature sensor of 0.1 × 0.1 × 0.03 cm, to be fastened onto a 6 Mylar sheet and contacted with AWG 50 Cu wires of 0.011-cm diameter. Final resistance values are 100 and 500 Ω in He; permissible power rating to avoid self-heating is 50 nW. Under these conditions quantum oscillations in Ga were observed below 4.2 K and in Bi below 3 K, and sinusoidal temperature fluctuations in superconductive He by mechanical pumping. The noise level from peak to peak was 1 μV.

The intrinsic time response of 0.1 W Allen-Bradley carbon resistors in the liquid He temperature range has been found to be of the order of msec[247] in agreement with theoretical estimates.[248] This implies that they are not in thermal contact with a solid surface. In this case their time response would be above 10 msec. Slicing the resistor into thin disks of 0.015 cm would reduce the time to about 0.1 msec.

Carbon film thermometers. In several investigations a step further was taken not only to further reduce the mass and size of the carbon sensor, but also to increase their flexibility in regard to shape and size. As early as 1936 Giauque[249] suggested the use of colloidal carbon films as thermometers, and Geballe[250] and co-workers followed this idea.

A comprehensive survey on prior and present activities has been given by Shen and Heberlein.[251] Colloidal dispersions of 270- and 700-Å graphite sold by Columbian Carbon Company under the name Aquablack B and M and of 20,000-Å graphite known as Aquadag from Acheson Colloids Co. were diluted 10 : 1 with water and sprayed onto a sample with two parallel Cu-Ni wires of 5.10^{-3}-cm diameter 0.05 cm apart. The formed carbon film must be dried either by baking or better by pumping in vacuum. The resistance of $0.05 \times 0.5\text{-cm}^2$ films made from Aquablack B was $>10^7\ \Omega$; from Aquablack M, $\sim 10^5\ \Omega$ at 1 K. By mixing equal portions of Aquablack B (27 μm) and Aquablack M (70 mμ), resistance values lower than those obtained with Aquablack M and higher sensitivity could be made. Variation of the mixing ratio permits one to match resistance and sensitivity to the desired values in a certain temperature range. Due to the sensitivity of these films to humidity, long-range stability and reproducibility is limited.

For pyrolytic carbon with a resistivity of ρ 293 = $1.22.10^{-3}\ \Omega$ cm under certain deposition conditions that were not specified, negligible magnetoresistance between 73 and 523 K was reported by Saunders.[252]

Since deposited carbon film resistors are based on pyrolytic carbon, possibilities to control the magnetoresistance exist.

A novel way to prepare carbon resistors for cryogenic measurements independent of commercial composition resistors was described by Mikhailov and Kaganovski.[253] Their method is similar to that used in making carbon brushes by mixing 77% coke with 23% pitch, both finely divided and molding parts of $7 \times 3 \times 1$ mm with 12,000 kg cm^{-2} that are then fired in a quartz tube filled with charcoal for 3 hr at 800°C. Then electrolytic deposition of Cu contacts on the ends of 0.02-cm-thick Cu wires were soldered to the bar and covered with a thermosetting material at 180°C. The temperature sensitivity decreases rapidly with increasing firing temperature between 750 and 810°C. The drift by cycling between 77 and 300 K is approximately 10 mK in the temperature range 2–4.2 K.

Carbon Bolometer for Low-Power rf Signals. An interesting use of carbon resistors has been proposed by Lalevic.[254] For very low rf power levels of 0.1–1 μW the signal-to-noise ratio can be improved by a bolometer operating at liquid He temperature. Allen-Bradley 12 Ω $\frac{1}{2}$ W resistors were chosen for this purpose. They were sliced into small disks $\frac{1}{32}$ inch thick with a mass of 4 mg and electroplated with Cu on both sides to attach 25-μm Monel wire. Their resistance at 1.2 K was 50 Ω. It was measured between 0.5 and 1 MHz at the end of a coaxial line with a sensitivity of 3 nW for power levels from 100 to 1000 nW at temperatures of 1.45–3.09 K.

A carbon resistor with a TC of $dR/R = 0.9\ K^{-1}$ has been used to stabilize the operating temperature of a superconducting bolometer near 3.8 K with an accuracy of 50 μK for several minutes.[255]

Carbon impregnation of porous alkali-borosilicate glass after leaching out the boron-rich phase resulted in a new type of carbon resistance thermometer with larger resistance ratios between 4.2 and 77 and 300 K (300 and 2, respectively) compared to commercial carbon resistors, but with twice their magnetoresistance at 4.2 K. The porosity of the glass units required encapsulation for better stability. Curve fitting permitted two-point calibration transfer.[674]

4.1.6. Superconductive Temperature Sensors and Tunnel Junctions

Giaever and Megerle[256] have used electron tunneling through an insulating layer of Al_2O_3 sandwiched between two metal films to determine the energy gap in superconductors as postulated by the BCS theory of superconductors.[257] It was observed that the tunneling current through the insulating film varies linearly with small applied voltages, if the metal films are in the nonsuperconductive state and the density of electron states in the two metals is constant over the applied voltage range. This principle has been applied to low-temperature thermometry by Bakker, van Kempen, and Wyder.[258] They measured current-versus-voltage characteristic across an Al-Al_2O_3-Ag junction with a resistance of 10–100 Ω mm^{-2} that has been deposited by consecutive evaporation of Al and Ag on the lower end face of a cylindrical synthetic ruby crystal, the Al_2O_3 layer being produced by partial oxidation of Al in pure O_2. Adiabatic demagnetization of the ruby with 0.3 at. % Cr^{3+} gives a temperature of 0.14 K. After going down to 0.09 K, it rises again slowly to 0.6 K after 2 hr. The I-versus-V curves are normalized by plotting $I(SIN)/I(NIN)$.* Fitting the measured I-versus-V curves with computed curves gives the value of the energy gap with an accuracy of 1%. The variation of the energy gap as function of temperature is compared with computer-generated curves according to the BCS theory and permits

* I(SIN) = current for superconductor — insulator — normal metal junction.
I(NIN) = current for nonsuperconductive — insulator — normal metal junction.

determination of the temperature with an accuracy of $\pm 6\%$. The magnetic effect on the energy gap limits this method to field-free applications. Giaever and Megerle[256] found a decrease of 20% for $\frac{1}{2}$ H_c = 52 Oe. A modification of this method has been suggested by Ignat'ev and Tarenkov[259]. The zero-bias conductance is a logarithmic function of the temperature, as first shown by Wyath[260] with the relation

$$\Delta g_{(0)} \propto g_0 \ln \frac{E_0}{kT}$$

where Δg_0 is the residual conductance and E_0 is the constant for each sample.

The sensitivity $d(\Delta g/g_0)/dT$ increases with decreasing temperature and is 26 K^{-1} for a junction Al-Al_2O_3-Al formed by vacuum evaporating an Al film onto an oxidized Al sheet, which is an order of magnitude higher than for carbon resistors.

Deviations from stoichiometry in Al_2O_3 act like paramagnetic impurities influencing the tunneling process. Monomolecular layers of aromatic organic compounds such as naphthalene or benzophenone increase the sensitivity and can be used to obtain a desired slope. The fast response (0.1 msec), small size (10^{-4} cm^2), and negligible mass (10^{-8} g) of this sensor, which responds to a power dissipation of $\sim 10^{-12}$ W, would be attractive for applications in the vicinity of critical temperatures. For superconductive quantum devices see Section 4.1.7.

4.1.7. Miscellaneous Low-Temperature Sensors

RADIATION-PROOF LT SENSORS. In order to meet the radiation conditions in nuclear rocket engines of the Nerva and Rover programs where fast neutrons up to 10^{18} nvt, En greater than 1 MeV and γ heating rates of 10 W g^{-1} are expected, a special radiation-intensitive thermometer has been developed by Tallman.[261] It uses amorphous Pd_{73} Si_{20} Cr_7 alloy as a disk-shaped foil of about 2-cm diameter and $5 \pm 2.5 \times 10^{-3}$ cm thick produced by rapid quenching (splat technique) of a liquid alloy globule and subsequent cold squeezing between a copper-faced anvil and fast-moving piston. For use in thermometer design the disks are cold rolled to 1.3×10^{-3} cm and attached to a 5×10^{-2}-cm-thick Ti beam, chosen for its thermal expansion matching to Pd and its low ratio of mass to strength. The thermometer design is shown in Figure 52. Below 60 K this alloy is a semiconductor. Though its negative resistance temperature coefficient is small, its sensitivity compared to Pt is higher, especially below 20 K (Figure 53), with good cycling stability down to 4 K. Besides its indifference to high radiation, a most interesting feature is its low sensitivity to strain.

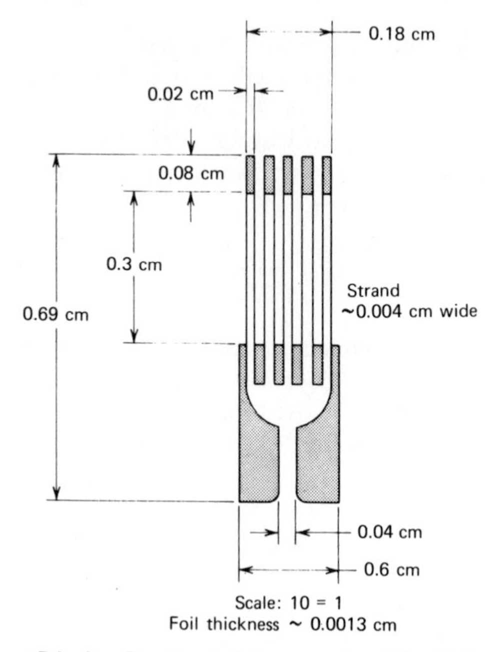

FIGURE 52. Die-cut Pd_{73} Si_{20} Cr_7 alloy foil thermometer. After Tallman.[261]

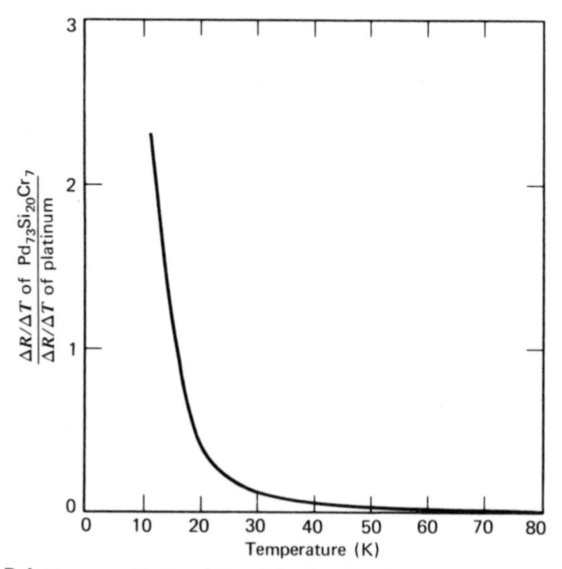

FIGURE 53. Relative sensitivity of the Pd_{73} Si_{20} Cr_7 thermometer compared to Pt between 10 and 80 K. Results of Tallman.[261]

The small radiation influence on the resistivity of this amorphous or at least highly disordered alloy (as shown in 50–75% of the foils by x-ray investigation) is easy to understand: It can be considered as defect saturated. Therefore additional lattice defects by irradiation with neutrons or heating by γ absorption have an insignificant influence (10 mK/error for 1 W g^{-1} γ heating), contrary to Pt where material with less than 10 ppm impurities is used to maximize the temperature coefficients and stresses are carefully avoided.

Therefore Pt is more susceptible to radiation-produced lattice defects. They can accumulate by prolonged use at low temperature where healing of the defects, which starts only at higher temperature than 100 K,[262] is impossible. Final performance evaluation of these sensors under high radiation is still pending.

The topic of cryogenic thermometers and radiation sensors related to superconductive quantum interference devices ("squid") is marginal to the main theme of this book. Therefore only two references are given,[263,264]

4.2. THERMISTORS FOR HIGH TEMPERATURES (>600 K)

4.2.1. Oxides

General Remarks. The upper temperature limit for each thermistor type is, at least theoretically, given by the chemical stability of the oxide phase used. Since the sinter temperatures are always much higher than the application temperature, the problem of material instability in use should in principle not exist, and this is certainly true for the basic chemical compounds in the thermistor composition such as NiO, CoO, Mn_3O_4, and so on.

However, sintered thermistor materials in general do not represent phase equilibria. Not only the sinter time for reacting the components, but also the cooling time to adjust to lower temperature equilibria are insufficient (for economical reasons, as pointed out in Section 5.3).

If such a nonequilibrium sinter body is heated to elevated temperatures exceeding 40% of the previous sinter temperature (in K), the following can happen:

1. Unreacted fractions of the composition continue to react with each other, although at a lower rate.

2. Even if the composition had enough time to react its components entirely, the defect concentrations at the sinter temperature are very different from those at lower temperatures. Therefore at each temperature a new equilibrium has to be established. The corresponding changes in the bulk grains and the interfaces produce a drift of the electrical properties.

Long-range stability of the electrical characteristics of the material can be enhanced by appropriate compositions in which divergent trends of individual components compensate each other. Although solid-state chemistry can give guidance in selecting the compositions, their ultimate quantitative definition requires empirical studies.

While these points are relevant to common thermistor compositions for moderate temperatures, another one has to be considered for higher temperatures: the decreasing temperature sensitivity of these materials. Since their B values are normally below 5000 (0.43 eV), their temperature coefficient, decreasing with $1/T^2$, is only $\sim 1.5\%\ K^{-1}$ at 600 K and less than $1\%\ K^{-1}$ above 700 K.

Therefore, not only chemical stability but also higher activation energy is necessary to make well-performing thermistors for higher temperatures. The materials meeting these conditions will be discussed in the next chapters.

Another crucial obstacle is the contact problem. It has two aspects (see also Sections 5.4 and 6.4).

1. The physicochemical and the mechanical stability of the contact: The contact metal not only should not oxidize at elevated temperature, but also not react with the thermistor composition nor migrate into it. Even if nonoxidizing noble metal contacts are used, the migration or contact corrosion problem still persists, especially under dc conditions. Also, slow reduction of the contact area by crystallization (grain growth) in contact layers and their evaporation over long periods can occur with corresponding increase of resistance for purely geometrical reasons. In case of Ag coatings, migration and even whisker formation have been observed.

2. It is quite obvious that different thermal expansion of thermistor and contact material can have a serious effect on the mechanical stability of contacts. This does not necessarily mean that contacts fall off, but they may become less adherent and thus increase the resistance of the entire units. This also applies to bead-type units with molded-in wires. Thermal stress, induced by the difference of expansion coefficient under static condition or nonuniform heating under dynamic conditions can slowly and irreversibly separate the contact wires from the surrounding thermistor material, often without resulting in a completely open circuit. The demand for high activation energies could be met in two ways:

a. Select a thermistor material from those transition metal oxides that reabsorb a minimum of oxygen neither during cooling from the sinter temperature, nor during prolonged heating at high application temperatures.

Mn and Ni oxides, pure or slightly doped with other metal oxides, could be used. For instance the resistivity of pure MnO prepared by sintering 5 hr at 1700 K in H_2 was found to be of the order of 10 Ω cm at 1000 C and could slightly be reduced by Cr doping within 1 at. % (~40%).[265] The maximal solubility of Cr (as oxide Cr_2O_3) in MnO between 1273 and 1473 K is of the order of ~1% and depends on the O_2 partial pressure in the sinter atmosphere. Assuming constant B at temperatures below 1273 K, resistivities of less than 1000 Ω cm could be expected at 1000 K, permitting production of thermistor disks of the order of 100 Ω (1-cm diameter, 0.2 cm thick).

b. Use oxides of a different transition element period, the rare earths. Most of these form only one type of oxide, M_2O_3 (Cerium, Praseodymium, and Terbium can have an 0 excess above this formula). They are especially attractive for high-temperature applications because of their high enthalpy of formation from the elements, which makes them chemically very stable at high temperatures. Table 24 makes this evident and shows for comparison some transition metal oxides normally used in thermistor compositions.

TABLE 24

	kJoule/mole		kJoule/mole
Y_2O_3	1906	Dy_2O_3	1866
La_2O_3	2255	Yb_2O_3	1814
Ce_2O_3	1820	NiO	240
CeO_2	975	MnO	362
Nd_2O_3	1783	Al_2O_3	1675
Sm_2O_3	1800		

As early as 1900 Nernst recognized the negative resistance temperature coefficient of some of these oxides and used them to develop a commercial electrical lamp. He also found that certain mixed compositions of such oxides with ZrO_2 had a lower resistivity than the components, a fact that much later was confirmed with the system La_2O_3–CeO_2.[266] Since this minimum was not found in oxide mixtures of equal valence state, such as La_2O_3–Nd_2O_3, the effect of controlled valence might be responsible for the minima. For room temperature up to ~500 K, most of these rare-earth oxides are practically insulating. Their dc resistivities have been measured with sinter bodies of equal density between 870 and 1570 K.[267–269] Since the geometry of the test specimen was not well enough defined, resistivity values in Table 25 have been rounded off.

The current–voltage characteristics in air and vacuum and the apparent absence of electrolytical polarization voltages and of oxygen discharge at the electrodes seem to support the concept of electronic conductivity.

TABLE 25[a]

Atomic number	57	60	62	63	64	66	68	70
Temperature (K)	La_2O_3	Nd_2O_3	Sm_2O_3	Eu_2O_3	Gd_2O_3	Dy_2O_3	Er_2O_3	Yb_2O_3
1070	7×10^4	6×10^3	2×10^4	1×10^5	4×10^5	6×10^5	7×10^5	1×10^6
1370	4×10^3	6×10^2	1×10^3	6×10^3	2×10^4	2×10^4	3×10^4	4×10^4
1570	1×10^3	2×10^2	4×10^2	2×10^3	4×10^3	6×10^3	6×10^3	7×10^3
B(1000–1570) K	13800	11500	13300	14300	16000	16200	17000	18000
Values by Rao et al.[271]	12200	13000	14800	15600	18200	—	—	18600

[a] Values in electron volts.

The early resistivity investigations of rare-earth oxides left some doubt concerning whether this criterion is conclusive enough to decide on an ionic or electron conduction mechanism. The situation is indeed ambiguous and both conductivity types can occur, depending on the atmosphere surrounding the test specimens.[270] Measurement of the emf of Gd_2O_3, Dy_2O_3, Sm_2O_3, and Y_2O_3 between Pt electrodes at temperatures from 940 to 1203 K at different O_2 partial pressures showed that at high O_2 pressures of 0.1 atm *p*-conduction is prevalent, while these oxides are pure ionic conductors below 10^{-5} atm ($\sim$8 mtorr).[270] In normal applications such low O_2 pressures would seldom exist.

The results of Noddack et al. indicate that the resistivity and the activation energy of the sintered rare-earth oxides fluctuate with the atomic number of the earth element from Nd_2O_3 to Yb_2O_3. More recent measurements between 670 and 1170 K[271] confirmed the general trends, though the numerical values differed in some cases.

For temperatures between 670 and 830 K the activation energies are lower and the log ρ versus $1/T$ curves change their slope between 770 and 870 K from 1–1.5 eV to 0.4–0.7 eV. The resistivity of all these oxides decreased with the O_2 pressure between 0.1 and 750 torr. They were *p*-type conductors with partial ionic component.

Recently thermistor disks have been made from yttria (Y_2O_3) and dysprosia (Dy_2O_3) and measured between 770 and 1470 K in air. The difficult contacting problem was solved by applying to both faces consecutive coatings of Pt paint until after firing at 1170 K a film resistance $\leq 1\ \Omega$ was obtained. Finally Pt foils were attached to the Pt film under slight pressure at 1870 K.[272]

Figure 54 gives the log ρ versus $1000/T$ characteristic. From these curves the B values can be calculated, with the results given in Table 26.

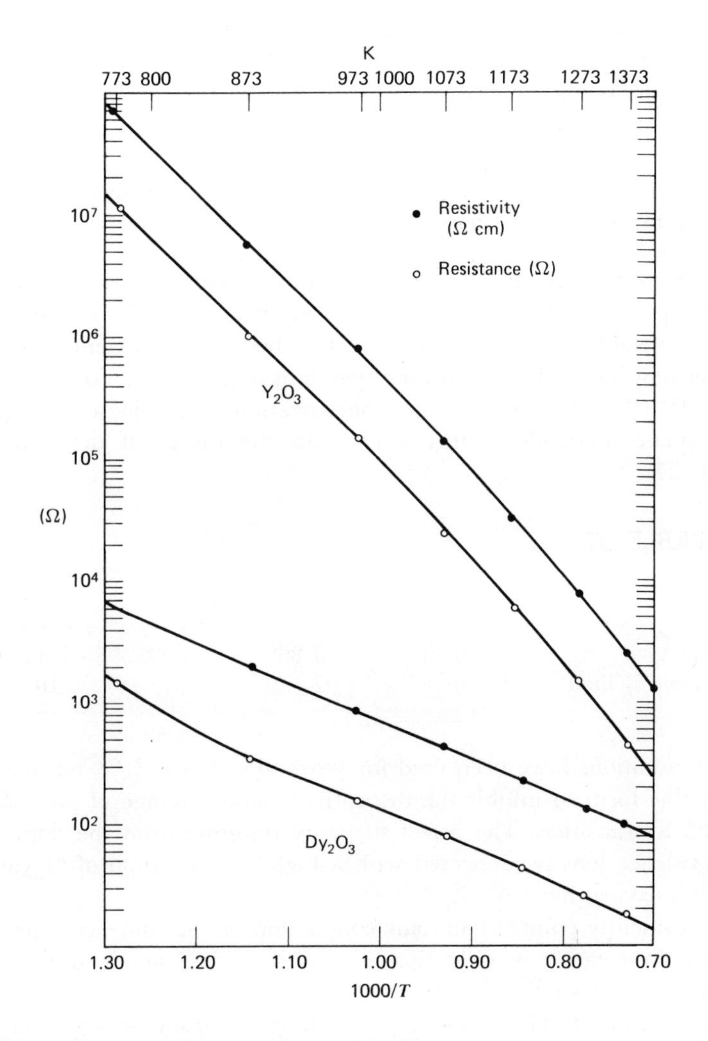

FIGURE 54. Resistivity of sintered Y_2O_3 and Dy_2O_3 as a function of temperature between 773 and ≈ 1400 K. Results of Hyde et al.[272] Courtesy of the Bureau of Mines, U.S. Department of the Interior (RI 7458).

TABLE 26 ***B* values corresponding to α % K^{-1}.**

	873–1073 K	1073–1373 K	at 770 K	1470 K
Y_2O_3	17,500	24,500	3.0	1.1
Dy_2O_3	6,900	8,900	1.2	0.4

The lower values between 873 and 1073 K indicate some impurity conduction. The temperature coefficients in the last two columns approximate those calculated from $\alpha = -B/T^2$. The dissipation constant at 770 K is 0.26 mW K^{-1} and the cycling stability $\pm 5\%$ after ten thermal cycles. Extended life tests are not reported. The oxides are sensitive to the oxygen potential in the surrounding atmosphere due to stoichiometric defects.[273,274]

MIXED SYSTEMS. As mentioned earlier, mixed systems of rare-earth oxides offer the possibility of obtaining lower resistivities, especially in compositions containing ions of different valence. The best-known system is based on ZrO_2 as host lattice doped with trivalent ions such as Sc^{3+}, Yb^{3+}, Y^{3+}, Sm^{3+}, and Gd^{3+}. The limiting doping concentration to retain a homogeneous cubic phase increases in this series with the radius of the doping ion (Table 27).

TABLE 27

	Sc^{3+}	Yb^{3+}	Y^{3+}	Sm^{3+}
r (Å)	0.81	0.86	0.92	1.00
mole % limit	6	7	9	10

Such additions have been used for years to stabilize ZrO_2 refractories in their cubic form to inhibit the disruptive volume change of pure ZrO_2 at its 1423 K transition. The defect structure resulting from the doping with lower valence ions is connected with a high concentration of O vacancies and high oxygen ion mobility.[275]

This evidently points to an ionic conduction mechanism—contrary to the concept of the earlier investigations with pure oxides—and is supported by a number of other studies.[276–279]

Doping with trivalent ions does not have as dramatic a concentration effect as alkali doping of the monoxides NiO, CoO, and CuO. This has been shown by Sc, Yb, Y, and Sm doping of ZrO_2[276] (Table 28).

ρ versus $1000/T$ curves for Yb-doped ZrO_2 from 900 to 1600 K are compared with those for Y-doped ZrO_2[280] and for La- and Y-doped ThO_2[281,282] in Figures 55–57.

If the temperature coefficient of Y-doped ZrO_2 is considered to be not large enough ($\sim 1.4\%$ K^{-1} at 1000 K), CeO_2 doped with 2.4–20 at. % Zr (2–15 wt % ZrO_2) would increase it considerably ($B = 21{,}000$ K). Beads with Pt leads sintered at 1670–1720 K in air for 2–3 hr and thermally stabilized at 1270 K for several hundred hours had resistance values of ~ 1 kΩ at 1273 K and $\alpha = -1.4\%$ K^{-1} ($\sim 2\%$ K^{-1} at 1000 K).[283]

TABLE 28 Resistivities (Ω cm) of doped ZrO_2 sintered at 2075–2275 K.

Doping Concentration Sc_2O_3 (mole %)	Temperature (K)				
	800	1000	1200	1400	1600
6	423	50	12	4.3	2.0
8	143	21	5.7	2.3	1.2
10	123	19	5.5	2.3	1.2
12	296	32	7.6	2.6	1.2
Yb_2O_3					
6	2,880	240	46	14	5.9
8	615	72	17	6	2.8
10	872	74	11	4.5	1.9
12	1,740	118	20	5.5	2.1
16	11,200	420	48	9.8	3.1
20	34,000	980	93	17	4.8
Y_2O_3					
6	7,600	760	166	56	25
8	5,500	630	152	55	24
10	910	74	14	4.3	1.8
12	4,600	240	32	8.3	2.9
16	19,000	560	56	11	3.0
Sm_2O_3					
6	4,700	350	62	18	7
8	10,000	620	93	23	8.8
10	2,600	170	26	71	2.6
12	8,700	320	48	13	4.6
16	1,950	700	80	17	5.2

THERMISTORS FOR 1000–2500 K. Early efforts to use alkaline porcelain compositions as sensors in jet engine exhaust ducts up to 1300 K were only partially successful.[284] Also alumina of different purity grades, magnesia (MgO), beryllia (BeO), and zirconia (ZrO_2) have been tried out at still higher temperatures (up to 1800 K).[285] The major problems are:

1. The resistivity of the material must be much higher than that of the surrounding air.
2. Application of stable and nonrectifying contacts.

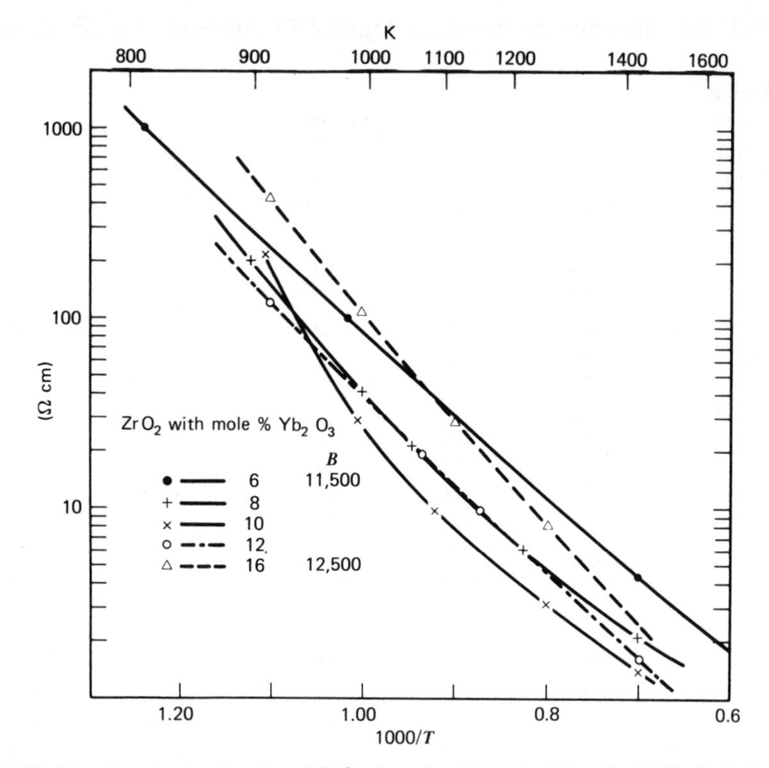

FIGURE 55. Resistivity of sintered ZrO_2 doped with up to 16 mole % Yb_2O_3 between 850 and 1600 K. After Strickler and Carlson.[276]

A critical study on the second point led to the conclusion that methods useful for lower temperatures, such as conductive cements, fail, and the safest solution is to insert the contact wire through a hole drilled transversely to the axis of an cylindrical sensor with a mechanically tied loop.[286]

Such sensors made from ceramic compositions containing 60% or 99.5% Al_2O_3 were tested between 1480 and 2030 K. Their temperature coefficient of resistance at 1900 K was 0.74 and 0.63% K^{-1} at 2030 K. Within the investigated temperature range the resistance is frequency dependent and drops from 20 to 1600 Hz approximately 10%, the major decrease occurring up to 500 Hz. Since the electrolytical conductivity of the materials requires ac measurements, test frequencies above 600 Hz are desirable to minimize the influence of frequency on the accuracy of temperature measurements. The repeatability of sensors over a 1000 hr calibration period was 0.6%, corresponding to ±1 K. The *B* values of these alumina-containing compositions differ considerably with the alumina content: 8000 for 60%, 55,000

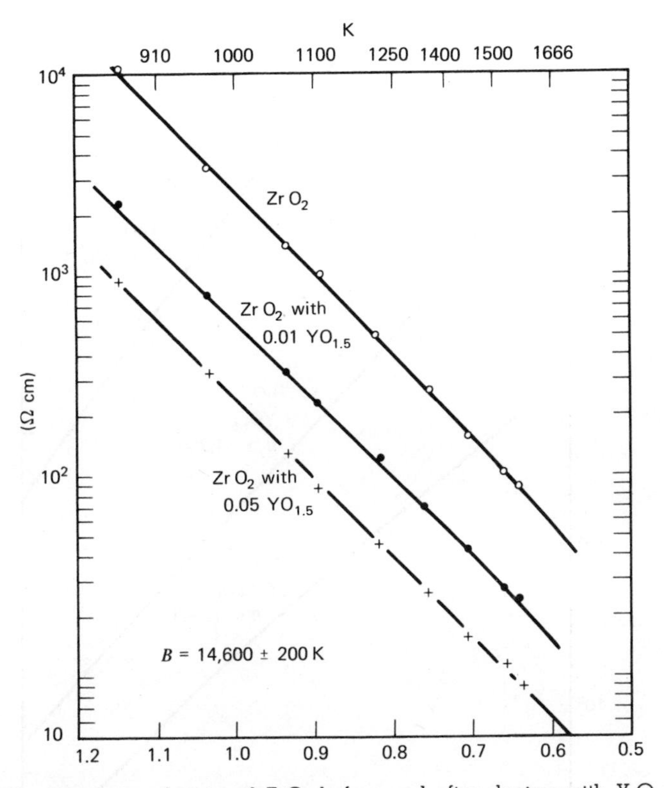

FIGURE 56. Resistivity of sintered ZrO_2 before and after doping with Y_2O_3 between ≈ 850 and 1700 K. Results of Noddack et al.[267]

for 99.5%. The latter value is somewhat higher than for MgO and BeO (40,000).[287–289]

An operational thermistor for use up to 2500 K was designed, made, and tested by Wolff.[290] It consists of spectroscopically pure grade ZrO_2 stabilized with 15 mole % Y_2O_3 (electronic grade.) Several methods of production were tried out:

Fusion. A pair of Ir–40% Rh wires were held in a double-hole MgO thermocouple insulator at distance of 0.15 cm protruding 0.8 cm. The wire "fork" was dipped into a premixed oxide paste with H_2O as plastifier and the formed bead fused in vertical position by oxyacetylene or octane-O_2 microtorch.

Slip Casting. The oxide paste was squeezed into a graphite mold holding the leads in grooves. After air drying for several days the green preform with the beads was slowly heated to 2000 K in air.

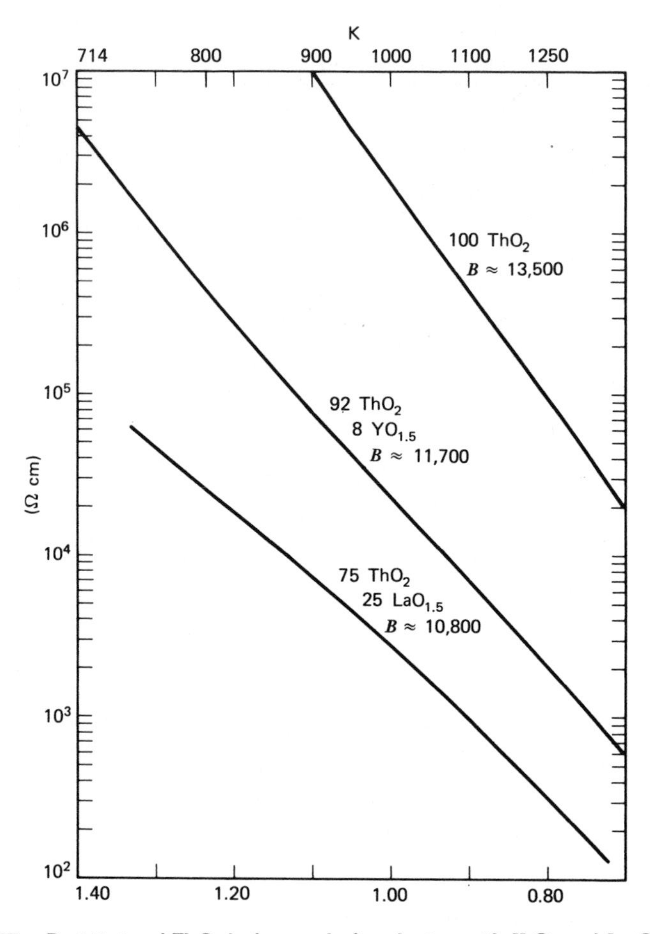

FIGURE 57. Resistivity of ThO_2 before and after doping with Y_2O_3 and La_2O_3 between ≈700 and ≈1300 K. After Hund.[281]

In the hot-pressing and press-forging process the oxide material was very dense, but lead attachment was difficult due to distortion of the wires though adhesion between contact metal and oxide was good.

As previously stated, yttria stabilized ZrO_2 has mainly ionic conduction[291] below 2400 K. Contrary to common convictions, the author considers this to be beneficial in making the conductivity insensitive to changes of O_2 partial pressure and eliminating a break in the log R versus K relationship observed if conduction at higher temperature would go from extrinsic to intrinsic. It might be somewhat controversial to trade in this benefit for possible electrolytical deterioration with resulting resistance drift. However, stability tests in air at about 1100 K and thermal cycling in vacuum between 1000 and 2500 K with cycling rates from 3 to 50 K min^{-1} have given very satisfactory results obviously since the resistance measurements were made with 1 kHz in the General Radio Type 1603-A bridge. However, a resistance drift occurs if a certain voltage threshold is exceeded (corresponding to the electrolytic polarization voltage). This threshold increases by a factor 40 between 1100 and 1800 K. There is also an initial (aging) drift by microstructural changes in the oxide. Also, vaporization losses exceeding 1% after 100 hr at 2200 K must be faced. The log R versus $1/T$ relation between 1660 and 2200 K is strictly linear, with a resistance change from 8.5 to 1.2 in this interval. (B ~17,000 K) (Figure 58). Since the resistance of the Ir-Rh leads increases with temperature, while the resistance of the oxide element decreases, an increasing error from the leads must be taken in stride. It corresponds to ~2.5 K at 2500 K. A linearization circuit was suggested.

Experiments with dc current densities up to 2 A cm^{-2} between 1600 and 2300 K have shown that cubic zirconia fully stabilized with Y or Ca is rapidly electrolyzed in air under formation of ZrN and oxygen-deficient ZrO_2. Although current densities in measurements are much smaller, the long-range effect of electrolysis should not be ignored. For partially stabilized ZrO_2, electrolytic effects tend to be smaller.[292]

In search for an electrode material meeting the stringent conditions of magnetohydrodynamic energy converters operating at at least 2200 K, semiconducting $LaCrO_3$, pure and doped with Ca, Al, Y, Ti, and Zr, was investigated after sintering in air, H_2, or vacuum. Between 1300 and 1730 K activation energies up to 0.50 eV were measured for air-sintered specimens with resistivities of ~10^3 Ω cm at ~550 K. Electrolytic effects were ~2% for a current density of 2 A cm^{-2}.[293]

Tin oxide-TiO_2 solid solutions have been suggested for temperatures up to 1200 K.[294]

Air-sintered compositions containing between 2 and 80 mol % TiO_2 and doped with small additions of Sb_2O_3, ZnO, and Ta_2O_5 have resistivities at 870 K between 10 and 10^3 Ω cm and B values up to 11,000. They have been suggested for thermistors in automotive antipollution devices.[295]

4.2.2. Nonoxidic High-Temperature Thermistors

Silicon Carbide and Boron. The highly refractory character of silicon carbide, elemental boron, and carbon makes them attractive as materials for high-temperature resistance thermometers. In the case of carbon, only diamond would be eligible, and only in its semiconducting natural (Type IIb) or synthetic form. It is treated in a special paragraph.

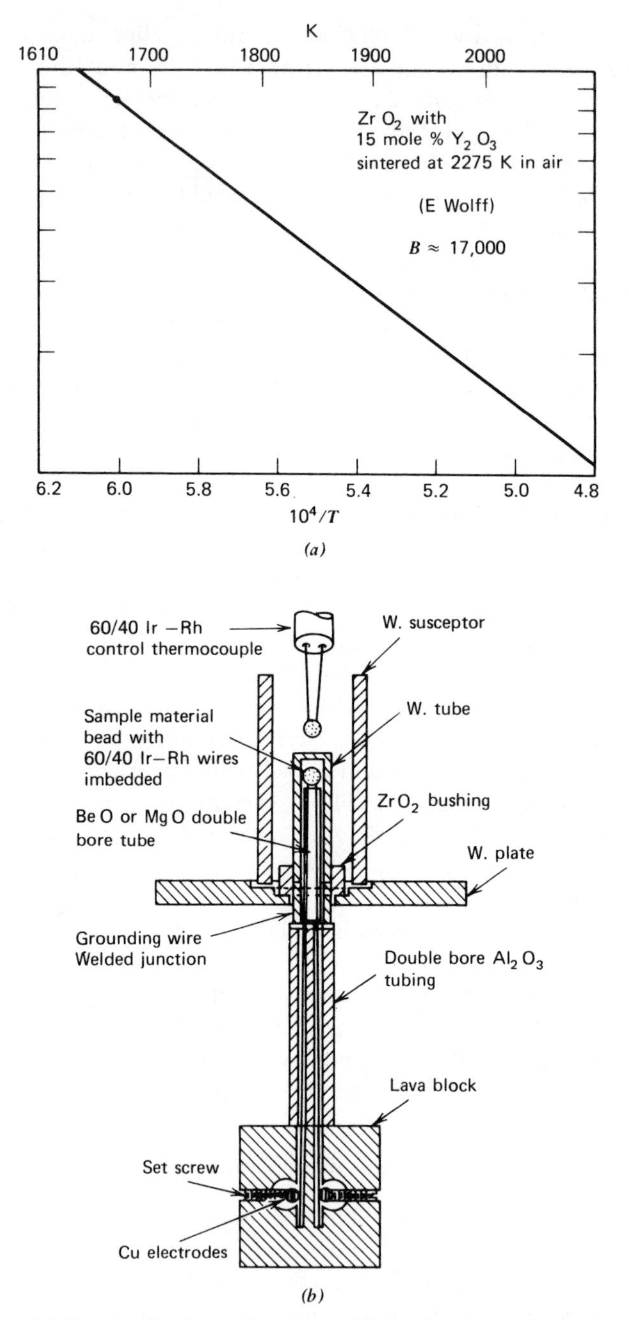

FIGURE 58. (*a*) Test jig for thermal cycling of ZrO_2 thermistor with 15 mole % Y_2O_3. (*b*) Resistance of ZrO_2 thermistor containing 15 mole % Y_2O_3 between ≈1600 and ≈2100 K. After Wolff.[290]

As previously stated, chemical stability and inertia are not enough to make a good material suitable as a high-temperature semiconductor. The activation energy (or the energy gap in the intrinsic case) must be high enough to guarantee sufficient temperature sensitivity, and at the same time the carrier mobility must be large enough to reduce resistivities to practically acceptable levels. A semiquantitative figure of merit for the refractory behavior of materials, expressed in their mechanical thermal properties as hardness, melting point, and thermal conductivity, is their Debye temperature, deduced from the temperature dependence of their specific heats.[296]

Ryan[297] plotted Debye temperatures of a large number of elemental and compound semiconductors as function of their intrinsic band gap (Figure 59). Two facts are evident from this plot:

1. The materials can be separated into three distinct groups: II-VI, III-V, and IV-IV compounds (such as SiC), the latter including elemental Ge, Si, and C (diamond) with increasing covalent bond character.

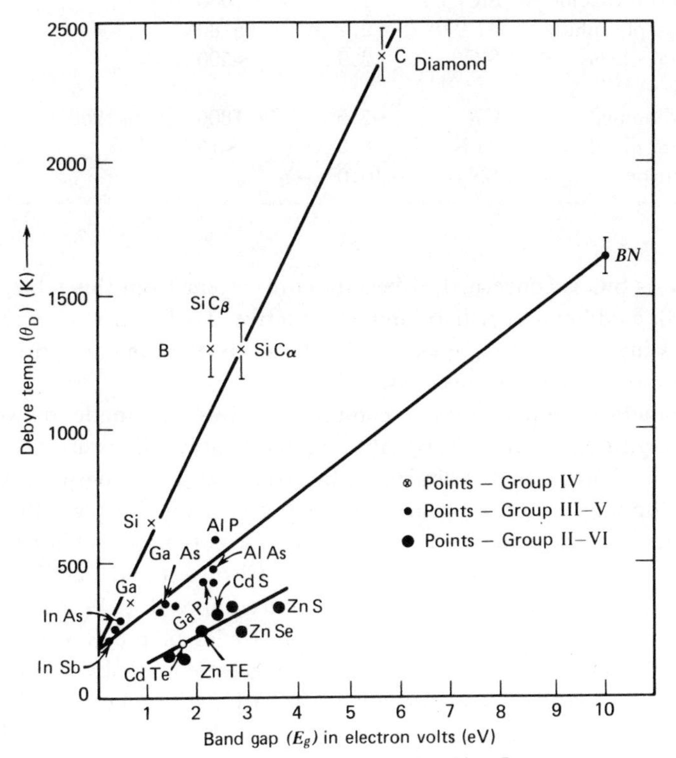

FIGURE 59. Debye temperatures versus band gap E_g. After Ryan.[297]

2. The band gaps in each group are approximately a linear function of the Debye temperature.

Boron seems to fall out somewhat from the general trend, which might be caused by measurements of impure material. A survey of all properties interesting for refractory semiconductors is given in Table 29 of reference 297.

TABLE 29 Properties of some semiconductors formed from elements B, C, N, Al, Si, and P.

Materal Units	Symbol	Energy Gap E_g (eV)	Mobility for Electrons (cm^2 V^{-1} sec^{-1})	Mobility for Holes (cm^2 V^{-1} sec^{-1})	Debye Temp. Θ_D (K)
Silicon	Si	1.1	1400	500	640
Boron	B	~1.5			1300
Boron phosphide	BP	2.0		~200	(1000)?
Cubic silicon carbide	SiC_β	2.3	>1000		1430
Aluminum phosphide	Al P	2.4	65–80		590
Hexagonal silicon carbide	SiC_α	2.9	~500		1200
Carbon diamond	C_D	~5.6	1800	1200	2200
Aluminum nitride	Al N	~6.0	~15		(1000)?
Boron nitride	BN	~10.0			1700

As far as SiC is concerned, it becomes quite clear from this table, that its cubic (β) modification will be more attractive. Its intrinsic energy gap of 2.3 eV is high enough for good sensitivity at elevated temperatures and its electron mobility favors lower resistivity.

For practical applications a compromise must be made between the desired high-temperature sensitivity and reasonably low resistivity. Therefore doping is necessary: *n*-doping with nitrogen or *p*-doping with Al. Since nitrogen-doping by diffusion into the solid SiC is much slower than by Al diffusion,[298,299] it is better done by pyrolytic formation of SiC in presence of nitrogen. As gaseous starting materials, Cl-substituted methylsilane-H_2 mixtures are used (dimethyldichlorosilane decomposed at 1473 K in H_2).[300]

For a recent study of the electrical properties of β-SiC with different nitrogen doping concentrations, the specimens were made by decomposition of methyltrichlorosilane and acetonitrile in an H_2 atmosphere[301] (Figure 60).

For heavy doping with carrier concentrations between $n = 2 \times 10^{20}$ and $n = 8 \times 10^{20}\ cm^{-3}$, the resistivity was nearly temperature independent.

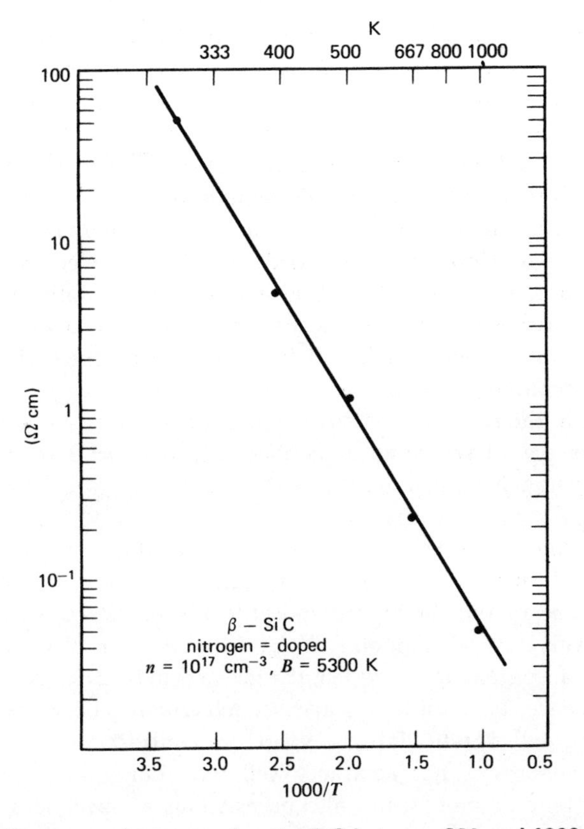

FIGURE 60. Resistivity of nitrogen doped β-SiC between 300 and 1000 K. Results of Golikova et al.[301]

Only for lighter doping with $n = 10^{17}$ cm^{-3} good semiconducting properties were obtained with B values of about 5500 K.

The difficulties to obtain reproducible B- or N-doped SiC crystals by other methods than pyrolysis for thermistor applications have been pointed out by Shaffer[302]; they relate to the formation of polytypes with variable energy gaps. The number of process variables such as temperature, pressure, and impurity concentration (and possibly first derivatives of the first two) are probably responsible for these difficulties. As Shaffer strongly suggests: "a completely different approach" is necessary to overcome these difficulties. The Russian publication previously mentioned does not reveal any problems with reproducibility.[301]

Sequential doping of SiC during sublimation growth for the vapor phase with N and B or Al results in the formation of an internal p–n junction with impedance values decreasing from 10^5 to 10^7 Ω at ~500 K to ~10^3 Ω at ~1000 K corresponding to B values of 7,000–10,000 K.[701]

BORON. The idea to use B as thermistor material dates back to 1959[303] and was followed up by development of sensors for nuclear energy. One of a pair of B thermistors with matched electrical properties is made of ^{11}B, the other of ^{10}B. Only the latter isotope reacts with neutrons according to $^{10}B + n = {}^{7}Li + {}^{4}He + 2.8$ MeV. This released energy in the ^{10}B thermistor unbalances the bridge and gives electrical signals for a flux $= 10^{11}\, n\ cm^{-2}\ sec^{-1}$. The preliminary work was done with boron layers on an alumina substrate.[304]

If B crystals or chips are to be used as sensors, the contact problem is most important, especially if the high-temperature stability of B is to be fully utilized. It has been found that Pt wire electrically heated to ~1280 K and brought in contact with B will form an eutectic melt with its surface, which after cooling is a good contact.[305]

Near 570 K the resistance-versus-temperature characteristic of β-rhombohedral boron can have some anoamlies, which depend on its degree of purity and its quenching state.[306]

SEMICONDUCTING DIAMONDS. Diamond is one of the best insulators over a wide temperature range, with an energy gap of ~5.5 eV. At room temperature resistivities of 10^{12}–10^{14} Ω cm and at 973 K of 10^{7}–10^{9} Ω cm have been measured with natural diamonds.[307] However, Custers[308] reported a large diamond, transparent to wave lengths greater than 2750 Å with unusual phosphorescence, low resistivity, and semiconducting behavior, which was subsequently called type IIb diamond, in contrast to type IIa, which neither phosphoresces nor conducts well. He found that semiconduction exists in all blue diamonds, but also occasionally in specimens that are not discolored. Further electrical and optical measurements with single specimen of IIb diamonds led to the conclusion that their p-type conduction with an activation energy of 0.4 eV is caused by a trivalent impurity similar to Ge doped with trivalent elements.[309–311]

Wedepohl[312] broadened these early data by a systematic study with six p-conducting diamonds. His dc resistivity and Hall-effect measurements between 200 and 800 K supplied data for the activation energy of the acceptors (~0.34 eV), the temperature dependence of the Hall mobility $\sim T^{-2.8}$), the effective mass of the carriers, and the concentration of donors and acceptors: N_D; 2–8 × 10^{15} cm^{-3}; Na, 10–63 × 10^{15} cm^{-3}. Raal[315] has determined trace elements in 25 diamonds of different sources by spectral analysis and found that the trace profiles in IIa (nonconductive) and IIb (semiconducting) diamonds do not differ drastically, though Al appeared to be dominant as an impurity in IIb diamonds. In a recent critical study Al has been replaced by B as the crucial dopant.[314]

A few years later Custer suggested the use of IIb diamonds as thermistors and pointed out their desirable features: corrosion and abrasion resistance, low specific heat, and excellent heat conductivity surpassing that of Cu at ordinary temperature, in addition to a rather high applicable temperatures (>600 K). A novel method to attach leads to diamond cylinders was developed.[313] A review of methods of bonding or electrical contacting diamonds concludes that solder alloys made of a group IVb or Vb elements (Ti, V, Nb, Tc) with a solvent such as Au, or Ag-Cu eutectic are most suitable for this purpose.[316] It embraces the initial methods of Custers and of Rodgers and Raal. It is assumed that these elements react strongly with the adsorbed oxygen on the diamond surface and thus clean it for the bond with the solvent metals. This principle is also mentioned in the Section 5.4.4.

The small abundance of natural semiconducting (IIb) diamonds limited the possibility of their application as thermistors. Therefore the synthesis of IIb diamonds with a high-pressure process was of interest.

During the pioneering work to produce man-made diamonds, crystals were made by adding the dopants to the diamond-forming catalysts, for instance Al to Fe catalyst, B to Fe-Ni, and Be to Ni.[317] This resulted in *p*-conduction with different activation energies, for instance, 0.32 eV for Al, 0.18 eV for B, and 0.2–0.35 eV for Be. It is not clear whether these differences are caused by different doping concentrations. The possibility of diffusing B into natural or man-made diamonds at high pressures and temperatures up to 60 kbar 2100 K for $\frac{1}{2}$ hr has been investigated. Drastic resistivity decreases have thus been accomplished (10^{10}–10^{4} Ω cm at 300°K). However, inhomogeneity of diffusion and trend to graphitization at higher temperature pose considerable problems.

A patent (U.S. 3,435,399, March 25, 1969) has been granted to General Electric Co. for producing diamond thermistors. It covers a wide scope of topics, starting from the doping of the diamonds with 0.001–0.15% B during their synthesis, describing the contacting procedure with a 61.9% Pd, 33.3% Ni, and 4.8% Cr alloy and simultaneous sealing into a glass envelope. These diamond thermistors are said to be superior to natural IIb diamonds, since they do not exhibit their resistivity reversal between 700 and 800 K.

In other studies it has been found that the resistivity is very sensitive against the doping concentration and decreases from 10^5 to 1 Ω cm between 5×10^{16} and 10^{18} B atoms cm^{-3}, while the color changes from light blue to black. Its temperature dependence exhibits two slopes; that of the B acceptor level has $Ea = 0.35 \pm 0.01$ eV. The *p*-conductivity, confirmed by thermoelectric measurements, is partially compensated by a deep *n*-level with $E_D \sim 36$ eV caused by nitrogen traces[318] (Figure 61). The ion injection method has been used to dope natural diamonds with B.[319] With doses of $\sim 10^{15}$ ions cm^{-2}, *p*-conductivity (Figure 62) with an activation energy of 0.3 eV between 600 and 900 K was produced, but only in a surface layer of 10^{-5}–10^{-4} cm thickness. Its resistivity was measured between 300 and

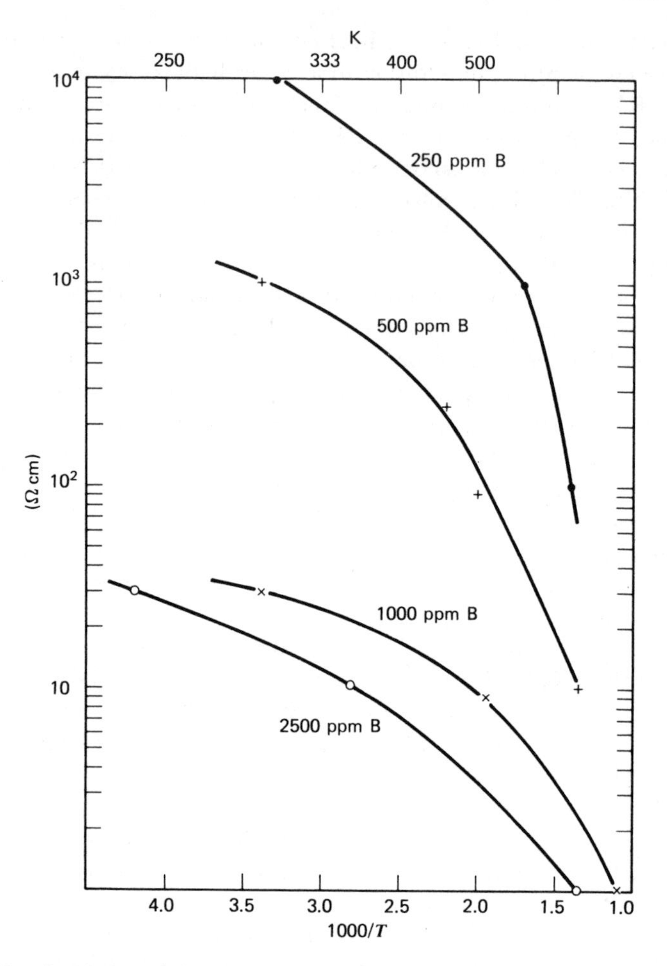

FIGURE 61. Resistivity of diamonds doped with up to 2500 ppm B between ≈250 and ≈700 K. Results of Bezrukov et al.[318]

1200 K and found to be ~5×10^5 Ω cm at 300 K, strongly dependent on annealings above 1400 K. Contact regions were made by (locally) 10–15 times higher injection doses. This method has potential interest to make film thermistors for high temperature and with small time constant, since the high heat conductivity (several times that of Cu at 300 K for synthetic diamonds) of the undoped diamond substrate represents an excellent heat sink.

Semiconductive diamond surface layers with an activation energy of 0.17 eV can be produced by implantation of Li^+ ions into nonconductive

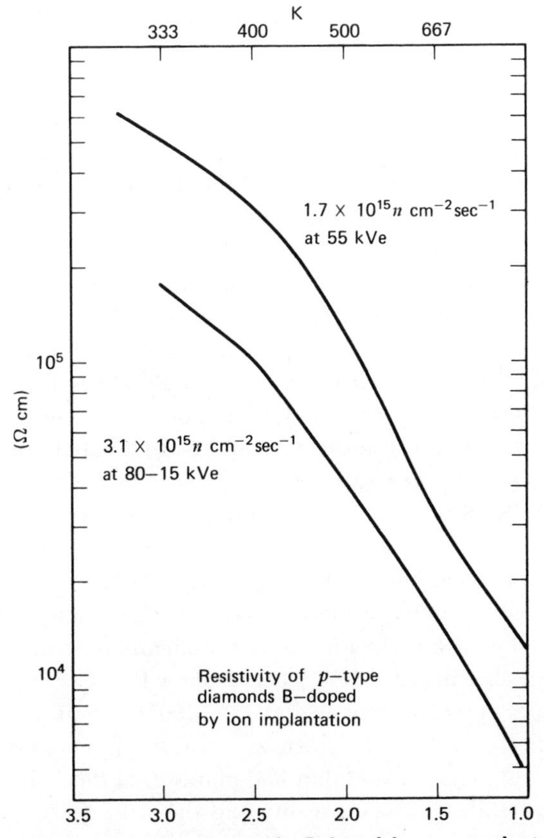

FIGURE 62. Resistivity of p type diamonds B doped by ion implantation. Results of Vavilov et al.[319]

diamond with doses of $<5.10^{16}$ ions cm^{-2} to minimize graphitization. Annealing at 1670 K and removal of the surface graphite leaves a conducting layer with $\sim 10^5$ Ω resistance at 300 K. This procedure would be of interest for making a temperature sensor with very good heat sink (extremely high dissipation constant), pending proof of their stability.[320]

Considerable insight into the effects of various dopants on the electrical and optical properties of semiconducting diamonds has been gained during the last 10 years. A treatment of these developments goes beyond the scope of this book. It should only be mentioned that nitrogen is the most common impurity in natural and man-made diamonds. In a substitutional lattice site it has one excess electron after using four valence bonds for the surrounding C atoms and therefore can act as donor capable to compensate acceptors. If the nitrogen donor concentration is smaller than the trivalent acceptor concentration, the compensation of electrons and holes is incomplete and semiconductivity can be observed.

Recalling that Custers was led to the discovery of IIb diamonds first by their unusual optical properties, their blue color, resulting from enhanced absorption in the red and infrared, can be considered as indicator for a non-compensated *p*-conduction. Raal has measured the absorption of the famous $42\frac{1}{2}$-carat blue Hope diamond and declared it to be the largest semiconducting diamond.

Even if a systematic control and production of semiconducting IIB diamonds would be possible, it remained to be shown that the initial high expectations as to their usefulness at temperatures above 600 K were fulfilled, especially after thermal cycling. The author has measured the electrical conductivity of a few natural IIb diamonds* between 200 and 800 K. The relatively small activation energy of 0.3–0.1 eV reduced the temperature sensitivity at 600 K to $<1\%\ K^{-1}$.

At 730 $\pm$ 10 K, the resistivity minimum, the temperature coefficient goes through zero and then becomes positive. During thermal cycling tests between 300 and 800 K, the resistance tended to drift slowly to higher values, obviously caused by deterioration of the silver contacts. This drift is reversible in H_2. The resistance minimum became more shallow after irradiation of the crystals with 0.6-MeV γ rays from a 10 mCi ^{137}Cs source.

At this time the expectations for IIb diamonds as high-temperature thermistors have been brought down to a realistic level. This was necessary in view of several press releases that had glamorized them. If the synthesis of reproducible crystals becomes a reality and the crucial contact problem can be solved, the high thermal conductivity of diamond (at 300 K, five times that of Cu) and its intrinsic chemical stability could open special fields of applications. Who would have dreamt of diamond heat sinks for diodes and lasers ten years ago?

Cubic boron nitride BN (borazone) has diamondlike hardness and corrosion resistance. Doped with Be, B, Si, Ge, S, or Se it can be made semiconductive with resistivities ranging from 10^{-2} to 10^{10} Ω cm. Its most interesting features is a strong trend to increasing B values with rising temperature. In a material with a resistivity of $\sim 3 \times 10^4$ Ω cm at 300 K, B increases from 500 (averaged between 90 and 300 K) to 1300 (averaged between 300 and 470 K) and to 3000 at 750 K. This is an unique case, in which for the entire temperature range from 90 to 800 K one thermistor

* Unpublished work by the author; crystals received by the courtesy of Industrial Diamonds, South Africa.

could be used with still reasonable resistivity at 90 K ($<10^7$ Ω cm) while retaining acceptable sensitivity above 600 K ($>0.6\%$ K^{-1})[321] (Figure 63).

The preparation of cubic BN at pressures of 120–150 kbar by pulse heating within milliseconds using a capacitor bank of over 8×10^4 μF has been described.[322]

A polycrystalline thermistor consisting of fine particles of diamond, cubic boron nitride, or silicon carbide hot molded in an externally heated pressure cell at temperature over 1570 K has been described. Molybdenum leads are molded in during the pressing. If B-doped diamond powder of micron-range particle diameter is molded, the resistance ratio between 300 and 1100 K is ~10 (compared to ~50 for synthetic diamond crystals.)[323]

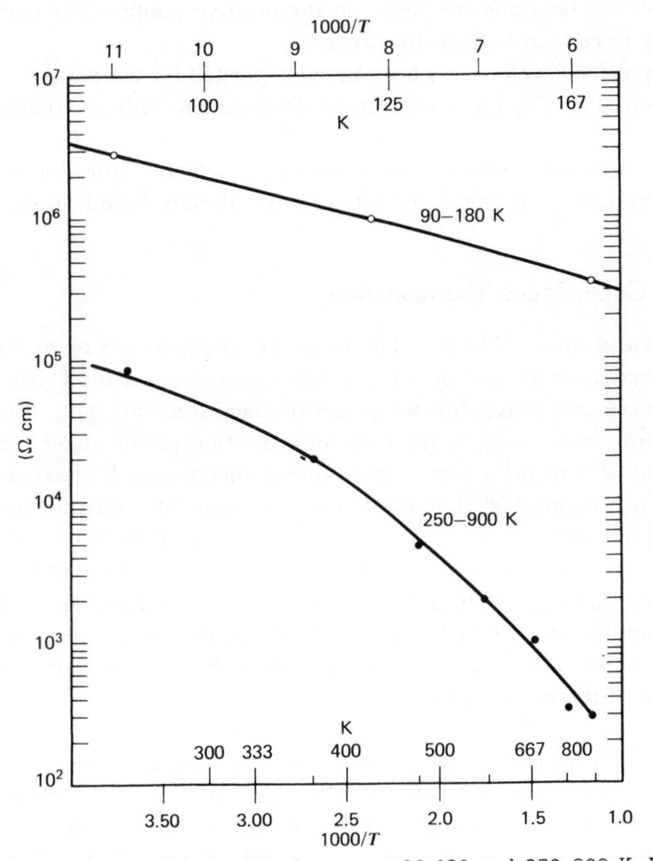

FIGURE 63. Resistivity of Borazon in the ranges 90–180 and 250–900 K. From U.S. Patent 3,435,398.

4.3. THERMOMETRY BASED ON NONRESISTIVE PROPERTIES

In principle, any temperature-dependent property of solid-state materials (including semiconductors) besides resistivity could be used for temperature measurements and secondary temperature scales. The general preference for resistivity is, of course, based on the simplicity of equipment for measuring it precisely and the ease of eliminating various side effects. This is especially true for semiconductors with temperature coefficients much larger than pure metals. Other electrical properties such as the dielectric constant and the dissipation factor have, in general, a much smaller temperature dependence. Ferroelectrical materials may be an exception, although even these, until recently, have hardly come into their own as capacitive temperature sensors. If properly doped for semiconductivity (PTC materials), large-scale applications are based on the positive temperature coefficient of resistivity in certain temperature regions.

The pyroelectric effect has been much neglected for temperature measurements until 1961.[324] The most recent development will be treated in later sections.

Electromechanical effects—resonance of piezoelectric vibration as functions of temperature—and associated effects have already found application.

4.3.1. Capacitance Thermometers

APPLICATIONS ABOVE 300 K. The large *TC* of capacitors made from ferroelectric materials for a wide temperature range around the Curie temperature is very attractive for temperature measurements and control. For temperature monitoring in the rotor of a traction motor under the working conditions of a train, a non contact thermometer in a *LC* passive resonant circuit has been applied. The Ba (Sr) titanate capacitor used for this purpose had sufficient sensitivity even with small coupling of the coils in the heterodyne detection circuit and remained stable within $\pm 3\%$ under severe test conditions, namely, 100 hr at 535 K, cycling between 220 and 500 K ($20\times$), 90% humidity and 1000 hr at 495 K. With the chosen bridge conditions the signal-to-noise ratio could be improved despite strong flashover and spark discharge at the commutator.[325]

APPLICATIONS BELOW 77 K. Recent dielectric measurements with monocrystalline specimens of Si, doped with B, P, B + P, or Sb show a strong temperature increase of the dielectric constant that can be shifted systematically to higher temperature between 10^3 and 5.10^5 Hz. This behavior is rather interesting since it would offer the possibility of adjusting the sensitivity of one sensor by selecting the optimal test frequency for each test temperature. The useful range would be 20 and 40 K for Si with a $1.5.10^3$-Ω cm resistivity.[326]

Crystallization of $SrTiO_3$ with perovskite structure in an aluminosilicate glass matrix produces a glass ceramic that has a sufficiently large dependence of its dielectric constant for applications as low-temperature sensor. Lawless[327,328] has reported on the production and the dielectric properties of this material over a temperature range from 25 mK to 300 K with special emphasis on the range below 72 K.

Since capacitance measurements are made with 5 kHz, the loss factor is also of importance because of possible self-heating. Although it increases from very low values above 200 K to 1–2% below 20 K with a peak of 2.5% at 40 K, it is small enough to limit self-heating power with 5-kHz and 7.5-mV test amplitude to picowatts. Response for nonencapsulated units approximates 70 K sec^{-1}. Two basic types have been developed for different temperature ranges.

1200 type:	60 mK–10 K
1100 type:	1.7–60 K

The largest sensitivity dC/dT is in linear region below 5 K with 180–300 pF/K and $d \ln C/dT = 1.5\ K^{-1}$ (Figure 64).

Curve-fitting studies for the *C*-versus-*T* relation were made for both types using series expansions. With fourth-order fits the residual standard deviations were reduced to −10 mK compared to 8–15 terms to obtain the same goal with Ge thermometers.[329]

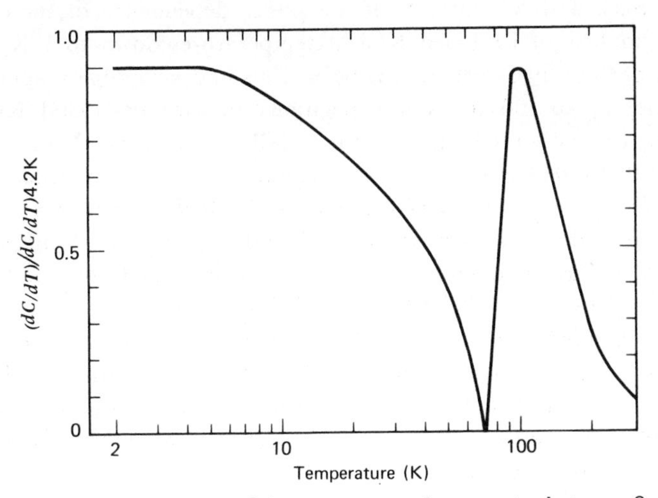

FIGURE 64. Relative sensitivity of the capacitance thermometer between 2 and 300 K (dC/dT at 4.2 K = 1). Results of Lawless.[333]

Both types have a capacitance minimum (at $\sim$0.11 K for type 1100 and $\sim$0.06 K for type 1200). Below these temperatures the capacitance seemed to increase with T^{-1}.[330,331] However, recent calibrations of the 1100 type with a nuclear orientation thermometer of ^{60}Co have shown that this relationship ceases below 25 mK.[332]

This glass-ceramic dielectric is prepared by melting 7 $SiO_2 \cdot 1Al_2O_3$ with 54 mole % $SrTiO_3$ and up to 2 mole % Nb_2O_5 at 1930 K in a Pt crucible. $SrTiO_3$ is crystallized from this glass by reheating to 1075–1575 K. For the preparation of cryogenic sensors, the annealing to crystallize $SrTiO_3$ was made at 1375–1475 K resulting in the two grades of different temperature dependence shown above.

Sheets of 25 cm² and 0.037 cm thick are ground and contacted with silk screened Au-Pt Multilayer elements with 51 capacitors are stacked together as with mica capacitors. Commercial units are marketed in encapsulated form.* Data for repeatability have only been given for unencapsulated units.

After the capacitance drift had been stabilized during the first 60-min immersion into liquid He, reproducibility during thermal cycling at He temperatures is better than 2 mK and after thermal cycling between room temperatures and liquid He ± 13 mK. The apparent maximal temperature error in magnetic fields up to 80 kG was of the order of 1 mK and not strongly influenced by the test frequency between 0.5 and 10 kHz[333] (Figure 65).

Another type of dielectric thermometer has been developed by a research group at Cornell University, Ithaca, New York.[334] Based on earlier observations by Sack and Moriarty[336] of the linear dependence of the dielectric constant on $1/T$ of Li-doped KCl at temperatures down to 1 K, a thermometer of the range 1–30 K was built. This had negligible magnetic field dependence up to 100 kG. The capacitance of a single-crystal KCl plate 1.9 cm² by 0.038 cm thick, doped with 3×10^{18} ^6Li ions cm^{-3} and contacted at the parallel 100 faces with evaporated gold, followed an equation $C(pF) = A + B/T$ with $A = 30.15(pF)$ and $B = 1574(pF)$ when measured with 10^5 Hz. The sensitivity increases from 110 mK at 30 K to 3.6 mK at 4.2 K and 1 mK at 1 K. The temperature range was extended down to 0.1 K by substituting Li with 0.009% OH or 26% CN as dopant,[335] but the insensitivity to magnetic fields was retained up to 50 K. In this case the frequency of 1 kHz is applied to limit dissipation losses that are larger than for Li-doped KCl (D at 100 KHz equals 10^{-8} W). Thermal cycling produced capacitance changes of a few tenths of one percent. Long-range stability of these capacitive sensors is still under investigation.

* Lakeshore Cryotronics, Eden, New York 14057.

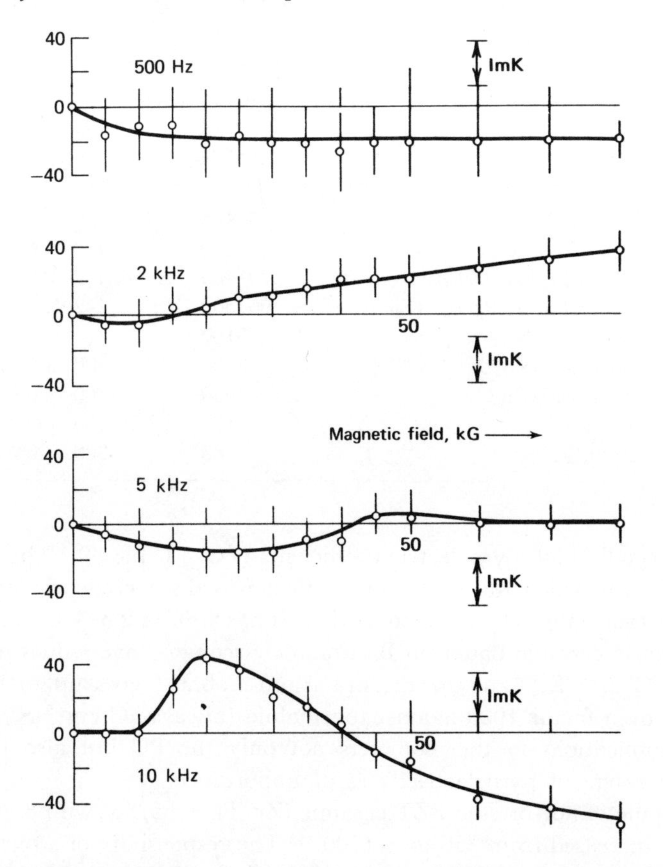

FIGURE 65. Influence of the magnetic field and test frequency on the sensitivity of a capacitance thermometer. Results of Lawless.[333]

75–200-Å thin films of TiO_2 or Nb_2O_5 sandwiched between 500-Å thick Al electrodes and supported by a thin mica sheet form a capacitive bolometer with a responsivity of $\geqq 2$ mV/μW and a time constant of 0.05 sec. With a detectivity of about 2×10^{10} cm $Hz^{1/2}$ W^{-1}, it is at least so good as other thermal detectors.[677]

4.3.2. Pyroelectric Thermometers

Since the early 1960s Lang[337] and some co-workers have pursued the goal of making pyroelectric thermometers practicable. The strong expansion of the ferroelectric research going on during this period has provided a choice of new materials for evaluation, since it was found that ferroelectric materials

TABLE 30

Material	Pyroelectric Coefficient at 300 K ($\mu C\ m^{-2}\ K^{-1}$)	Range (K)
Tourmaline	2.3–4.4	20–648
$BaTiO_3$ ceramic	180–200	4.2–396
$Pb(ZrTi)\ O_3(PZT_4)$	370–400	4.2–348
$Pb(ZrTi)\ O_3(PZT_5)$ (Figure 66)	400–500	4.2–523
$Sr_{0.75}Ba_{0.25}Nb_2O_6$	3100	
$Sr_{0.73}Ba_{0.27}Nb_2O_6$	2800	300–500
$Sr_{0.48}Ba_{0.52}Nb_2O_6$	650	300–500
$LiTaO_3$	176	300–500
$LiNbO_3$	83	300–500

have especially high pyroelectric coefficients (PC) (Table 30). The classical material with which the pyroelectric effect was discovered accidentally in 1703 was tourmaline crystal from Ceylon. It has a PC of 2.3–4.4 $\mu C\ m^{-2}\ K^{-1}$. Ferroelectric ceramic based on Ba-titanate-zirconate have values from 300 to 500 $\mu C\ m^{-2}\ K^{-1}$ and strontium-barium-niobate* goes up to 3100, an increase by a factor 1000 against tourmaline (always $\mu C\ cm^{-2}\ K^{-1}$).[338–341]

For applications in thermometers not only the PC but also the temperature range of pyroelectricity is of importance.

By doping of hot-pressed PZT ceramic (Zr/Ti = 65/35) with 8 at. % La, its PC is increased from 350 to ~1700.[342] The responsivity of a pyroelectric detector is given by $V/W = PC/\rho \cdot Cp \cdot a(C_c + C_0)$, its noise equivalent power (NEP) in watts multiplied by the ratio rms noise voltage to responsitivity. ρ, Cp, and a are the density, specific heat, and thickness of the detector plate. C_c and C_a are the capacitances of the detector and the amplifier. In most cases $Cc \gg Ca$. ω is the angular frequency. The NEP for the materials mentioned above is $<10^{-10}$ W $Hz^{-1/2}$. All pyroelectric detectors are limited by their thermal noise caused by ac conductivity. Signal and noise decrease with increasing frequency. Since the signal decreases faster than the thermal noise, the detectivity is better at lower frequencies. Based on a PC of 200 $\mu C\ m^{-2}\ K^{-1}$, a $BaTiO_3$ disk of 1-cm^2 cross section and 0.1-cm thickness, 1 K temperature change produced an electric change of 2.10^{-8} C. Assuming a dielectric constant of 1400, the capacitance is 10^{-9} F and the resulting voltage across the disk is 20 V, compared with 20–200 $\mu V\ K^{-1}$ for metal or semiconductive thermocouples. At first glance this appears to be rather

* $Sr_{0.75}\ Ba_{0.25}\ Nb_2O_6$.

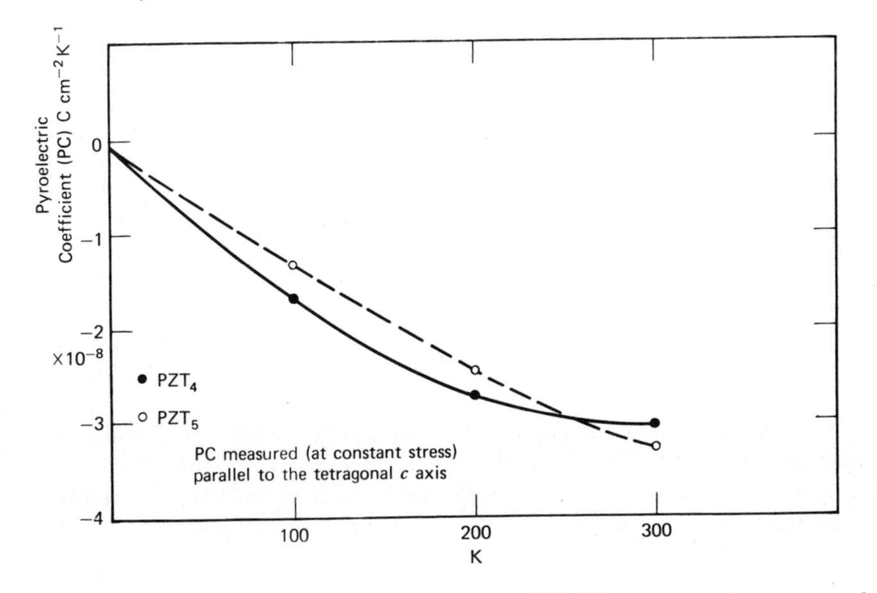

FIGURE 66. Pyroelectric coefficient of PZT-4 and of PZT-5A at constant stress measured in a direction parallel to the tetragonal c axis. After Lang.[324]

favorable. There are some drawbacks, however. First of all the pyroelectric voltage signals appear at a source with very high internal impedance and cannot be measured with a normal electromagnetic voltmeter. Secondly random fluctuations produce a noise in the thermometer that limits the sensitivity. Lang et al. have analyzed these factors and proposed several solutions for the optimal use of pyroelectric sensors.

1. The pyroelectric voltage should be modulated by periodically short circuiting the pyroelectric sensors or the voltage measuring instrument at its input. This minimizes the NET.
2. When measuring temperature rates, no modulation should be applied, Response to high-frequency temperature fluctuations is improved by use of a smaller shunt resistor or a sensing material with a lower dielectric constant.
3. Occasional repolarizing of the pyroelectric ceramic by a large dc field is desirable.

A commercial windowless detector* based on a strontium-barium niobate crystal has the specifications given in Table 31 and can be operated between 253 and 363 K (below the Curie temperature of the detector crystal).

* Molectron Corporation, Sunnyvale, California 94086; approximately $1000.

TABLE 31

Load resistance (Ω)	Frequency range	Rise Time	Responsivity (50-Ω Output) Resistance
10^9	70 Hz	5 msec	125 V/W
10^8	700 Hz	500 μsec	15 V/W
10^6	70 Hz	5 μsec	150 mV/W
10^4	7 MHz	50 nsec	1.5 mV/W
10^3	70 MHz	5 nsec	0.15 mV/W

The change in spontaneous electric polarization produced by the absorbed heat radiation is balanced by current in an external circuit. The voltage drop across a load resistor (given in the first column of the Table 31) resulting from this current is measured and used to compute responsivity according to the formula

$$Rp = \frac{V}{W(\omega)} = \frac{p(T)\, R_L}{\rho c_p d[(1 + \omega R_L)\,(C_L + C_C)]}$$

V = voltage signal
$W(\omega)$ = absorbed radiation of angular modulation frequency
$p(T)$ = pyroelectric coefficient at temperature T
RL = effective load resistance
d = distance between crystal electrodes
c_p = specific heat
ρ = density
C_L = capacitance of external load circuit
C_C = crystal capacitance = $\varepsilon_r \varepsilon_0 \cdot A/d$
ε_r = relative dielectric constant
ε_0 = permittivity of free space
A = electrode area

Increasing R_L increases also R_p but also decreases the range of flat frequency response as shown in Figure 67. Responsivity and flatness are optimized by increasing R_L and holding R_L $(CL + C_c)$ below unity. This means reduction of the electrode area of the crystal and minimizing the external capacitance of the load circuit. An infrared bolometer using the pyroelectric effect of triglycine sulfate (TGS) has been suggested for thermal imaging, laser calorimetry, and multielement line scanning.[343]

Oriented films of 10–20-μm-thick polyvinylidene fluoride have pyroelectric coefficients of about $3nC$ cm^{-2} K^{-1}, corresponding to signals up to 47 V W^{-1}

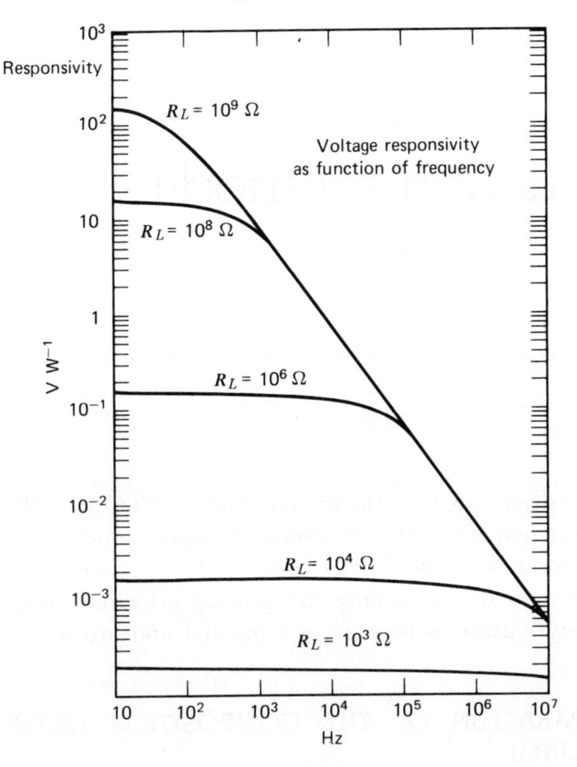

FIGURE 67. Voltage responsivity of pyroelectric sensor as function of frequency for different effective load resistance between 10^3 and 10^9 Ω. After Molectron, Sunnyvale, California.

with a NEP = $5.7.10^{-9}$ W $Hz^{-1/2}$ at 90 Hz. Though less sensitive than crystalline pyroelectric materials, they have the advantage of low price and large receiving areas.[344]

A rather novel thermometric principle was recently described by Hirschkoff and co-workers.[345] Its material base is neither "fish nor fowl", neither semiconductor nor dielectric. During the static magnetization of copper nuclear spins with the superconducting quantum interferometric device (SQID) magnetometer in a dilution refrigerator, a temperature-dependent part of the magnetization was found as a background obeying a Curie law for values $H/T < 1000$ Oe K^{-1} and saturating for larger values. This effect was traced to the organic (linseed and tung oil base) insulation of the niobium wire of the magnetometer. This insulation contains small amounts of Mn or Cr naphtanate, which are probably causing this magnetic effect. The authors believe that this enamel might be useful as a magnetic thermometer in the mK region if large time constants are tolerable.

5

Production methods

This chapter deals only with the manufacture of NTC and PTC thermistors. For the production of Ge, Si, and compound semiconductors such as GaAs an abundant literature exists that is outside of the scope of this book. The following chapter describes only the general principles, not technological details. Many variations have been suggested and are used.

5.1. PREPARATION OF THE COMPOSITION (BLENDING AND MILLING)

5.1.1. NTC Materials

In most cases the thermistor compositions are only blended and milled. Solid-state reactions are left to the sinter process. Dry or wet milling is done in ball mills of porcelain, mullite, or sillimanite (minerals of the general composition $xAl_2O_3y\ SiO_2$); for smaller wear resistance and less contamination, rubber-lined mills or those made of agate are recommended. The usual milling time ranges from 4 to 48 hr and conditions are normally chosen in the shop to optimize economy and quality. To obtain rapid comminution to micron dimensions, vibration mills with accelerations of up to 10 *g* are useful.*

5.1.2. PTC Materials

After blending plus wet milling up to 6 hr, the slurry is dried with or without previous filtering, then dried again and reacted for 2–4 hr at 1270 K. The

* Vibratom Oscillating Mill, N.V. Tema, The Hague, Holland.

calcination product is again milled up to 6 hr dry or wet (in the latter case redried) and blended with a binder or lubricant (polyvinylalcohol), stearic acid, or another wax free of inorganic impurities.

A study of the mechanism of the $BaTiO_3$ formation, based on geometrical considerations of optimal reaction surface between the component particle, has indicated that $BaCO_2$ particles of 1 μm diameter mixed with TiO_2 particles of 0.3 μm diameter would be the optimal condition for the formation of $BaTiO_3$.[346]

For real production runs, one has to rely on the possibility of approximating this goal by trial and error. A wet process is based on the earlier method[74] to precipitate Ba-titanyl oxalate tetrahydrate for conversion to high-purity $BaTiO_3$. Its application to PTC materials required studies to determine the amount of rare-earth dopant[347] La or Nd[348] that was co-precipitated.

In order to avoid considerable losses of dopant, the precipitation has to be done under controlled pH conditions.

Another wet method is based on simultaneous hydrolytic decomposition of barium bis-isoprooxide and titanium tetrakis tertiary amyloxide. The produced particle size range is 50–150 Å, the purity 99.98+%.[349]

5.1.3 Organic Thermistors

It has always been taken for granted, that thermistors and other semiconducting components are to be of inorganic nature. The reason for this opinion was obviously the conviction that organic materials would have extremely high resistivity, poor stability, and lack electronic conduction. In recent years a new look developed, based on the increasing knowledge of organic semiconductors. The expected potential of easier fabrication and better control of the manufacturing process with resulting improvements in reproducibility stimulated a number of patents and publications on a global scale. The scope of this book does not permit to go in details. (see U.S. Patent 3.239,785, French Patent 1.443.121 plus Additions 88.337 and 89.238, German Patent 1.954.255 in list of Patents.)

Nylon thermistors for electrical blankets based on Nylon compositions with semiconductive compounds have been suggested.[710,711]

5.2. COMPACTING AND FORMING

5.2.1. Disks and Washers

The raw materials, mainly oxides, often with added dopants and lubricants, are ball milled to a homogeneous mixture, as mentioned above. In some cases an organic binder is blended in during or after the main milling operation. Automatic molding requires a free flow of powder from the "filling shoe" into the mold. Fine powders often do not meet this condition. Therefore a "pelletizing" or granulating process is necessary after ball milling.

Numerous binders and lubricants have been described and used to mold parts with good green strength at high rates. Isobutyl-methacrylate applied as diluted solution in organic solvents (xylol) has been often mentioned in the literature besides less sophisticated additions like "fish oil" and turpentine. The advantages which a binder and lubricant offer in the molding process must be carefully balanced against their obvious disadvantages. These are:

1. Possible chemical interaction with the oxides and changing their oxygen-to-metal ratio. It is possible that in most cases slight reducing effects can be offset by later heating in oxidizing atmosphere. However, in view of the normally short sinter times there is always the possibility that the compositions might "remember" an earlier partial reduction by the effect it might have on the sinter kinetics.
2. Inorganic traces in form of alkali or alkali earth elements might dope or compensate intentionally doped compositions (see Section 3.5.2).

In general, it will always remain a desirable goal to abstain from organic binders and lubricants whenever possible. Often the right amount of humidification ($\sim$0.5–3%) is good enough to make organic binders expendable. Inorganic binders, if necessary, are always a permanent constituent of the composition. Glass powders have been often used in amounts of 2%. Also in this case the obvious advantages in obtaining mechanically strong units can result in undesirable excessive capacitance values or alkali doping. SiO_2 and B_2O_3 ,devoid of cations and used in small amounts, act mainly as binders with no doping effects.

The molding pressure depends on the nature and composition of the oxide mixture and the added binder lubricant and may vary between 70 and 900 kg cm^{-2} (1000–12,000 psi). Too high pressures can produce laminations perpendicular to the molding direction, especially with fine powders. In general, molding dies are tapered to facilitate ejection of the molded compact from the die without cracking. The compact under pressure within the die has still some elasticity and tends to expand when pushed from the die by the lower plunger. This sudden release of stress can lead to cracks in the compact that often do not heal during sintering.

5.2.2. Rods

By adding to the ball-milled oxide mixtures organic binders (natural resin tragacanth) or synthetic plastifiers as used in the ceramic industry, plastic putties can be prepared that are extruded through cylindrical dies as rods, tubes, or other profiles, dried, and then cut to length. Here again not all plastifiers suitable for ceramic materials can be used. Reducing action by the plastifier itself or its carbonization products and inorganic impurities or constituents have to be taken into consideration.

The plastifier is driven out by slow heating, to avoid cracking or gas bubble formation. Final sintering requires uniform heat to prohibit warping and bending of the extended pieces. For technical details see references 350 a and b.

5.2.3. Beads

There is a lower limit in size and mass for die molding powders, either determined by their ability to flow freely or by the tendency of the tiny plungers to fracture. On the other hand many applications require thermistors of fast response and small power rating and therefore small size. This prompted development of a new technique using a slurry of thermistor material. Lead wires of 25–125 μm (1–5 mil) diameter with higher melting point than the sinter temperature of the thermistor composition are held in a jig under slight tension parallel to each other in a distance determined by the desired size of the bead, for instance 3–16 mil (80–400 μm). The oxide slurry containing an organic volatile binder is applied between the wires at regular distance and then dried and sintered in controlled atmosphere at a temperature sufficient to obtain density for holding oxide bead and lead wires together. After sintering and cooling, the string of beads is cut to provide different lead arrangements. For details see U.S. Patents 2,552,640, 1951; 2,973,283, 1961; 3,016,506, 1962; 2,897,584, 1959; 2,977,558, 1961; 3,068,438, 1962; 2,739,212, 1956.

To improve the long-range stability of beads and to protect them against chemical attacks, they are normally coated with a thin glass film or hermetically sealed into the tip of a thin-walled glass tube. A number of modifications of this encapsulation method exist and have been described. One of the most interesting is suggested in U.S. Patent 3,430,336, which claims to attain minimal time constant using a thermal gradient to the environment in the direction of current flow by making the outer enveloping surface conductive and using it as electrode.

For beads of sufficient size, leads have been successfully molded in by using special tools.

A novel method of producing a bead thermistor suitable for temperatures above 600 K has been suggested in U.S. Patent 3,479,631, November 18, 1969. Ilmenite ($FeTiO_3$) ground to <100 mesh (<0.015 cm) is heated in air at $\sim$1590 K (2400°F) and then melted at $\sim$1800 K (2800°F) to form a glass. Dipping suitable metal wires into the melt, even in reducing atmosphere to prohibit wire oxidation, produces beads. By annealing in air at temperatures between 470 and 970 K, the electrical resistance can be adjusted as desired over a range of four orders of magnitude. Sealed in glass envelopes, they remain in stable resistance and slope after 120 hr at 770 K (~930°F) and after passage of dc current corresponding to ~17,000 C.

$PbO \cdot PbCrO_4$ can be melted in air at ~1200 K in a Pt crucible. Rapid withdrawing a pair of Ag, Au, or Pt wires from the melt produces a bead, which can be aged at lower temperatures. Doping with trivalent rare earth (Nd, Eu, Tb, Dy, or Y) reduces the resistivity.[675]

GLASSY THERMISTORS. The possibility of making thermistors from a melt is intriguing. It would reduce the production process to a simple dipping of a pair of leads into a melt, possibly followed up by an annealing to reduce stresses and to adjust the electric properties. Most glasses are ionic conductors and have very high resistivities in the temperature range usual for thermistors. There are few exceptions:

1. Phosphate glasses containing Ti, V, Mn, Fe, Co, Ni, Cu, Mo, and W. Lowest melting points (1170 K), resistivities, and activation energies can be obtained with 50–88% mol. % V_2O_5 (10^4–10^5 Ω cm at 300 K and $B \approx 4000$).[351]

Chalcogenide glass bolometers[352] as 10-μm-thick sputtered film have a NEP of 10^{-9} W $Hz^{-1/2}$, but inherently slow response (~1 sec). Glasslike microthermistors of the V_2O_5-P_2O_5 ZnO-TiO_2 system have been described.

2. Chalcogenide glasses: These nonoxidic semiconductors based on the general formula $Me_xX_y \cdot Me$ can be metals and X: S, Se, Te, As, Sb. In general they are crystalline, but can be prepared as amorphous (disordered) phase. One example is Tl_2 Se As_2 Te_3 with the resistivity of 300 Ω cm at 300 K and an activation energy of 0.35 eV (B = 4000 K). Its low softening temperature of 360 K simplifies hot pressing of sensors or bolometers, but also limits its upper application temperature (see also bolometers).[353]

5.2.4. Flakes and Chips

For infrared detectors, thermistors of small mass but larger ratio of area cross section to thickness are desirable. Flakes can be made by coating smooth substrate plates (glass or metal sheet) with a thermistor slurry, drying it on the substrate, removing the film, and cutting it into flakes of desired size and shape and sintering it on a smooth ceramic plate. Conventional sizes of flake thermistors range from 0.5 × 0.5 to 3 × 3 mm.

5.2.5. Sheets and Strings

As early as 1968 a press release stirred attention to a plastic thermistor with sufficient flexibility to be used in blankets and developed by Matsushita Electric Industry Co., Osaka, Japan. A technical report sometime later indicated that a truly organic semiconductor and not a composite of an inorganic semiconductor in a plastic material was developed. It can be molded into wires or sheets and detect average temperatures over large areas with twice the sensitivity of conventional thermistors.[354]

5.2.6. Films

The general trend to microcircuitry did not stop at semiconducting temperature sensors. The first step in this direction, the bead and the chip, was followed up by thick-film (~1–10 μm) technology, to produce chips and flakes in quantity and with uniformity. The techniques of doctor blading are described on page 150 and silk screen printing have been used or suggested. The earliest efforts in this direction are to be found in a Western Electric flake patent.[355]

Since then several methods, mainly in patents, have been suggested and tried to produce film thermistors. It is not only impossible to strive for completeness in reporting them, it is also doubtful whether they are all hold enough promise for reliable units.[356]

Although there have been indeed serious efforts to market film thermistors,* a major breakthrough to large-scale production has not occurred. The most promising approach has still been the thick film based on the technology of flake production. In this case even unsupported films can be made, while thin films (<0.5 μm) always require mechanical support by an insulating substrate, often also suggested for "thick films". In the following pages a few characteristic examples for thick-film methods are given.

1. $Co(NO_3)_2$ + Mn-acetate in molar ratio 14:1 are coprecipitated as hydroxydes and then reacted 3 hr at 1323 K (1920°F). The oxide powder mixture fraction with <0.5 μm particle size is separated by sedimentation at pH ~3. A suspension of ~50% of this oxide mixture in 5% polyvinyl alcohol–water is applied to polyethylene foil and dried at 323 K. The dried oxide layers are cut into rectangular specimens 0.5–2 mm wide and 2–4 mm long, preheated 3 hr to 723 K (750°F) and then rapidly (~300 K hr^{-1}) heated to the sinter temperature of 1323 (1920°F). After 3 hr the units are cooled in the furnace to 873 K (1115°F) and then rapidly to room temperature.[357]

* Thinistors by Victory Engineering Co.

The translation of common thermistor compositions on Mn-Co and Cu oxides basis into thick-film printing has been suggested in a number of patents of international scope as shown in the following Table 32.

Thin-film technology requires more expensive equipment and is less adaptable to mass production. There are two approaches open: evaporation or sputtering.

Vapor deposition of semiconducting films lends itself mainly to elemental semiconductors such as Si and Ge or compound semiconductors of the III-V (GaAs) or II-VI (CdTe) type, when both components can be co-evaporated within a reasonable margin of temperatures. Experiments to evaporate transition metal oxides in presence of O_2 have been disappointing and led to oxide films of poor reproducibility and stability. This contradicts claims by Miura that mixed oxide films of Mn, Ni, Co, and Cu evaporated on fused quartz had the same electrical characteristics as the bulk material and could be used as infrared bolometers.[358] Also, experiments to obtain useful thin film thermistors by controlled oxidation of evaporated about 1000-Å-thick metal films on mica or other insulating substrates were not very successful.* The necessary oxidation temperature promoted grain growth and the films, though apparently free of nonoxidized metal (Co, Ni, Cu, separately or mixed), had rather small temperature coefficients. Similar effects have been found by Lark-Horowitz and co-workers,[359,360] even with vacuum evaporated or chemically deposited (by dissociation of GeH_4) Ge films that differed very much from bulk material. They had a rather small temperature coefficient of resistivity and much lower hole mobility and were obviously highly disordered. Deposition of oxide films by reactive sputtering in Ar with variable percentage of O_2 holds more promise to obtain films of transition metal oxides. Mixed oxide films from prealloyed or geometrically split cathodes can be obtained with desired metal ratios.

For the preparation of thin-film thermistors with PTC properties, several methods have been suggested. Spreading a slip paste plastified by addition of a few percent of polyvinylalcohol with a steel blade over a flat glass plate ("doctor blading") resulted in mechanically uniform 50-μm-thick films, if the spacing between blade and plate was controlled by a micrometer head. The films required firing at 1670 K in oxygen atmosphere on setter plates of prefired undoped $BaTiO_3$, $Ba(Sr)TiO_3$, or Pt to obtain high PT coefficients of 8–10% K^{-1}. Reactive sputtering of a vacuum sintered cathode made from Ba, Sr, Li, and Ti hydrides with subsequent oxidation in air or direct sputtering of PTC material in Ar has been tried with variable success. By using a pyrolytic graphite substrate, unsupported PTC films ($Ba_{0.600}Sr_{0.397}La_{0.003}TiO_3$) 5–25 μm thick have been prepared. After separation from the

* Unpublished by the author.

TABLE 32 Patents.

(a) System Mn-CoO Oxide with Pd + Ag	Components (mole %)	(a) Substrate and (b) Vehicle	Sinter Temperature in Air
U.S. 3,408,311, Oct. 29, 1968	85–40 Co_2O_3	(a) Al_2O_3	923–1093 K
	15–60 MnO_2 to 1 part of oxide mixture 3–6 parts Pd 0–3.6 parts Ag	(b) Decanol, terpene or esters of higher alcohols with 1–4 parts glass binder	(1200–1515°F)
1 kΩ at 300 K	$\alpha = 0.6\%\ K^{-1}$		
(b) System Mn-Co Oxide	(mole %)		
U.S. 3,469,224, Sept. 23, 1969	40 MnO 40 CoO	(a) Oxidized Ni with NiO 33 μ thick 100 per comp.	1633 K (2475°F)
	20 Cu_2O	(b) 23 *n*-butyl Carbitol 2 Et-cellulose	15 min Ag electronic Screen printed
1.2 kΩ at 300 K			
(c) System Mn-Co Oxide	Parts		
Brit. 1,226,789, March 31, 1971	33 Co_3O_4 27 MnO_2	(a) Al_2O_3 (b) Diethylene glykol Monobutyl Ether Et-cellulose Glass part	1023 K 1400°F Contacting with Pd-Ag conductive LnK
300 Ω at 300 K			
(d) System Mn Oxide			
Span. 363,566, Jan 1, 1971	Mn oxide	(a) Au-Pt electrode as base Pt-Au cover electrodes	Together with electrodes
		(b) Et-cellulose Terpineol	1273–1373 K (1820–2010°F)

substrate, sintering at 1670 in Ar and cooling in air a *TK* of ~6% K^{-1} was observed.

From 0.1- to 1-μm-thick $BaTiO_3$ films have been made by simultaneous evaporation of its constituent oxides by electron bombardment and reacting then on a Pt, Rh substrate at temperatures above 1070 K. This method results in films with nonstoichiometric ratios Ba/Ti, hardly inviting its extension to PTC films.[361,362] The preparation of thick-film PTC thermistors by silk screening of a paste and subsequent firing of the film substrate combination has inherent short-comings, partially caused by the different thermal expansion of film and substrate. It has been suggested to prepare a porous wafer with ~50% theoretical density by prefiring of the basic Ba(Sr) composition at temperatures below 1570 K, soaking it with the desired rare earth or other dopants, and then firing it at 1670–1725 K for 1 hr.[363]

Sputtered VO_2 films on Al_2O_3 substrates with semiconductor to metal switching temperature of 361 K have been produced by rf sputtering[364]; see Section 2.3.

5.2.7. Indirectly Heated Thermistors

So far only thermistor units of all sizes and shapes have been described. They did not include heated units that are assemblies of the temperature-sensitive semiconductor and practically temperature-independent resistance heaters. The first assemblies of this type were made by Osram (Germany) in the early 1930s as a semiconducting sensor tube of partially reduced $MgTiO_3$ radiation heated from an axial *W*-heating coil. The small capacity of 2 pF between heater and sensor made these assemblies very suitable as hf regulators. Since they required about 4 W heating power, the further trend went toward smaller assemblies. A typical example is shown in Figure 68 where an approximately 175μm bead thermistor is surrounded by a heating coil and thermal contact between both is made with an electrically insulating cement of low thermal resistance.

Commercial types of indirectly heated thermistors are now available with maximum power rating for the heating coil between 30 and 200 mW and a heater efficiency up to 98%. The insulation resistance between heater and thermistor is $>10^6\ \Omega$ and the capacitance is 1–2 pF. They are produced by a number of companies: Philips (Valvo), Standard Telephone, and Siemens with various thermistor resistance and power ratings. For details one must be referred to sales pamphlets of these companies.

An ingenious method for making an indirectly heated thermistor as an integrated element from *one* material for sensor and heater has been suggested in a recent patent. It has been found that the activation energy of a vacuum evaporated Ge film depends very much on the temperature of

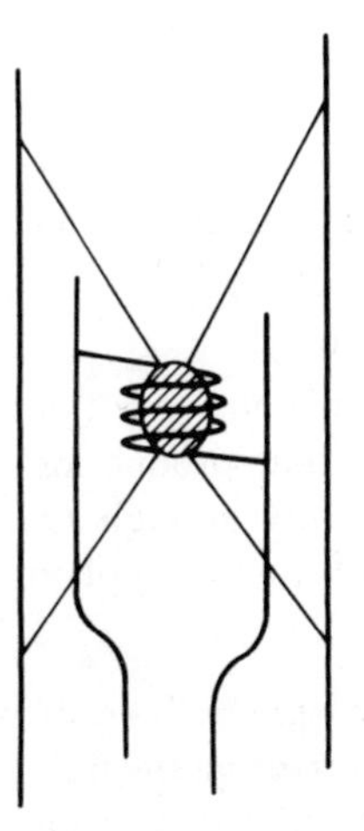

FIGURE 68. Construction of indirectly heated thermistor (Schematic). After Scarr and Setterington, IRE Transactions on Component Parts, 6-22 (March 1961).

substrate on which it is deposited. For example it decreases for a chosen Ge grade from 0.36 eV ($B = 4160$) to 0.095 eV ($B = 1100$) and to ~ 0, if the substrate temperatures of 500, 620, and 635 K are used. Simultaneously the resistivity starting with $10^3\ \Omega$ cm drops considerably. Sequential evaporation, first with high, then with low, substrate temperature, produces at first the heater and then the sensor film. Sandwiched between both is a ≈ 5-μm thick SiO film, produced by vacuum deposition between Ge evaporation. Appropriate masking in each step prohibits short circuits. Resistance ratios of ≈ 100 can be attained with ≈ 25 mA heater current.[365]

A three-terminal thermistor ("Varitherm") has been suggested for use in nonlinear networks and ac voltmeters.[366] Its third lead acts like a control electrode similar to transistors. The single voltage–current characteristic of a normal two-lead thermistor is spread into a family of curves.

5.3. SINTERING

The sinter process for disks, washers, and rods is similar except for the sometimes different preheating procedure for extruded parts prior to the main sintering. The molded units are packed into boats and moved slowly into the preheated furnace (batch-type sintering) or pushed through the furnace. It is often practical to presinter the molded disks, washers, or rods at a temperature a few hundred degrees lower than the final sinter temperature. This prepares a reserve pool of units milled and molded under the same conditions. It is then possible to adjust the electrical properties of the end product by slight sinter variations in the molded batches. A number of factors determine the sinter condition:

1. Sinter time, which is identical with the "soaking" time in ceramic industry during which the sinter batch is held at a certain maximal temperature ($\frac{1}{2}$–2 hr).
2. Heating and cooling rate (normally less than 50 K min^{-1}).
3. Atmosphere. The chemical nature of the earlier thermistor types (see Chapter 1) required a reducing atmosphere, mostly dry H_2. The production of CuO thermistors of superior properties called for sintering at high O_2 pressure to suppress partial dissociation into $Cu_2O + \frac{1}{2} O_2$ at the applied sinter temperature.

Later the trend was toward an air, O_2, or N_2 sinter atmosphere. Thermistors based on the transition metal oxides of Mn, Ni, Co representing the bulk of the commercial output during the last decades, are sintered either in air or O_2. For compositions containing larger amounts of Fe oxide, N_2, pure or with less than 1% O_2, is used (U.S. Patents 2,492,543 and 2,616,859). This applies also to thermistors with low B values, based on highly reduced iron oxides. PTC units based on doped Ba-Sr-titanate, can be sintered in air. Increasing O_2 partial pressure promotes a higher resistance ratio. Pretreatment in N_2-air mixtures with about 10^{-2} atm O_2-partial pressure followed by sintering in O_2 has been found to be useful for obtaining optimal electrical properties.[367]

Since the sinter process is the most crucial phase of the thermistor production, and nearly as decisive as the chemical formula, the basic principles involved are treated at length in the following paragraph.

Sintering is generally defined as densification and bonding of molded parts at temperatures well below the melting temperature. In some cases the densification feature is less pronounced if the molding process has already produced parts with a high relative density. It is impossible even to quote the immense literature on sinter phenomena and theories in this context. Therefore reference is made only to some recent reviews which can be used as base for further search.[368–372] Since sintering of thermistors in most cases implies solid-state reactions of several components, the processes involved are often more complex than those described in textbooks and review articles, though certain analogies and principles will be found there for practical estimates and approximations. Since Aristotle[373] it has seemed to be a dogma that solid materials could not directly react with each other if no liquid or gaseous phase participated. "*Corpora non agunt nisi liquida*"* was an axiom of

* J.A. Hedvall, the pioneer in solid state chemistry deserves the merit to quote the original statement of Aristotle's: "Ta hygra meikta malista ton somaton."

chemistry for many centuries (since 350 BC) until less than about 100 years ago, when certain reactions between solids Rinman's green ($Co_3O_4 + ZnO$) and Thenards blue ($Co_3O_4 + Al_2O_3$) had interest for analysis and others for production.

There was indeed some truth in the ancient dogma, if perfect crystals of the components are brought together. The facts that many solids either are in defect state or can become defective by impurities, thermal transitions at elevated temperatures, or disorder at higher temperatures by increasing lattice vibrations, have contributed to break down the dogma, especially by Hedvall's numerous investigations.[374]

Hüttig and his co-workers in a large number of publications have studied solid-state reactions between oxides and have tried to identify certain intermediate stages of higher reactivity by x-ray diffraction, surface activity, and magnetic susceptibility.[375]

These investigations cannot be used for studies of the sinter kinetics, since they have been made (nearly without exception) with powder mixtures, often with the objective of studying the properties of mixed oxide catalysts. Therefore only studies of precompacted powder mixtures are meaningful for evaluating the factors that determine the reaction and sinter rate. They are the following:

1. The free surface energy of the reacting powder particles.
2. The specific powder surface per gram, which is a function of the submicron particle size. In general the microscopically visible (secondary) powder particles are clusters of primary particles (crystallites) with linear dimensions between 300 and 3000Å.

In many cases it is naive to assume that standard sieve analysis in the range above 1 μm is capable of characterizing powders for their reactivity in solid state. This method defines only the size of the secondary cluster particles. Typical examples to illustrate this misconception are iron oxides (Fe_2O_3) prepared either by dissociation of sulfate or oxalate. Although the first one has the larger "particle size," it is more reactive due to its higher defect concentration compared to the oxide from oxalate.[376]

Since reaction between several components starts at grain boundaries to progress further by volume or grain boundary diffusion, the first two factors are decisive for the initial reaction rate.

All these factors have not received the attention they deserve. Obviously, methods to determine submicron particle size, microstrain, and defect concentration have not been familiar to technologists engaged in making thermistors. Sometimes the BET method, based on monomolecular N_2 adsorption, is used to characterize the specific surface of ceramic or thermistor powder compositions. Preferences for certain grades of raw materials to obtain optimal sinter and electrical properties implicitly pointed to favorable combinations of several factors mentioned above. The lack of quantitative characterization by one or several of these factors made many formulas a "black art."

Since most of the common thermistor compositions after sintering form spinel phases, the mechanism of spinel formation is of vital interest. For the formation of composite ionic compounds according to the reaction $AO + B_2O_3 = AB_2O_4$, Wagner[377] made the following simplifying assumptions:

1. The compact boundary where the reaction between both components takes place has no defects that are not determined by thermodynamic conditions (e.g., temperature), which means that no active centers are frozen in.
2. The reaction rate is mainly determined by the diffusion of the component ions into the reaction zone or across the reaction product.
3. There is no interaction between the ionic motions within the reaction zone except for the condition that electroneutrality must be conserved. This implies that cations A and B diffuse in opposite direction (toward each other).

Based on this concept, a number of investigations were stimulated that confirmed the validity of a parabolic reaction rate. This means that the reacted fraction at the interface increases with the square root of the elapsed time. Schmalzried[378] has studied the reaction rates for the formation of the spinels $NiAl_2O_4$, $NiCr_2O_4$, $CoAl_2O_4$, $CoCr_2O_4$, and Co_2TiO_4 at $\sim$1700 K, taking in account the volume expansion of the reaction zone caused by the transfer diffusion of the reacting ions. He used 6–10-μm Pt-wires as markers between the initial unreacted oxides NiO or CoO and Al_2O_3, Cr_2O_3, or TiO_2. His results can be condensed in the following manner:

1. The reaction rate can be influenced by the gas atmosphere. For the formation of $CoCr_2O_3$ and Co_2TiO_4 it is about 4 times smaller in N_2 than in O_2; however, it is equal for $Co_2Al_2O_3$.
2. It increases in O_2 by a factor of 10 for Co_2TiO_2 between 1490 and 1660 K, for $CoCr_2O_4$ between 1630 and 1760 K, and for $CoAl_2O_4$ between 1620 and 1760 K.
3. The reaction rates are mainly determined by the transport of the Ni^{2+} or Co^{2+} ions, while the oxygen necessary to form the spinel phase is partially transported through the atmosphere. This implies that electrons diffuse along with the cations to preserve electroneutrality. This is accomplished in N_2 atmosphere by migration of the Ti^{4+} ions in the opposite direction.

The previous considerations dealt with stoichiometric ratios of the oxide components. In nearly all practical cases of ceramic sinterings with solid-state reactions, this condition is not prevailing. This is especially true for oxide thermistor compositions, which often require large deviations from the spinel stoichiometry to meet electrical specifications.

Investigations with nonstoichiometric spinel systems $(1 - x)$ NiO+ x Al_2O_3 and $(1 - x)$ NiO+ x Fe_2O_3 sintered 24 hr in O_2 at 1873 and 1623 K, respectively, have shown that the porosity of the sinter product increases rapidly with surplus of the second component ($x > 0.50$).[379,380] This effect is explained in detail by the different vacancy flux emitted by pores using simple theoretical diffusion models and the basic Kelvin equation or the increased vacancy concentration at the surface of a pore against bulk. Anion-deficient materials sinter better than those with cation deficiency (surplus of Al or Fe). Similar effects of reduced sinterability have been found for the systems

$$NiO + x\ Li_2O + y\ Fe_2O_3$$

$$NiO + x\ Li_2O + y\ Al_2O_3$$

where drastic changes of porosity appear already with concentration of less than 1% y. For NiO-Al_2O_3 the change from O_2 to N_2 atmosphere has not much influence on porosity, confirming the previous results on reaction rate (Schmalzried). Optimal diffusion of vacancies can be expected for the condition

$$DcCc + DiCi \approx DoCo$$

Dc, Do, Di, and Co, Cc, and Ci are the diffusion constants and concentrations of cations, anions, and interstitial cations, respectively.

Although these examples do not represent usual thermistor compositions, such as Ni(Co) Mn spinels, their principles can well be applied to them.

For temperatures of ~1700 K reaction rate constants of spinel formation are of the order 10^{-12} cm^2 sec^{-1}. Using the parabolic rate equations $x^2 = 2Kvt$, it follows that after 1 hr the reaction between the components has been penetrated to a depth x of ~1 μm. If the primary particles are <1 μm, complete reaction can be expected at this temperature. However, for more common sinter temperatures between 1300 and 1500 K (~1870–2250°F) the reacted zone, at least theoretically, is much thinner. A certain compensation can be expected under practical manufacturing conditions by the preceding milling of the powder composition, which not only promotes homogenization, but also offers the opportunity of creating microstrains and defects beyond the thermodynamical equilibrium concentration. Both contribute strongly to the activation of the sinter process.

Finally, not only the processes at the maximum sinter temperature but also during the cooling period determine the structural and electrical properties of the sinter product:

1. The transition from a structural phase stable at high temperatures to that stable at lower temperatures can be retarded or occur with different rate at various sites (interior or surface) resulting in an inhomogeneous product.
2. The mutual solubility of coexistent phases normally changes with temperature. Rapid cooling can freeze in the solubility ratio for the higher temperature; however, subsequent long time annealing (in life test) would tend to restore this equilibrium for lower temperatures.
3. While the preceding phenomena can also occur in ideal nondefect systems, in real systems with defect structures cation or anion vacancies can also be frozen in. Lacking chemical equilibrium with the surrounding atmosphere promotes this kind of nonequilibrium condition (see also Section 6.4.2).

5.4. CONTACTING

5.4.1. General Considerations on Making Contacts to Semiconductors

It is desirable that the resistance of NTC or PTC thermistors be determined mainly by the resistivity of the bulk material and that the influence of contact resistance be minimized. This condition is not always fulfilled. It has been found that the contact resistance of commercial NTC thermistors was up to 20% of their specified nominal resistance, at least at temperatures around ambient. To keep the contact resistance small has been over the years a question of trial and error by using different commercial silver pastes. For PTC units this problem is much more severe. If a conventional silver contact is fired onto a sintered PTC unit, its resistance is so high that only a few percent of it represents bulk resistance, the remainder being contact resistance.

Table 33 below gives observed resistance values for one type of PTC disk with different contacts applied. It is obvious, and it can be shown by four point probe or van der Pauw potential measurements, that the bulk resistivity essentially determines the resistance values below 5 Ω, while all higher resistance values are due to contact resistance.

It seems surprising that noble metals with negligible tendency to oxidize in air (Pt, Au, Ag) produce the highest contact resistance. On the other hand, less noble and especially Al, notorious for its spontaneous and permanent surface oxide film, make low resistance contacts.

Therefore there has to be something more complex than the naive oxide layer concept to explain these data.

The foundation for understanding these phenomena and to make good ohmic contacts has been laid by the introduction of the enrichment and depletion layer model in semiconductor surfaces. This concept has been most successful in all theoretical treatments of semiconductors diodes and transistors. In the case of sintered oxide semiconductors this model, normally applied to single crystals, is probably too primitive for quantitative considerations, although qualitative trends might be predictable.

THE SCHOTTKY BARRIER LAYER MODEL. In a contact metal and semiconductor, separated from each other, certain electron or hole concentrations exist, determined by the nature, defect structure of the material, and last but not least, by temperature. Also, it is known that for each material, metal or semiconductor, a critical energy value can be defined that permits emission of an electron into vacuum; it is called the *work function* of the material. The relation between work function of different metals and their contact resistance on well-defined semiconductors such as Ge or Si is treated extensively in several textbooks and in the literature. Therefore a brief description of the conditions at the interface between a semiconductor and a contact metal has to be sufficient. If, in the case of an *n*-semiconductor all donors are dissociated and have supplied one conduction electron per donor, the electron concentration is given by $n_e = Nd^+$. In the contact metal another electron concentration exists that is determined by the Fermi level of the metal[381–383]

If metal and semiconductor are brought into contact with each other within a distance of atomic dimensions, their different electron concentrations induce a current flow until a new equilibrium is reached, resulting in a concentration gradient of electrons from the interior of the semiconductor to the contacted surface. The positively ionized donors cannot neutralize the in-flowing electrons since they are basically localized within the lattice and can only move by slow diffusion processes. The electron concentration in the semiconductor layer adjacent to the metal is determined by

1. The work function of the electrons in the contact metal.
2. The chemical bond of the electrons in the metal and the semiconductor.
3. The concentration of ionized donors.
4. Surface conditions on metal and semiconductor (adsorbed impurities, gases, or vapor, or in extreme cases, mono-, or multimolecular films of reaction products, oxide for metals, or stoichiometric deviations in the semiconductor surface.

Electrons can easily flow from a contact metal into an n-type semiconductor and no appreciable contact resistance develops. In contact with a p-type semiconductor the situation is different.

The electrons from the metal flowing through the contact interface neutralize the holes in the semiconductor, thus depleting the contact surface layer of the semiconductor of carriers. Even if undissociated acceptors are still abundant in the interior of the semiconductor, they cannot freely and rapidly diffuse to the interface to restore the equilibrium disturbed by the in-flowing electrons. As a result a depletion layer is formed that represents a high contact resistance. In diodes and transistors these depletion layers perform a vital role in the function of these devices. In thermistors they are undesirable.

Figure 69 shows the effect of enrichment and depletion of carriers in contact zones. The electron concentration at the interface the semiconductor metal junction is given by

$$n_s = N_c \cdot \exp\left[-\frac{(\varphi \text{ metal–semiconductor})}{kT}\right]$$

N_c is the effective density of states in the conduction band and is not determined by material constants of metal or semiconductor. However,

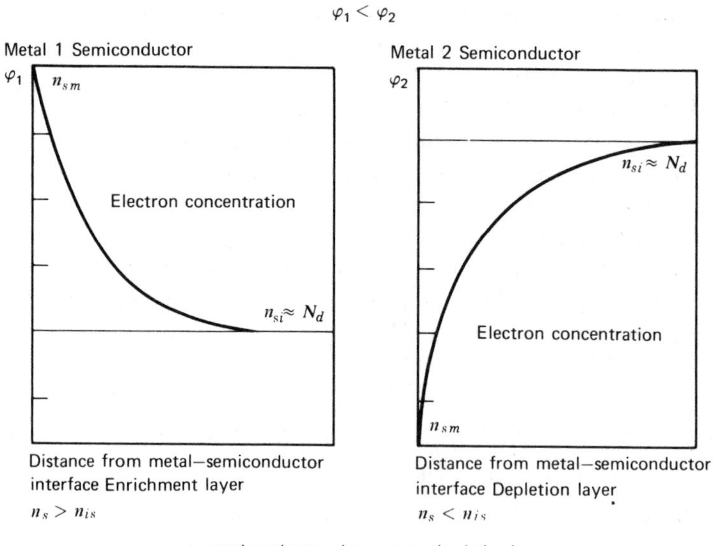

FIGURE 69. Concentration gradient in n-semiconductor in contact with metals of different work function.

knowledge of the work function of metals in contact with the semiconductors is vital to calculate or at least to estimate the probability of contact resistance. Schottky and Spenke[383] have shown that the work function φ_{ms} metal–semiconductor for electrons (n) and holes (p) is determined by the relation:

$$\varphi_{ms(n)} + \varphi_{ms(p)} = E_c - E_v = E_g$$

where the lower edge of conduction band is E_c, the upper edge of valence band is E_v, and $E_c - E_v$ is the band gap E_g. This follows from the definitions

$$\varphi_{ms(n)} = E_c - E_F$$

$$\varphi_{ms\,(p)} = E_F - E_v \; (E_F = \text{Fermi level})$$

Most important for practical consideration is the possibility of deriving the work function metal–semiconductor from the normally well-known value metal–vacuum.

According to Spenke[385] φ_{ms} is a linear function of φ_{mV}:

$$\varphi_{ms(n)} = \varphi_{mV} - E_c$$

E_c has the same meaning as in the preceding equations. A dielectric theory of the carrier height at metal–semiconductor and metal–insulator interfaces was presented by Hirabayashi.[386] The differential resistance at $V = 0$ at the interface metal–semiconductor,

$$R_0 = C \cdot \exp\left[\frac{\varphi_{ms\,(n)}}{kT}\right]$$

or for a given temperature,

$$R_0 \sim C \cdot e^{\varphi_{m\,V}}$$

For p- a contact metal–p-semiconductor, the relationship is inversed:

$$R_0 \sim C \cdot e^{-\varphi_{m\,V}}$$

Therefore one can spell out the following practical rules:

Semiconduction type	contact resistance
n-type	increases with work function of contact metal
p-type	decreases with work function of contact metal

A semiquantitative confirmation of these relations for p-conductive Se dates back to 1939, when Schweickert[385] showed experimentally that the maximal barrier resistance of an Se rectifier decreased by 5 orders of magnitude between K ($\varphi \sim 2$ eV) and Au ($\varphi \sim 5$ eV).

It is now easy to understand why the contact resistance of doped PTC materials, which are n-semiconductors, also varies so much with the contact

TABLE 33

Contact Metal	Work function (eV)	Resistance at 300 K (Ω)
Al	3.4–4.3	2–4
Zn	3.4	3–4
Sn	4.37	50–100
Ag	4.5	1,800
Pt	5.4	25,000

metal. Table 33 correlates work function of the contact metal with the observed resistance value for PTC units of identical size and shape.

It is of course naive to assume that the qualitative trend between work function of contact metal and contact resistance can easily be developed into a quantitative formula. Even contact interfaces on monocrystals often do not match Schottky's model especially if reactions between contact metal and semiconductor are possible, or more trivially, if it is difficult to keep the surfaces to be brought into contact clean.

The difficulty of keeping surfaces, especially of metals, clean could be a blessing in disguise. If the metallic or nonmetallic but conductive films on a contact metal can change their work function into the right direction, this could be beneficial to reduce contact resistance. Efforts of this kind have indeed been made, often only on an empirical basis and retained as proprietary knowledge. Siemens has disclosed that a special treatment of the contacted surface of doped $BaTiO_3$ PTC units has reduced the barrier resistance to Ag electrodes considerably. The decrease of the work function of Pd, Au, Ag, and Pd-Ag by surface contamination has been systematically investigated.[700]

Since thermistor compositions are often modified by doping, the influence of the doping concentration on the work function of the semiconductor is of considerable interest. Direct measurements with NiO, CuO, and ZnO doped with Li^+-Na^+ and Cs^+ indicated an appreciable reduction of the work function within the first few percent of dopant concentration (Table 34).

The effect of small additions of ThO_2, Fe_2O_3, Cr_2O_3, or MgO to NiO, CuO, and ZnO is ambivalent. Changes of work function between ± 0.4 eV were observed.[386]

Following Langmuir's[387] classical work to reduce the work function of Th by a monomolecular layer of Cs, many new experiments in this direction have been made during the last decades; a critical study was published 1962.[388] The fact that fractional coverage of a monomolecular layer can result in a substantial change of the work function opens many possibilities

TABLE 34 Decrements of work function in eV; concentration in wt %.

Semiconductor	Dopant	1	2	4	6
NiO	Li	0.25	0.31	0.38	0.42
CuO	Li	0.38	0.50	0.62	0.70
ZnO	Li	0.43	0.48	0.63	0.70
	Na	0.83	0.95	0.96	0.97
	Cs	0.57	0.72	0.65	0.58

to advance good "contacteering" from a black art to a scientific objective. It is, for instance, of considerable interest that the work function for Cr, Mo, and W is about 1 eV higher than for their silicides,[389] suggesting a beneficial effect of a superficial treatment with Si.

In conclusion, some realistic reservations should be made as to the validity of theoretical concepts. Most theoretical considerations leading to the interaction of depletion layer and contact resistance are based on a model of a continuum of metal and semiconductor, both homogeneous and monocrystalline. Sintered oxides, whether with NTC or PTC characteristics, are far from matching this model. Even if well sintered, they still have porosity and they consist of individual grains. Often the grains differ in composition. This may occur if sinter time is insufficient for complete reaction or if, in equilibrium, homogeneous phases were impossible based on phase diagrams. The structure of PTC units with bulk and grain boundaries differing in composition adds to the complexity of the problem. Finally, the effect of the chemisorbed humidity or other atmospheric components has to be considered, especially in stability tests, which often reflect more the behavior of contacts.

Investigations of the influence of adsorbed gases and vapors on the work function are numerous. One with transition metal oxides V_2O_5, NiO, CuO, and ZnO found an increase of φ with O_2, and decrease with H_2, CO, CO_2, C_3H_8, and isopropyl alcohol, if adsorbed. This also would indicate opposite effects when adsorption changes the contact resistance.[390] Data based on the adsorption effect on photoelectronic emission of Cr_2O_3, NiO, and ZnO support this picture.[391]

Evidence exists that films have lower φ than compact metals. Annealing removes the cause: high concentration of defects.[392]

For *n*-type semiconductors (such as TiO_2-doped Fe_2O_3), the work function increases with increasing relative humidity. This effect is slightly larger in O_2 (115 mV) than in Ar (85 mV). With low relative humidity ($>15\%$) Ar increases and O_2 decreases the work function.[393]

5.4.2. Normal Contacting Procedure for NTC Thermistors

In the early phases of thermistor production, metal coatings were applied by flame spraying (mainly Cu), brush painting, or spraying of conductive suspensions (mainly silver). A number of proprietary suspensions some based on the old Cary Lea process,* others on suspensions of decomposable noble metal compounds have been used in large-scale operations. The "ancient" art of making silver and gold ornaments on ceramic and glass branched out in numerous new formulas for metallizing of ceramic parts for the electronic industry and produced as a spin-off a number of compositions especially suitable for contacting of thermistors where performance conditions were more stringent than for capacitors or inductors due to higher operating temperatures, temperature gradients, and frequent temperature changes. It was therefore necessary to develop formulas that met these conditions and in addition formed metal layers with a sufficient margin of safety when wire leads were attached by soldering or other bonding methods.

Metallizing preparations based on organic Pt, Au, and compounds, dissolved in organic liquids, give bright and well-adhering films on glass and nonreducible ceramics, but could reduce oxide thermistors, at least superficially, thus producing interlayers of different (in most cases higher) resistivity.

In Table 35 a number of well-established and time-tested conductive preparations are compiled without claim for completeness.

The upsurge of microelectronics has not only stimulated systematic work on metal inks and paints but also development of new methods of their application. The need for uniformity and reliability in mass production has led to printing processes. One of them is the screen printing method, which has also become important for thick-film resistor fabrication (squeegee process). The conductive paste is squeezed through a fine screen onto the substrate either by gravity, dead weight, or air or coil spring pressure. A minimum clearance between the bottom of the screen (0.3 cm) and the substrate is desirable for proper snap-back of the coating material.

The wide application of the screen printing method, at first an art, has led to systematic technological studies on the variables responsible for a good performance. As result of these strong efforts, it is now possible to print resistors within 10% of their intended values, which means $\pm 5\%$ of a thickness of 25 μm. The requirements for conductive contact layers on thermistors are less stringent. Here it is sufficient that the entire conductive area represents a plane of equal potential and furthermore that it has sufficient

* Resulting in the formation of a colloid silver paste.

TABLE 35

Metal		Brand		Method of Application	Firing Temperature (K)
Silver		DuPont USA	7313	Dip, brush, or spray	870–970
		do.	7345	Squeegee	870–970
Pd-Ag		do.		Dip, brush, or spray	800–1080
Pt-Au		do.	7553	Squeegee	1030–1100
Au-Ag		do.	6976	Squeegee	1000–1050
Pt fluxed		do.	6855	Dip, brush, or spray	800–1080
Pt unfluxed	mg cm^{-2}				
	15 ± 5	do.	7919		
	mg cm^{-2}	Degussa West-Germany	Argalvan		
Ag	15 ± 5	do.	103	Dip, brush, or spray	1030–1130
Ag	12 ± 4	do.	178 L	Dip, brush, or spray	1030–1130
Ag	12 ± 4	do.	281 L	Dip, brush, or spray	
Ag	15 ± 5	do.	303	Squeegee	1070
Ag	15 ± 5	do.	309	Squeegee	1070
Pt	10	do.	175	Brush; very resistant to soft solder	1070
		DODUCO W. Germany	Argonor		
Ag		do.	N		830–975
Ag		do.	N2		1025–1075
Ag		do.	NB3		775–875

thickness to permit wire attachment. The physical performance of a screen printer has been analyzed with special emphasis on the squeegee pressure as most critical factor.[394] Since substrates are not always perfectly flat, and this applies still more to sintered thermistors without additional surface grinding, a cambered surface can result in variations of the film thickness. In addition to a number of screen printers that are able to compensate tolerance in surface flatness and to cope with other interfering variables, a torsion bar squeegee head has been designed and found to meet the following conditions:

1. Enable the squeegee to float up and down over surface irregularities.
2. Keep squeegee pressure on the substrate constant at all points, at all times at a preset value.

It is very important to know the rheological behavior of printing pastes to judge or predict their performance. To adjust pastes by simple viscosity measurements using the Rotovisco viscosimeter can be misleading. A screen viscosity index, which is the ratio of the paste viscosity during screening and subsequent leveling, has been determined for a number of Ag, AgPd, Au, and AuPd pastes and related to their practical performance.[395]

5.4.3. The Contact Problem with PTC Thermistors

The theoretical considerations on the electrical characteristic of PTC materials deal with the intergranular boundary layers and the barrier zones at the surface of each grain. There is still another boundary effect of great importance. the potential barrier between the PTC material and a contacting metal. As early as 1956 Sauer and Flaschen[64] found in their study of $BaTiO_3$ and $Ba_{(1-x)}Sr_xTiO_3$ systems with $x \leqq 0.3$ doped with one to three milliatoms La per mole an enormous influence of the contacting metal on the resistance of their PTC disk units. similar to the results shown in Table 33 (Section 5.4.1).

Although this sequence of contact "quality" needed some later revision, it is in principle still valid. There is no doubt that the noble metals Ag, Au, and Pt make poor contacts while In and In alloys are still the best contacting materials, at least as far as their electrical contact resistance is concerned. As shown in Section 5.4.1 the work function of the contact metal in relation to that of the semiconductor and its conductivity type is important for the size of the contact resistance. Since the latter decreases with increasing temperature this trend bucks the positive resistance characteristic of the bulk material (in this case bulk to be understood as sum of intra- and intergranular resistance.. While the intergranular barrier junctions largely disappear in single crystals of doped $BaTiO_3$ and similar ferroelectrics, surface barrier junctions with Sm-doped $BaTiO_3$ crystals persist as shown by Goodman.[42] This was also investigated with Ce-doped $BaTiO_3$ junctions as a function of temperature between 270 and 400 K at zero bias,[396] with a method described before.[397]

For Cu contacts with an area of $\approx 1.22 \times 10^{-8}$ m^2, a barrier capacitance of ≈ 200 pF was found, somewhat dependent on the phase transitions at 280 and 395 K. With an effective $\epsilon_r = 1000$ the resulting thickness of the carrier-depleted barrier junction would be

$$t = \frac{1000 \cdot 8.854 \cdot 1.22 \times 10^{-8}}{200} = 54 \times 10^{-8} \text{ m} = 5400 \text{ Å}.$$

Assuming that the resistivity of the depletion zone is 1000 times higher than that of the bulk crystal ($\sim 10\ \Omega$ cm), the contact resistance per square centimeter would be $R = 10^4 \times 5 \times 10^{-5}$ or $5 \times 10^{-1}\ \Omega$. This is much too small to explain the contact resistance values observed with polycrystalline sintered material. In reality, a donor gradient, resulting in a diffusion potential and interfacial space charge, can complicate this model.[689,690]

In and In alloys, especially with Hg and Ga as second component, have been used extensively for experimental studies, since their contact resistance is virtually zero as proven by four-probe measurements. For large-scale production, however, other contact metals that are cheaper and easier to handle had to be sought. It has been found that Pb alloys with down to 15% In can be used successfully with very small contact resistance. Also, interlayers sandwiched between the ceramic PTC body and any other contact metal that is strong enough to serve as contact plate with lead connection have been used.

Flame-sprayed Al layers are nearly as good as In alloys, obviously due to the low work function of Al. In Table 36 the resistivities for four different PTC units measured "contactless," using the Van der Pauw method, are compared to those found after contacting with flame sprayed Al.*

The four units investigated were chosen with different dopant concentrations, which explains the wide scope of their resistivity. The experimental error of the Van der Pauw method, described in detail on p. 173, could be $\pm 5\%$ since it is based on the successive measurement of values that cumulate their individual error.

It is hard to imagine that the two-terminal resistivity that always contains the contact resistance should be smaller than the "contact-free" resistivity. Therefore systematic errors of the Van der Pauw method must play a role. Nevertheless, it has been proven that Al contact has no appreciable contact resistance. Another interesting experiment has been made that clearly

* Unpublished work by the author.

TABLE 36

Method	Unit	1	2	3	4
			(Ω cm at 298 K)		
A. Van der Pauw resistivity		13,960	8,120	6,150	555
B. Al-contact		15,500	8,800	6,000	460
Ratio B/A		1.11	1.08	0.98	0.83

shows the depressing effect of a less satisfactory contact on the PTC characteristics. A highly doped $(Ba_{0.85}Sr_{0.15})\ TiO_3$ with a resistivity of 100 Ω cm was measured between 298 and 623 K simultaneously by the four-probe method and with the conventional two-terminal contacts, in this case using Al coating.

In Table 37 the relative resistance against $R_{298} = 1.00$ is given for a series of temperatures. It is self-evident that the relative resistance between the sondes in the four-probe method reaches much lower values in the NTC range below 373 K and much higher values in the PTC range up to the turning point at about 498 K.

Upon cooling, erratic readings were observed. It was found that (1) the peak is reached in the 423–498 K region, and (2) above the peak temperature range the resistance drops faster for the "sonde" region, although absolute values remain higher up to 620 K.

TABLE 37 Comparison of four-probe measurements with two electrode values. Relative resistance at elevated temperatures of $Ba_{0.85}Sr_{0.15}$ (+ 0.01 Nd) based on 1.00 at 298 K.

K	Relative Resistance Total	Relative Resistance Sonde Region	Resistance Ratio Total/Sonde Region
298	1.00	1.00	1.00
373	0.86	0.36	2.30
398	45	32	1.40
423	73	474	5
498	64	474	7.5
523	48	310	6.5
548	36	181	5
573	24	97	4
598	18	67	3.7
623	17	53	3.1

Electroless (often called autocatalytic) plating. This method always requires five basic components: the metal salt that provides the ions for deposition, a complexing agent preventing uncontrolled precipitation, a reducing agent, a pH adjuster for best performance, and a stabilizer. Besides Ni, also Co, Cu, Pd, Ag, and Au can be deposited.[398] Nickel contacts formed by electroless plating have been suggested for various semiconductors including PTC materials.[399] They are made in a wet process in which Ni^{2+} ions are reduced by hypophosphite according to the reaction:

$$Ni^{2+} + (H_2PO_2)^- + H_2O = Ni + 2H^+ + H(HPO_3) \quad (1)$$

$$+ (H_2PO_2)^- + H_2O = H(HPO_3)^- + 2H \quad (2)$$

$$(H_2PO_2)^- + H = H_2O + OH^- + P \quad (3)$$

The elemental phosphorous thus formed reacts with the concurrently deposited Ni, resulting in Ni electrodes containing variable amounts of P, increasing with decreasing pH from 10 to 2, for instance[400]:

pH	12	10	6	4	3	2
P (%)	0.3	3.4	6.8	12.8	14.3	16.8

The initially nonohmic contact has to be heated to 440 K to become ohmic. Further improvements to eliminate the heat treatment and to achieve ohmic contact as deposited have been made.[401] As in other electroless processes, the surfaces to be metal coated require a predipping with $SnCl_2$ (1%) and $PdCl_2$ (10^{-4} %), apparently to promote formation of metallic nuclei.

Even considering the improvements made with electroless metal-coating, its application to sintered oxide ceramics still has some drawbacks:

1. The sinter bodies are porous and can absorb small quantities of the aqueous solutions used in this process. The removal of electrolytes from capillaries is notoriously difficult, especially after a large part of the surface has been coated with metal.

2. Even if the electrolyte has been washed out from the capillary pores (possibly supported by electrophoresis), the necessary drying of the parts can be a nuisance, especially if vacuum drying must be applied to avoid undesirable heating effects.

3. Finally, the parts must be surface ground to remove the nickel deposit where it shortens out the resistance.

5.4.4. High-Temperature Contacts

1. Silver, gold, and Pt pastes: poor to acceptable after drying and firing.
2. Ag oxide contact: very good mechanically, but liable to diffuse and migrate at temperatures > 850 K.
3. Molybdenum powder paste (fired only in reducing atmosphere).
4. The Mo-contact has been modified for metal-ceramic sealings by adding 25% Mn powder (150 mesh) to 200 mesh Mo powder and suspending this mixture in pyroxylin (DuPont 5511) with a 1 : 1 mixture of amyl-acetone–acetone as thinner.[402]
5. Brazing with metals by dissociation of hydrides in vacuum or in highly purified H_2, A, He, N_2 after O_2 gettering

a. TiH_4 in vacuum on ceramics[403]

b. ZrH_2, superior wetting characteristics, only tank N_2 necessary.

c. TaH_x.

d. NbH_x.

Procedure. Waterpaste or solution in nitrocelluse binder for uniform coatings are used. Eutectic Cu-Ag alloy (mp: 1052 K) or pure Ag (mp: 1133 K) are placed on the hydride coating and heated in vacuum of $\sim 10^{-4}$ torr to the melting temperature of the brazing alloy or silver. Pure Al (933 K) is especially suitable to bond an Nb and Ta coating formed by hydride dissociation.

All these metals and their hydrides are extremely sensitive to oxygen, to a lesser extent nitrogen, which can form nitrides especially with Ti.

Wetting and bonding also can be accomplished by direct action of certain brazing alloys containing Zr prepared by vacuum melting of the components.[404] Table 38 gives a number of compositions in wt. %.

Silver with 15% Zr can be applied in forepump vacuum or in oxygen-free nitrogen.

TABLE 38

Base Metal Component	Additions	Wets and Bonds
Silver	10–25	Ceramics, carbides
	optimal 15	Sapphire, diamond
Silver	26 Al 10 Zr	Oxide ceramics
Al	10 Nb	Oxide, ceramics
Pt	0.05–5 Zr	Porcelain

The color of these alloys goes from light orange to darker tones for higher Zr content if brazing is done in A with 0.4% N_2. In general it has been found that Zr-containing alloys had better wetting and bonding properties compared to Ti alloys. Direct bonding of gold electrodes eliminating gold paste with its possible detrimental effects of its fluxing additions has been suggested by pressing gold foils of 50 μm at elevated temperatures onto the semiconductor surface.[405]

5.4.5. Novel Contacting Methods

TRANSFER TAPES. In recent years thin coatings of numerous materials have been produced by a dry method. It is based on the simple principle that a flexible tape covered with an adhesive containing the desired coating material is pressed against the surface to be coated. The coating metal is transferred from the flexible plastic tape to any substrate. This "green" coating is subsequently heated to temperatures above 1000 K to remove the organic adhesive binder by decomposition, evaporation, or oxidation, depending on the material to be coated and to sinter the applied film. Metals, ceramics, or glass coatings can be made with a green thickness ranging from 5 to 125 μm using transfer tapes.[406]

Transfer tapes are available or can be prepared for the following metals: Pt, Pt-Au, Pd, Pd-Ag, Au, Ag, Mo-Mn, Mo-Ti, and Ni. This wide variety makes it possible to select the metal most suitable for low contact resistance. At the same time it imposes restrictions on the thermistor materials to be coated. Whenever a coating material requires firing in a nonoxidizing or reducing atmosphere, the possible reducing effect on the thermistor composition must be taken in consideration and minimized. Aftertreatments in mildly oxidizing atmosphere could be a solution.

It has been found that polyacrylic-type adhesives decompose almost completely to their gaseous monomers and leave no solid residue which might slowly react with the oxide surface. Ettre[407] has described an automatic tape transfer machine and made a cost comparison of its efficiency with silk screen spray and paint operations taking full account of necessary capital investment and lifetime of the equipment. A substantial cost reduction per unit is evident even for hand feeding. Of equal importance is the clean metallizing for small parts with 1.2-mm diameter.

ACTIVATED SOLDERS. Activated solder alloys containing small concentrations of strongly reducing elements (for instance, ~0.02% Li) can be bonded by pressure to sintered oxides at temperatures below their mp, if the crack-free oxide surface is ground or polished.[408]

5.4.6. Contacting by Thermolysis

Coherent metallic films of the transition elements Cr, Fe, Co, and Ni can be produced by thermal dissociation of their carbonyls ($Ni(CO)_4$ $Mo(CO)_6$) at relative low temperatures (<400 K) and less expensively than by evaporation and sputtering. Their application to sintered oxides is limited by the degree of porosity since deposition would occur also within the pores and thus shorten parts of the semiconductor body. The carbonyl chemistry has developed a large number of derivatives that are easier to handle than the low-boiling carbonyls themselves. This has opened new possibilities for thermolytic contacting, as shown in the dissociation of diacetyldihydrazon carbonyls of Cr and Mo to form 400–2000-Å thick metallic films on Si by decomposition in Ar/O_2 atmosphere between 520 and 670 K. The carbonyl compound can be applied in liquid solution (German Offen. 2,012,031, September 23, 1971). In another thermolytic process Ag-Al or Ti-Al alloy can be produced by decomposition of $Ag(AlH_4)$ or $Ti(AlH_4)$ at temperatures between 470 and 570 K. Such contacts have been suggested for dense semiconductor bodies of Si.

Metallic laminates consisting of multiple vacuum deposited layers (sequence of Ge, Au, Ni, Pd, and Au) have been suggested as solderable coatings with low resistance and good mechanical strength on La-doped $BaTiO_3$.[409]

The application of ohmic contacts to Ge, Si, and III-V semiconductors has attained a high level of reliability and perfection. It is often based on the thermolytic deposition of dopants which produce surface layers which are degenerate semiconductors. The immense patent literature on this topic is beyond the scope of this book.

5.4.7. Characterization of Metal–Semiconductor Contacts.

When is a contact "good"? Even if the rules given in the general considerations on metal–semiconductor contacts have been followed and the technology to apply them been successful (proper choice of deposition conditions), there is always need for contact quality control. Several methods will be described:

Current-versus-Voltage Characteristics. The simplest approach is to check whether the contact is "ohmic," that is nonrectifying and without voltage dependence. The current-versus-voltage characteristics is measured, if necessary with short (1–50 msec) current pulses to prohibit self-heating. The current voltage characteristic is linear only if the semiconductor zone adjacent to the contacting metal has no carrier depletion zone.

Stöckman[410] has shown that not only for depletion, but also for enrichment, layers a nonlinear potential distribution through the semiconductor can occur. In the second case majority carriers are injected from the boundary zone and penetrate the semiconductor, thus reducing its resistivity, until a lower limit is attained when the initial carrier concentration has been swamped over by the injected majority carriers.

SEPARATION OF BULK RESISTIVITY AND CONTACT RESISTANCE.

Four-Probe Method. This classical method to measure resistivity while eliminating contact resistance is well known and much applied. Accuracy is determined by the precision of measuring the probe distance and length and the uniformity of the current flux pattern. The possible sources of error and their correction have been discussed by several authors.[411,412]

Van der Pauw Method. A real innovation in measuring the resistivity and Hall coefficient of flat samples of arbitrary shape has been based on measurements of potential differences between four alternate points at the periphery of a flat plate and the corresponding current values between these points.[413] Its usefulness is restricted to two-dimensional specimens (disks, bars, chips, films, or rings). The resistivity is

$$\rho = \left(\frac{\pi d}{\ln 2}\right) \frac{R_{MN,OP} + R_{NO,PM}}{2} f$$

The two resistance terms are defined by (Figure 70)

$$R_{MN,OF} = \frac{V_P - V_O}{i_{MN}}$$

$$R_{NO,PM} = \frac{V_M - V_P}{i_{NO}}$$

The factor f is a function of the ratio $RMNOP/RNOPM$ and is still 0.8 for a ratio 5. Only for extreme ratios of 40 does f drop to 0.5.

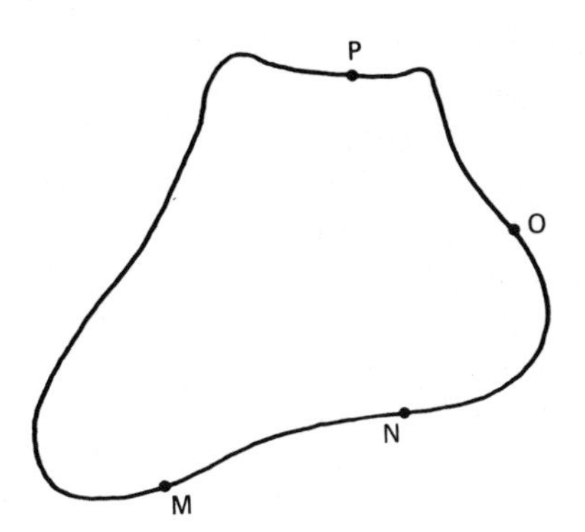

FIGURE 70. Van der Pauw resistivity test set-up.

The Spreading Resistance Probe Method. The need to measure resistivity profiles and inhomogeneities in semiconductors led to the development of the spreading resistance probe.[414]

The ideal relation between the measured resistance of a point contact probe with an effective radius a of the "contact ball point" and the in-depth resistivity of a semiconductor is given by

$$R = \frac{\rho}{4a}$$

where R is measured in ohms, Ω, a in centimeters, and ρ in ohm-centimeters.

The difficulties of making the point contact geometrically reproducible are considerable and have stimulated extensive studies on this topic. On the other hand, it is generally taken for granted that the contact of an ideal geometric point is ohmic and nonrectifying, in other words that the ratio R/ρ is only a function of the radius of the point probe. Since this is not true in most cases, a correction factor G has been used, which varies with the material and has to be determined by calibration with a second method (four-probe or van der Pauw). Normally tungsten carbide probes with ≈ 15 μm radius are used under a load of 10–20 g to minimize mechanical deformation of the contact area, though some investigators also used hardened steel needles.

For W probes with 10 μm point radius on n-type Ge, reproducibility up to 50 g load has been found.[415]

Published data deal mainly with n- or p-type Si and Ge. Measurements with sintered oxides complicate conditions and their interpretation, since second-order effects caused by surface roughness, grain boundaries, and chemisorption of polar molecules such as H_2O enter into the picture. Also, the current density at the contact point has to be held below a level to ensure reproducible and identical IV characteristics in the forward and reverse direction, which is normally defined by a power dissipation at the contact of about 10–100 μW. It is, of course, desirable to minimize barrier resistance by choosing a contact point material with a work function favorably matching the semiconductor type. Since most transition metal oxides in thermistors are p-type semiconductors, materials with high work function should be chosen that simultaneously have to be hard and nondeformable. Among the high-melting, conductive, and hard borides and carbides, tungsten carbide with $\varphi = 4.6$ eV is indeed a good choice. Re, Os, Ir, Pt, and some of their alloys have a higher work function, but are not so hard as tungsten carbide. Os-Ir probes have also been used.[416]

For n-type PTC materials based on doped $BaTiO_3$, Ti-Zr or Hf-carbides might be best suitable. Ultimately the availability of tips with well-defined contact radius will limit the choice.

The previously mentioned methods lose accuracy if the contact resistance is less than 3% of the total resistance. In this case another approach is possible. A pulsed resistance bridge permits to detect resistance change of ~0.1% caused by nonlinear behavior of a semiconductor contact.[416] The resistance of the specimen is measured with short pulse of 3–5 μsec at a repetition rate of 30–200 Hz as a function of the applied voltage. The resistance unbalance is given by

$$\frac{\Delta R}{R_0} = \beta E^2.$$

R_0 is the resistance for the electrical field approaching zero. The constant β is of the order of 10^{-4} cm² V² and its sign depends on the energy gain or loss of the carrier in the field. Plots of $\Delta R/R_0$ versus E^2 for good contacts are straight lines, independent of polarity. For poor contacts each polarity has a different curve and linearity is lacking. This is shown in Figure 71 for two InSb samples, one with a good contact, the other with a poor contact. Another method to separate bulk and contact resistance uses the frequency dependence of the total impedance. It has been successfully applied to CdS photoconductors.[417] The interpretation of the measurements requires the

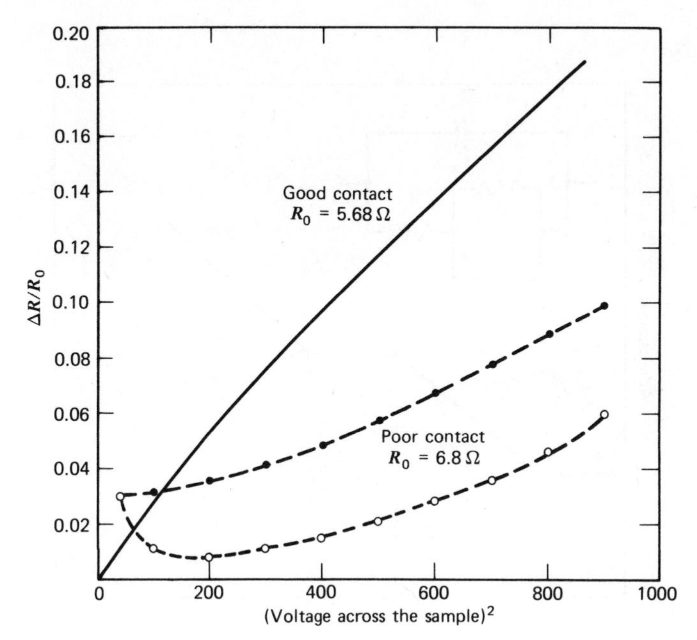

FIGURE 71. Characterization of good and bad contacts in pulsed resistance bridge. (Open dots negative, full dots positive polarity.) After George and Bekefi.[416]

proper choice of an equivalent electrical circuit. Assume the simple case of a bulk semiconductor R_b in series with a contact resistance R_c, both shunted by either the capacitance C_c of the contact or C_g, the geometrical capacity of the bulk unit due to the contact electrodes. The admittance of the entire circuit remains constant until C_c shunts R_c. For $R_c \gg R_b$ and $C_c \gg C_g$ the frequency dependence of the entire admittance is shown in Figure 72.

Consecutive measurements of the frequency dependence give clear evidence of contact deterioration. Prior to this work, Heywang used ac measurements to separate contact, barrier layer, and bulk contribution to the ac impedance.

Increasing contact noise can also be used to characterize the deterioration of semiconductor contacts. With good contacts on crack-free sintered semiconductor bodies noise amplitudes of $<3\ \mu V$ per applied volt are measured. Thermal cycling or mechanical vibration can raise the noise to $>10\ \mu V\ V^{-1}$ as consequence of contact deterioration.

HF Methods. During the last 20 years a number of so called "contactless" methods to determine the resistivity of semiconductors, thus separating bulk and contact resistance, have been published. Each of them has certain shortcomings, even if applied to monocrystalline samples. For polycrystalline sintered oxides with grain boundaries, all hf methods have serious limita-

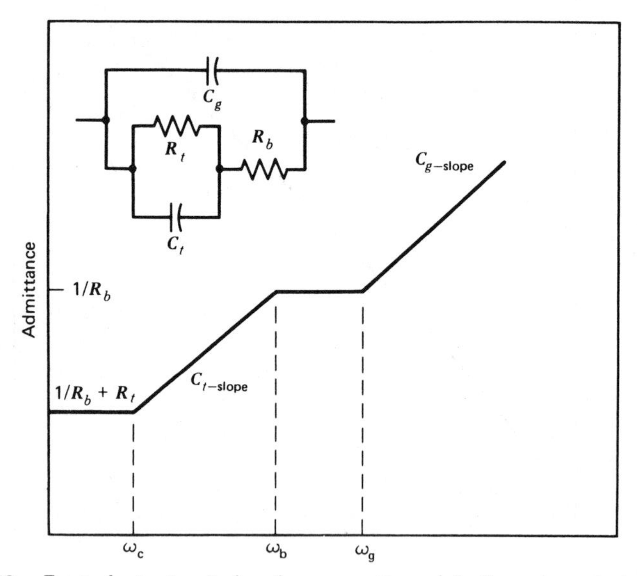

FIGURE 72. Equivalent circuit for the separation of bulk and contact resistance. After Wagner and Besocke.[417]

tions. Most work has been done with Ge and Si for resistivities of 10–10^4 Ω cm.[418–420]

It has been found that the microwave conductivity of *n*- and *p*-type Ge and Si measured by the reflectivity of the circular $TE_{0.1}$ mode (the circular electric wave with the lowest cutoff frequency) at 48 GHz follows the same temperature dependence as thc conductivity in the temperature range 300–180 K. With increasing *p*-type resistivity of the Ga-doped Ge- and B-doped Si, the deviations from the dc value (resistivity increments) become larger.[421]

Contact-free resistivity measurements in hf- fields could give results that are irrelevant for dc or low frequencies. To overcome this difficulty the principle of inducing a torque on a cylindrical specimen suspended in an uniformly rotating electric or magnetic field could be applied.[691] Resistivities between 50 and 10^3 Ω cm have been measured, giving at 2–20 Hz an upper sensitivity limit of $\sim 10^4$ Ω cm.[692]

5.5. MEASURING, ADJUSTING, AND STANDARDIZING RESISTANCE VALUES

When measuring the resistance value of the finished thermistor, its high-temperature sensitivity must be taken in consideration. For a specified tolerance of $\pm 1\%$, its temperature must be defined to be less than 250 mK. This condition can be easily met in a liquid bath, but becomes more critical in air (see Section 6.3). As a general rule, the power input during measurement should be less than 0.1 mW. A bath liquid must meet the conditions to have high resistivity at temperatures up to 450 K, not to react with the sensor material, and to be easily washed out from the pores. Replacing the liquid bath by a fluidized bed of sand or alundum (Al_2O_3) obsoletes some of these problems and at the same time expands the usable temperature range considerably (up to about 1300 K). The scatter of the resistivity and material constant B in each production batch is a more serious obstacle to the narrow tolerance. Resistivity fluctuations from unit to unit can be compensated by mechanically adjusting their size or the extension of their metallized electrode areas. Various grinding or lapping operations have been successfully used to manufacture large quantities of disk or washer units with limited resistance tolerance. This method is not applicable to beads nor does it permit the adjustment of B values. In these cases heat treatment at moderate temperature (<600 K) in controlled atmosphere can be successful. It has been suggested to compensate the inevitable self-heating during the grinding by bringing the nonstandard thermistor in good thermal contact with a standard thermistor with the desired resistance value. Both thermistors are in a bridge circuit that is brought into balance when both thermistors have equal resistance. The inevitable self-heating error would be the same for both units.[422]

5.6. PROTECTIVE COATING AND ENCAPSULATION

The main purpose of coating and encapsulation (canning) has always been to preserve the initial resistance-temperature characteristic over long time periods by minimizing the detrimental effects of humidity and atmospheric contaminations, in addition to protecting against corrosion and mechanical damage. To accomplish this, it always was necessary to take in stride a larger time constant and a smaller dissipation constant, and a great deal of effort has been made to minimize these detriments by sufficiently good internal heat sinking of the sensors.

THICK COATING AND POTTING. Glyptal and later silicone, Teflon, and enamel coatings have been used. The problem here is not so much only to obtain a protective coating, but to retain it without cracking in thermal cycles. Silicone rubber coatings of the RTV type, initially very promising in cryogenic applications, have failed badly after a large number of quenching cycles. For the benefit of this material, it should be added that the cycling conditions were very severe (repetitive dunking in liquified gases). Cooling in air to the same low temperature drastically reduced the failure rate. Heating cycles for Teflon- and enamel-coated units would be similarly severe, although in this case the coating does not have the additional handicap of brittleness as in the cryogenic case. Fluidized bed coating with epoxy resins has been very successful in making durable and attractive coatings on single thermistors or assemblies for use at temperatures up to $\sim$450 K ($\sim$370°F). Glass coatings on bead thermistors have to be chosen for lowest melting temperatures and expansion coefficient matching those of the bead materials and possibly also the lead wire. Glass-covered beads and thermistors sealed into a glass envelope can be used as probes to $\approx$800 K ($\approx$1000°F); for maximal stability, however, only up to $\approx$600 K ($\approx$600°F).

The possibility of thermal resistance drifts and contact deterioration during the sealing process suggests the use of relatively low melting glasses. The devitrifying solder glasses developed by Corning permit hermetic sealing at temperatures as low as 720 K ($\sim$840°F) in 60 min and 760 K ($\sim$900°F) in 5 min (Glass codes 7572 and 7583). Glasses 8407 and 8661 developed for ferrite or barium titanate sealing are in the same bracket. The widespread use of sealing glasses in the electronic industry has given a strong impetus to new developments and simultaneously expanded basic knowledge about the physical properties of glasses. Basically, lower melting points mean larger expansion coefficients; but by using thermosetting or devitrifying glasses, the expansion can be lowered by 20% for the same softening or melting point, and lower thermal expansion means less breakage in thermal cycles. A critical review on this topic has been given by Rabinovich.[423] For a novel group of still lower melting glasses based on the system Tl-As-Fe-Se-S and As-S-I, applications for encapsulating electronic devices below 550 K ($\sim$530°F) have been suggested by Flaschen and co-workers.[424]

A link between these compositions and sealing glasses are the tellurite glasses with melting temperatures down to 680 K.[425]

Encapsulation. The sales bulletins of thermistor manufacturers contain numerous types of encapsulations into evacuated air or He-filled containers made of glass, stainless steel, or other metallic materials and in a great variety of designs. One of the smallest and fastest responding types is the microminiprobe (Fenwal) with a time constant of $\approx$1.4 sec and a dissipation constant of 0.15 mW K^{-1} in still air. It contains a 0.028-cm-diameter bead in a thin-walled glass tubing of 0.050 cm o.d. The majority of sturdy probes uses Cr- or Ni-plated metal tubings made of Al, stainless steel, or alloyed Cu with o.d.'s of $\approx$0.25 cm, 1–16 cm length, and with the sensor mounted (sometimes potted) at the closed end. For measuring surface temperature, surface sensor assemblies exist with a sensor mounted to a larger metal plate to optimize heat exchange with the object to be measured.

For specific information on assemblies and housings, commercial pamphlets are available from:

Fenwal Electronics Inc., Farmingham, Mass. 01701
Keystone Carbon Company, St. Marys, Pa. 15857
Victory Engineering Corp., Union, N.J. 07083
Siemens America, Iselin, N.J. 08830
Siemens West-Germany, WWB, Munich 8, West-Germany
Yellow Springs Instrument Co., Yellow Springs, Ohio 45387
Standard Telephones and Cables Ltd., Footscray, Sidcup, Kent, U.K.
Ohizumi Ltd., Tokyo, Japan
Toa Electronics Ltd., Tokyo, Japan

As a spin-off from the general trend to improve reliability of electronic components, hot or cold weldable small cans offer new possibilities for economical and efficient size reduction of encapsulated sensors. A forerunner of this development was the He-filled can housing a Ge thermometer. Cryogenic thermistors for the He and liquid H_2–range (as reported in 4.1.1.) have also gained in reliability by canning.

6

Basic characteristics of thermistors

6.1.1. RESISTANCE-VERSUS-TEMPERATURE CHARACTERISTICS

As shown in Chapter 2, log R is in the first approximation a linear function of $1/T$. For small temperature intervals this dependence permits reasonable interpolation of temperatures since other factors contributing to resistivity, such as mobility, have a smaller temperature dependence. For larger temperature ranges more sophisticated equations are necessary. The empirical formula of Bosson and co-workers[426] provides a good approximation for two thermistor types over a temperature range from 250 to 400 K: $R = R_0 \exp [B/(T + \Theta)]$. Θ is a characteristic of the thermistor; its physical interpretation is not clear. It seems more meaningful to introduce a temperature dependence into the factor preceding the exponential term, as suggested prior to Bosson by the Bell Telephone team[447] or to consider a temperature dependence of the activation energy B.

For the first case it was suggested to write the resistance temperature equation

$$R = AT^c \exp\left(\frac{B}{T}\right)$$

with c a small positive or negative number and A a true material constant.

A new phase not only in describing but even in linearizing temperature characteristics of thermistors started with the systematic work by researchers at the Yellow Spring Instruments Co. Biological and medical applications require a wide range linear response of thermistors, and this demand was met by studies of Trolander and Harruff.[427]

A short time later Steinhart and Hart[428] investigated curve-fitting techniques for the calibration of thermistors. The results are similar. But whereas the first technique uses explicit functions of resistance, the second uses functions explicit with temperature.

Although the functions explicit in resistance appeal more to the physicist, practical considerations should remain the governing factor for a choice. A detailed description of the first approach is given. The exponent in the basic relation

$$R_T = A \exp\left(\frac{B}{T}\right)$$

is expanded as polynominal resulting in

$$R_T + A \exp\left(\frac{B}{T} + \frac{C}{T^2} + \frac{D}{T^3} + \cdots\right) \tag{4}$$

This new expression lends itself to computer programming and was calculated to the third degree. An explicit example of the procedure has been recently presented by Trolander, Case, and Harruff.[429] The increasing accuracy in describing a measured R-versus-T characteristic by expanding the number of terms in the exponential can be clearly seen from Table 39, which expresses deviations ΔT in K. The Bosson formula reduces them nearly by a factor 10 compared to the single exp (B/T) relation. By adding the terms C/T^2 and D/T^3, further improvements are attained up to a factor of 3–4.

Trolander et al. have rewritten Steinhart and Hart's equation, which is explicit in temperature,

$$T^{-1} = A + B \log R + C \log R^3$$

into a form explicit in resistance. They then inserted their data used for the polynominal equation of Table 39 and obtained a fit still better than that of their column based on three exponential terms. This is evident in the last column of Table 39.

The temperature coefficient $(dR/dT)\,100/R$ was calculated for Equation 3 (two exponential terms) by first determining the B and C values using a computer and then solving for R and its first derivative with respect to temperature. The sensitivity values thus obtained were used to calculate the apparent temperature deviations in Table 39. The detailed computer program of this procedure was given by Trolander, Case, and Harruff.

The higher-order terms in the exponential could be attributed to a temperature dependence of the activation energy. This is obviously an oversimplification, since the mobility of current carriers is also a function of temperature. Some basic data are known on the temperature dependence of the activation energy. According to Bube[430] dE/dT is negative and of the order of -5×10^{-4} eV K^{-1} for temperatures between 90 and 400 K and

TABLE 39 Resistance values computed by each of four equations for the temperature span −20°C to 120°C (253 to 393 K), and corresponding temperature deviations from a prototype thermistor.

Temp. (°C)	Temp. (K)	Exponent: Prototype (Ω)	B/T Equation 1 (Ω)	ΔT	$B/T + \theta$ Equation 2 (Ω)	ΔT	$B/T + C/T^2$ Equation 3 (Ω)	ΔT	$B/T + C/T^2 + D/T^3$ Equation 4 (Ω)	ΔT	Steinhart and Hart ΔT
−20	253	21859.9	23060.0	−0.94	21976.3	−0.091	21945.4	−0.067	21828.1	+0.025	+0.008
−10	263	12464.6	12747.9	−0.42	12489.0	−0.036	12482.9	−0.025	12459.5	+0.008	+0.002
0	273	7359.85	7359.85	0	7359.85	0	7359.85	0	7359.85	0	0
10	283	4486.07	4417.22	+0.32	4482.33	+0.017	4483.30	+0.013	4486.96	−0.004	−0.003
20	293	2815.90	2745.09	+0.56	2812.79	+0.024	2813.54	+0.019	2816.28	−0.003	−0.002
30	303	1815.69	1760.33	+0.72	1813.97	+0.022	1814.35	+0.016	1815.74	−0.001	0.000
40	313	1200.00	1161.33	+0.80	1199.41	+0.012	1199.54	+0.009	1200.00	0	0.000
50	323	811.425	786.132	+0.82	811.425	0	811.425	0	811.407	+0.001	0
60	333	560.399	544.765	+0.77	560.623	−0.011	560.572	−0.009	560.372	+0.001	+0.001
70	343	394.648	385.662	+0.67	394.926	−0.021	394.868	−0.016	394.648	0	0.000
80	353	282.952	278.420	+0.50	283.228	−0.030	283.185	−0.025	283.021	−0.007	−0.007
90	363	206.367	204.638	+0.27	206.515	−0.023	206.493	−0.020	206.411	−0.007	−0.007
100	373	152.911	152.911	0	152.911	0	152.911	0	152.911	0	0
110	383	114.975	116.010	−0.32	114.847	+0.040	114.866	+0.034	114.937	+0.012	+0.010
120	393	87.637	89.260	−0.69	87.411	+0.098	87.446	+0.083	87.573	+0.028	+0.023

about 10^{-3} eV K^{-1} for temperatures >400 K. A decrease in band gap can in the first approximation be explained by changes in energy levels caused by thermal expansion. It has been shown[431, 432] that this can only account for one-quarter of the observed effects. The broadening of the energy levels at the bottom of the conduction and the top of the valence band due to collision between electrons and phonons, though less effective in nonpolar crystals, would produce in ionic crystals, such as oxides of thermistor materials, a temperature effect dEg/dT of -5×10^{-4} eV K^{-1}, in good agreement with experimental data as estimated by Radkovsky.[433] Recently, Swartzlander, Jr. has compared eight different model equations with regard to their maximal and average error and their suitability for computer routines to fit polynomials.[706,707]

6.1.2. Linearization of Thermistor Characteristics

The nonlinear resistance-temperature characteristic inherent in its exponential relationship has always been considered the main disadvantage of thermistors. Therefore efforts were made earlier, to linearize their characteristic.[434–438] It is often forgotten that even the classical resistance thermometer using platinum (and, for that matter, other pure metals such as Ni or Cu) is plagued by nonlinearities. They are not as dramatic since the basic relation between resistivity and temperature is in first approximation linear. However, the greater demands on accuracy made on the Pt resistance thermometer require corrections for their nonlinearity. The resistance can be described by the following empirical equation:

$$R_T = R_0(1 + AT + BT^2 + CT^3)$$

where R_0 is the resistance at 293.15 K and the factors A, B, and C are:

	A	B	C
Ni[1]	6.0177×10^{-3}	6.535×10^{-6}	1.09×10^{-8}
Pt[439]	3.975×10^{-3}	-5.85×10^{-7}	0

The small negative second-term nonlinearity of Pt can be suppressed by connecting a platinum and a nickel sensor in series. By using a resistance ratio Pt/Ni of about 19, a linear 100-Ω sensor with a sensity of 0.4056 Ω K^{-1} can be made (hybrid linearization). Other methods of linearization including electronic techniques as well as the use of particular Pt alloys have been suggested. Their description goes beyond the scope of this book and only references to recent publications are given.[440–443]

Returning to the main topic of this chapter, the work done during the last decade in linearization of thermistor characteristics is reported. The first method of compensating the nonlinear R-versus-T characteristics of thermistors was the passive shunt. Their intrinsic high sensitivity permits their use with low parallel and high series resistance, thus attaining a reasonable linear approximation.[444] Making the shunt or the series resistor active, that is temperature dependent, is an additional means for attaining a linear output. Two examples have been given by Kimball and Harruff[443] to show the effect of single and double active shunting (Tables 40 and 41); Figures 73 and 74).

Double shunting reduces the nonlinear output to the equivalent of ±0.04 K. The analytical forms for these networks are:

$$R_x(\text{at T}) = 2768.23\ \Omega - 17.115\ \Omega\ \text{K}^{-1}\ (\text{T} - 273.16)$$

$$R_y(\text{at T}) = 1850.03\ \Omega - 9.2471\ \Omega\ \text{K}^{-1}\ (\text{T} - 273.16)$$

TABLE 40 Performance data for two thermistor network of Figure 73.

When $R_1 = 3200\ \Omega$, $R_2 = 6250\ \Omega$, $T_1 = 30\text{k}\ \Omega$ at 298.16 K, and $T_2 = 6\text{k}\ \Omega$ at 298.16 K.

Temperature (K)	Required R_T	Generated R_T
273.16 + 0	20,510	20,325
273.16 + 5	16,600	16,685
273.16 + 10	13,790	13,895
273.16 + 15	11,684	11,740
273.16 + 20	10,040	10,046
273.16 + 25	8,719	8,702
273.16 + 30	7,643	7,611
273.16 + 35	6,741	6,713
273.16 + 40	5,980	5,962
273.16 + 45	5,326	5,320
273.16 + 50	4,759	4,763
273.16 + 55	4,262	4,271
273.16 + 60	3,824	3,836
273.16 + 65	3,435	3,445
273.16 + 70	3,086	3,092
273.16 + 75	2,773	2,772
273.16 + 80	2,448	2,483
273.16 + 85	2,227	2,221
273.16 + 90	1,991	1,984.5
273.16 + 95	1,777	1,771
273.16 + 100	1,578	1,580

2768.23 and 1850.03 are the resistance values at 273.16 K. It is necessary to take into account the basic concept that thermistors are always part of a network including a resistance bridge, an input source, and an output meter. Common bridge connections already have an intrinsic nonlinearity. Therefore, the linearization problem is that of the entire network. It has been treated by Trolander, Case, and Harruff for several networks with a number of thermistors using Thevenin equivalent circuits (Figures 75*a*–75*d*).

For the output 3–4 to be linear with temperature requires that e_x also meets this condition. This is certainly not the case for e_x across the thermistor R_x.[445]

A network of N thermistors, each in series with fixed resistors $R_1, R_2, \ldots, R_N$ as shown in Figure 76(*a*) for resistance and in Figure 76(*b*) for voltage

TABLE 41 Performance data for three thermistor network of Figure 74.

When $R_1 = 2160\ \Omega$, $R_2 = 4031\ \Omega$, $R_3 = 9025\ \Omega$, $T_1 = 45\text{k}\ \Omega$ at 298.16 K, $T_2 = 15\text{k}\ \Omega$ at 298.16 K, and $T_3 = 3\text{k}\ \Omega$ at 298.16 K.

Temperature (K)	Required R_T	Generated R_T
273.16 + 0	12,392	12,375
273.16 + 5	10,579	10,581
273.16 + 10	9,164	9,171
273.16 + 15	8,034	8,035
273.16 + 20	7,108	7,106
273.16 + 25	6,336	6,334
273.16 + 30	5,684	5,679
273.16 + 35	5,122	5,118
273.16 + 40	4,638	4,633
273.16 + 45	4,211	4,208
273.16 + 50	3,837	3,833
273.16 + 55	3,502	3,500
273.16 + 60	3,206	3,203
273.16 + 65	2,935	2,935
273.16 + 70	2,694	2,692
273.16 + 75	2,471	2,472
273.16 + 80	2,270	2,271
273.16 + 85	2,085	2,086
273.16 + 90	1,914	1,915
273.16 + 95	1,756	1,758
273.16 + 100	1,612	1,612

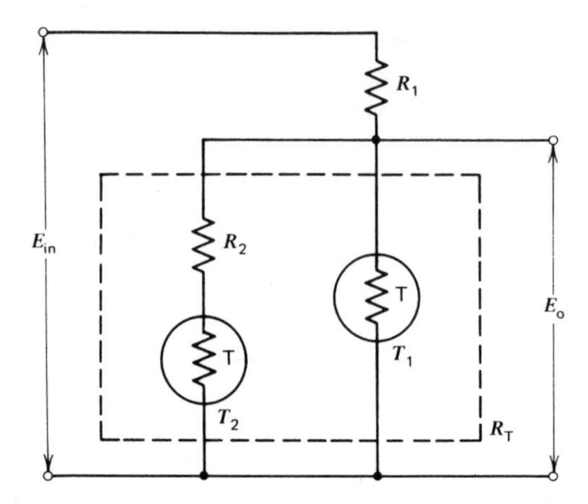

FIGURE 73. Thermistor shunting another thermistor to improve linearity. Courtesy of Kimball and Harruff.[443]

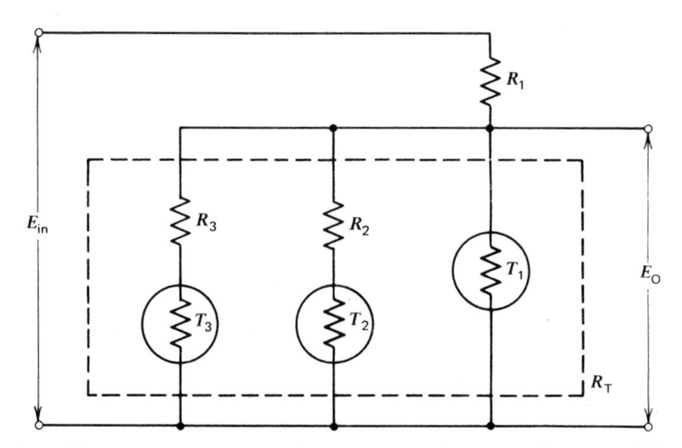

FIGURE 74. Three thermistor network to improve linearity. Courtesy of Kimball and Harruff.[443]

output would result in a resistance R_x given by the equation

$$\frac{1}{Rx} = \frac{1}{R_1} + \frac{1}{R_{T1}} + \frac{1}{R_{T2} + R_2} + \frac{1}{R_{T3} + R_3} + \frac{1}{R_{T4} + R_4} \cdots + \frac{1}{R_{TN} + R_N}$$

The temperature dependence of each thermistor is given by:

$$R_T = A \exp\left(\frac{B}{T} + \frac{C}{T^2} + \frac{D}{T^3} + \frac{E}{T^4} \cdots + \frac{X}{T^{x-1}}\right)$$

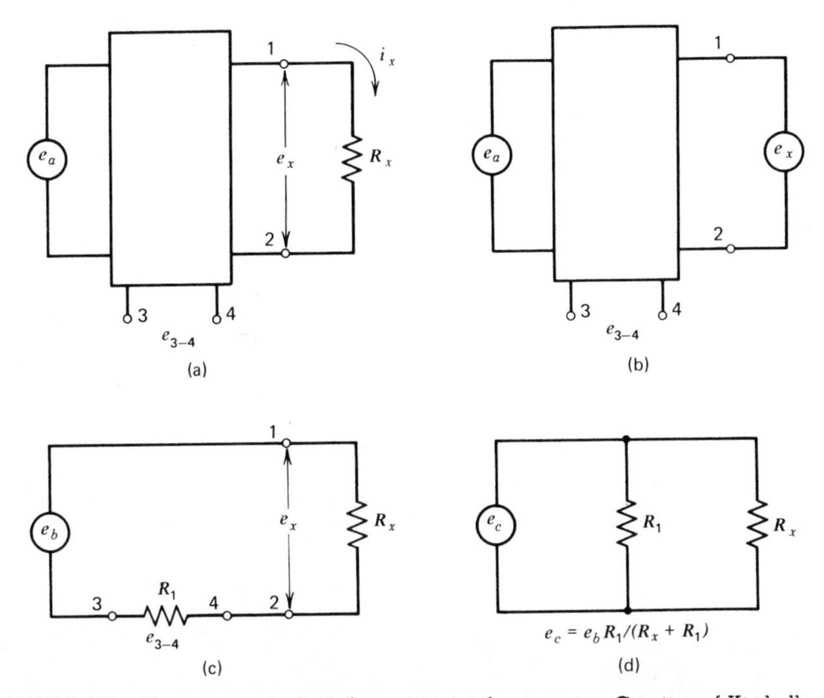

FIGURE 75. Thevenin equivalents for resistance thermometer. Courtesy of Kimball and Harruff.[443]

This leads to an equation that is rather cumbersome and difficult to solve explicitly. After assuming certain reasonable values for a desired temperature range (Figures 77 and 78), iteration by computer has been successful for a network with three thermistors over the temperature range 273–373 K and resulted in the following equations:

$$R_x = 1839\ \Omega - 9.17(T - 273.15)\ \Omega$$

$$e_0 = 1.057\ V - 5.31 \times 10^{-3}\ \mathrm{V\ K^{-1}} \times (T - 273.15)$$

The accomplished linearities and the relative contributions made by each thermistor to the output of the network in Figure 78 are shown in Figure 79.

Multiple-thermistor networks are able to meet the requirements for a linear characteristic and still have sufficient sensitivity in intervals exceeding 100 K. There are, of course, limits as to their practical and economical feasibility. One is the volume of space for which the temperature is uniform and is to be monitored. This limits the number of thermistor shunts in the network. Since they are active components, they have to be in the sensing volume, while the fixed resistors R_1, R_2, . . . , R_N of the network can be

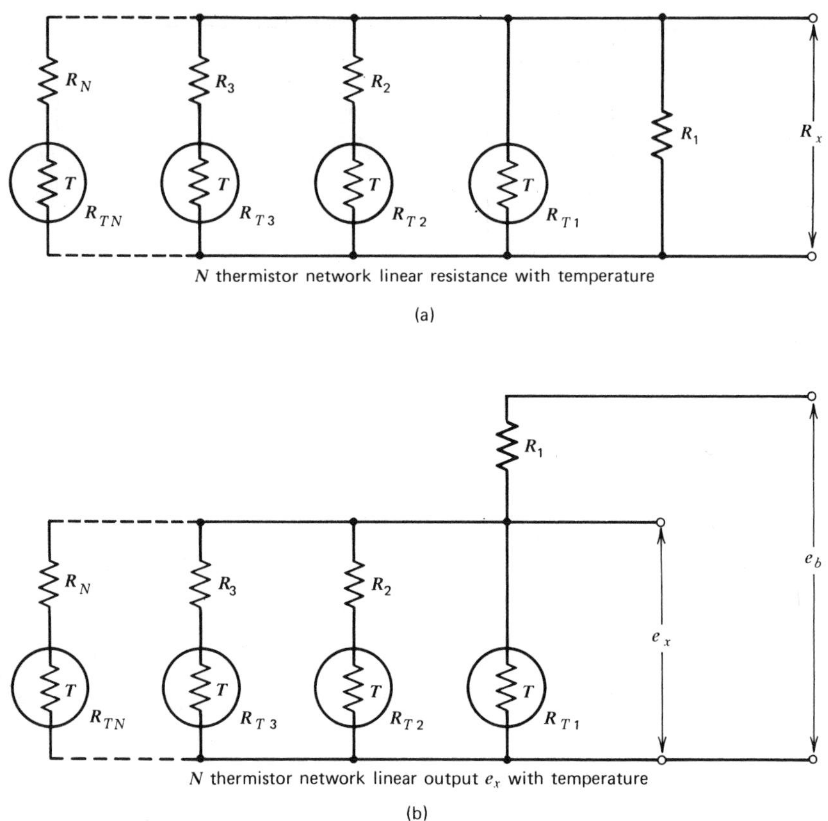

FIGURE 76. Generalized multithermistor network. (*a*) Linear resistance with temperature. (*b*) Linear output e(*x*) with temperature. Courtesy of Kimball and Harruff.[443]

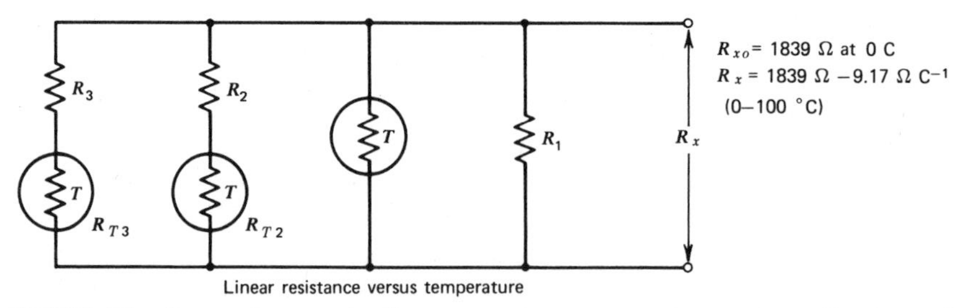

FIGURE 77. Practical example. Linear resistance versus temperature. Courtesy of Kimball and Harruff.[443]

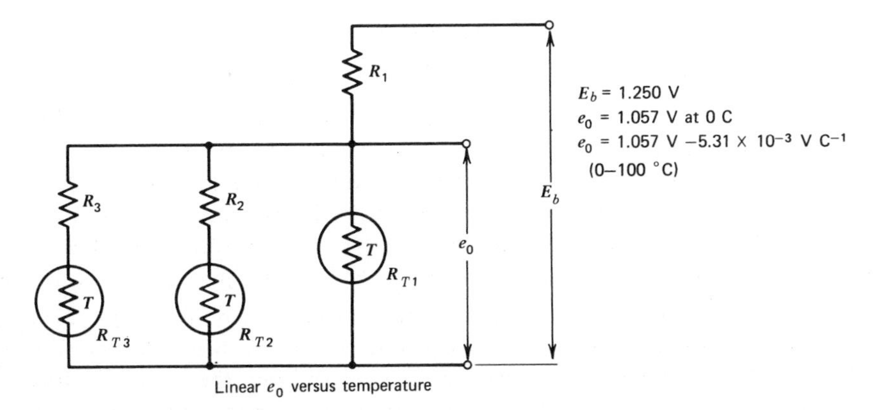

FIGURE 78. Linear voltage output versus temperature network. Courtesy of Kimball and Harruff.[443]

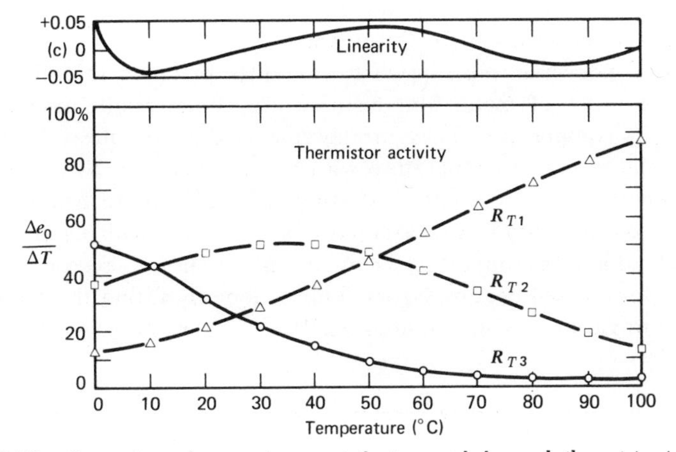

FIGURE 79. Linearity and percentage contribution made by each thermistor to network output versus temperature in Figure 78. Courtesy of Kimball and Harruff.[443]

somewhere outside. There is also the problem of uniform adaptation of the network thermistors to the temperature in the test volume, which is related to their time constant. If one or several of the network thermistors have different time constants, the entire linearization process becomes erratic and highly dependent on the time rate of temperature variations. If the sensor network controls the input of a heater system, these differences can be minimized by increasing the distance between sensor and heat exchanger. Kimball and Harruff have given two examples:

1. *Linearized Psychrometer*: This measures the temperature difference between a wet and a dry bulb (the latter has the ambient temperature). The relative humidity determines this difference and can be read from psychrometric tables. A constant temperature sensitivity is necessary for calibration at a convenient temperature valid for the entire temperature range. Figure 80 shows the circuit with T_D representing the dry and T_M the wet thermistor.
2. *Linear Bolometer*. When measuring radiant energy by absorption in a thermistor of suitable design (see Section 7.1.7), the ambient temperature fluctuations are troublesome, first of all by shifting the zero-radiation resistance of the sensor but also by the variation in temperature sensitivity for different radiation inputs. The linearization network as shown for the psychrometer could serve the same purpose.

Temperature compensation of the Clark-type polarographic oxygen electrode current that has a positive but nearly linear temperature coefficient is another example for this method.

Integrated networks consisting of two or three precision thermistors and the same number of metal film resistors are commercially produced as thermolinear components. They are specified either for linear voltage or linear resistance versus temperature with a sensitivity of ≈ 20 mV K^{-1} for temperature ranges of 80–140 K starting from 218 up to 373 K. Their deviation from linearity can be tailored to specification with a minimum of ± 30 mK and a maximum of ± 1.1K, with restrictions imposed on the input voltage to curtail self-heating errors. This is done by fitting the thermistor composite to various resistor composites.[446]

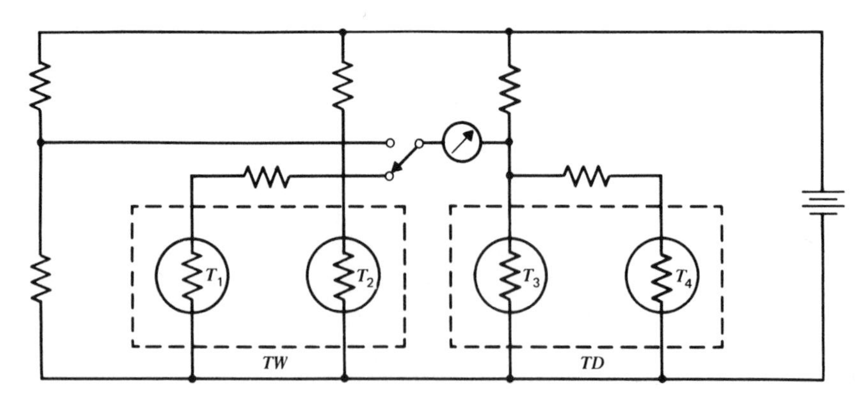

FIGURE 80. Linearized wet bulb circuit. Courtesy of Kimball and Harruff.[443]

6.2. THE STATIC VOLTAGE-CURRENT CHARACTERISTIC

The interesting properties of a thermistor when self-heated by a current have been presented and explained in the basic publication by the Bell Laboratories quoted earlier.[447] Typical voltage-current characteristics are shown in Figure 81 for a PTC and an NTC unit. For an NTC unit each point of this static state curve is measured by allowing sufficient time for the voltage at each current setting to attain a steady value in still air. As long as the self-heating power VI is smaller than the heat dissipation, the resistance is constant and given by the slope of the straight line starting from the origin. When self-heating becomes appreciable, the voltage increase lags against the increment of current until, for a current I_m, a voltage peak V_m is reached where a further increase of current causes proportionally the same decrease in resistance. For still higher currents the resistance drops faster than the current increase, resulting in a falling voltage-current region with negative resistance. Only at very high current values the decreasing temperature coefficient of resistance forces the characteristics to reverse into a positive region. This often cannot be realized in practice if the produced temperatures lead to degradation of the material or other irreversible resistance changes.

It has been shown that the influence of various factors on the V-I characteristics, such as initial (cold) resistance values, B constants, and different dissipation conditions and ambient temperatures, can be better displayed by log-log plots of voltage versus current. This not only has the advantage

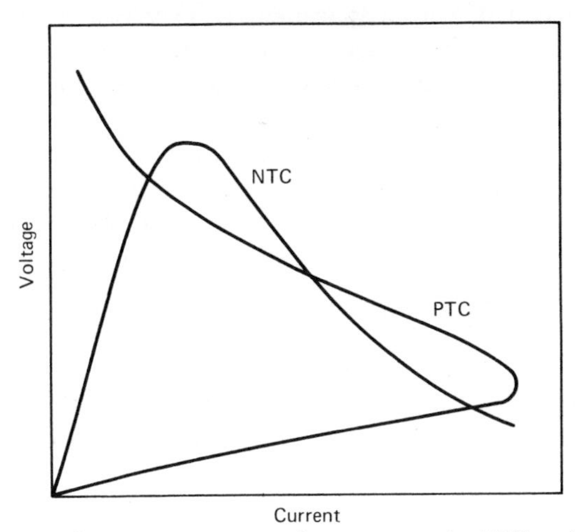

FIGURE 81. Static voltage versus current characteristics for NTC and PTC units.

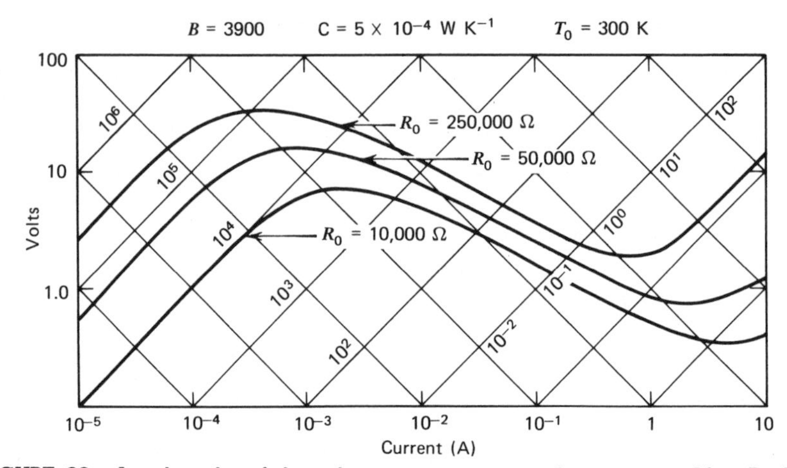

FIGURE 82. Log–log plot of the voltage versus current characteristic. After Becker, Green, and Pearson.[447]

of covering a wider range of current and voltage values but also of normalizing the initial slope in the positive resistance region to 45° for all resistance values. For different thermistor resistance values, a parallel shift of the curves occurs. While a positive 45° slope represents a constant resistance value, a negative 45° slope will represent a constant power input (Figure 82).

This would permit one easily to see how the power input increases with increasing current. This is of interest if the thermistor is used as a power meter. In the following figures, examples of various V-I characteristics are given as a function of various parameters. These are computed idealized curves, in which only one factor was assumed to be changing. In reality, some parameters change simultaneously with temperature, for example, B as well as the dissipation constant while the ambient temperature can be held rather constant, if the device is in contact with a heat sink. Becker and co-authors have also calculated the temperature of a thermistor at the voltage peak by differentiating its two basic equations of state with respect to current and setting $dV/dI = 0$. This leads to the relation:

$$T_m = \frac{B}{2}\left[1 - \left(1 - \frac{4T_0}{B}\right)^{1/2}\right]$$

For $B = 4000$, $T_m \sim 320$ K; for $B = 2000$, $T_m = 370$ K; for $B = 1600$, $T_m \sim 400$ K; and for $B = 1200$, $T_m = 600$ K. The peak disappears at B values of less than 1200 (Figure 83).

By inserting T_m peak temperature into these basic equations—one for the temperature dependence of resistance, the second for the dissipation equi-

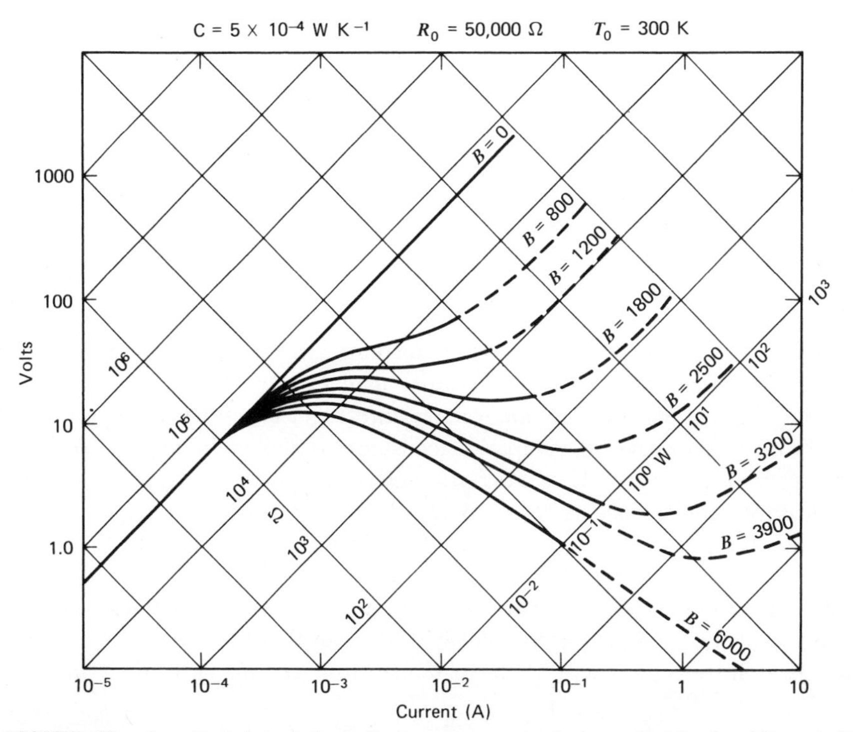

FIGURE 83. Log–log plot of the voltage versus current characteristic for different B values. After Becker, Green, and Pearson.[447]

librium—the power input, resistance, current, and voltage at the peak can be calculated.

$$P = D(T_m - T_0)$$

$$R_m = R_0 \exp\left[B\left(\frac{1}{T_m} - \frac{1}{T_0}\right)\right]$$

$$V_m = \left\{D \cdot R_0(T_m - T_0) \exp\left[B\left(\frac{1}{T_m} - \frac{1}{T_0}\right)\right]\right\}^{1/2}$$

$$T_m = \left\{\frac{D}{R_0} (T_m - T_0) \exp\left[-B\left(\frac{1}{T_m} - \frac{1}{T_0}\right)\right]\right\}^{1/2}$$

Figure 83 also indicates that in a certain range of B values a region of horizontal V versus I characteristics is favored that is useful for voltage regulation. A maximum extension of such a region is very desirable. This is achieved to some degree by higher dissipation losses at higher temperatures (i.e., current values). In this case the negative resistance region is lifted to an approximately horizontal line. Thermistors operating at temperatures where radiation losses start to play a role in the energy balance are likely to meet this condition. A more systematic approach is to arrange the thermistor in series with a fixed resistor having the same positive resistance as that of the thermistor in the negative region. The compensation of these opposite trends is, of course, limited by the extension of the region with nearly constant negative resistance. This is determined by the constant B and its first derivative dB/dT and also by design features, such as heat leaks and radiation losses that have already been mentioned.

Figure 84 shows that with increasing ambient temperature the voltage peak decreases and shifts to higher currents and the negative resistance slopes for different ambient temperature tend to converge. In this case the self-heating temperature dominates any differences in ambient temperature. The effect of increasing dissipation constants is shown in Figure 85. The apparent similarity of V-versus-I characteristics that have a voltage peak ("turnover point," according to Burgess) and a negative resistance region as in reverse-biased Ge rectifiers and thermistors led Burgess[448] to a quantitative analysis of these phenomena assuming different functional relations between resistance and temperature. The thermal nature of the resistance reversal at the peak is in agreement with the fact that the "turnover power" at this point increases with ambient temperature according to the equation

$$P = \frac{1}{D}\left[\frac{B}{2} - T_a - \left(\tfrac{1}{4}B^2 - BT_a\right)^{1/2}\right]$$

where D is the dissipation constant and T_a is the ambient temperature.

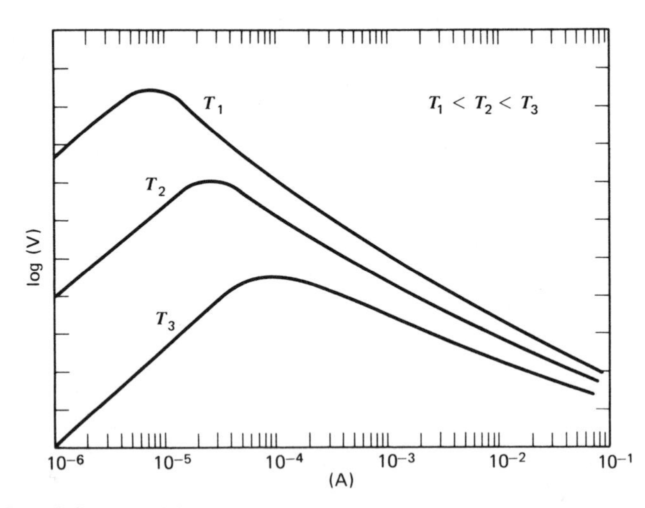

FIGURE 84. Influence of the ambient temperature on the voltage current characteristic. After Becker, Green, and Pearson.[447]

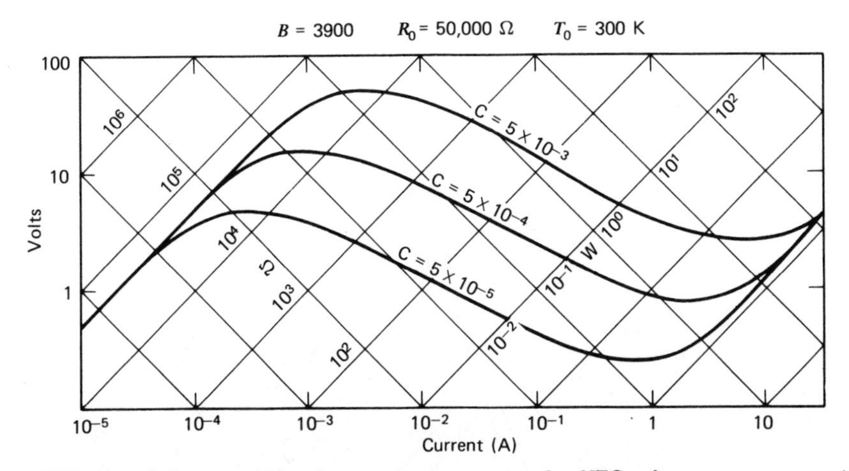

FIGURE 85. Influence of the dissipation constant on the NTC voltage versus current characteristic. After Becker, Green, and Pearson.[447]

For example, the relative increase in turnover power between 300, 400, 500, and 600 K is given by the ratios 1:2.5:4.18 for $B = 4000$. The peak voltage, however, decreases with ambient temperature as shown previously. This can be very useful to switch safety devices that prohibit the overheating of rooms or equipment (see Section 7.2.3).

The inverse resistance temperature characteristic of PTC thermistors results in a corresponding displacement of their *V-I* characteristic relative to NTC thermistors as shown in Figure 81. It has become common practice to plot PTC characteristics with the voltage as abscissa. Analogous to the voltage peak in the NTC case, a current peak occurs. Its amplitude and location in the *I-V* diagram is, as in the NTC case, determined by the heat dissipation conditions, the ambient temperature, and the initial cold resistance. These effects are shown in the Figures 86–88.

6.3. DISSIPATION AND TIME CONSTANT

DISSIPATION. The dissipation constant is defined as the power input necessary to raise the temperature of the sensor 1 K above ambient temperature. Therefore

$$D = \frac{P}{\Delta T}\ (\mathrm{W\ K^{-1}}) \quad \text{or} \quad \Delta T = \frac{P}{D}.$$

It is normally measured by determining the resistance as a function of

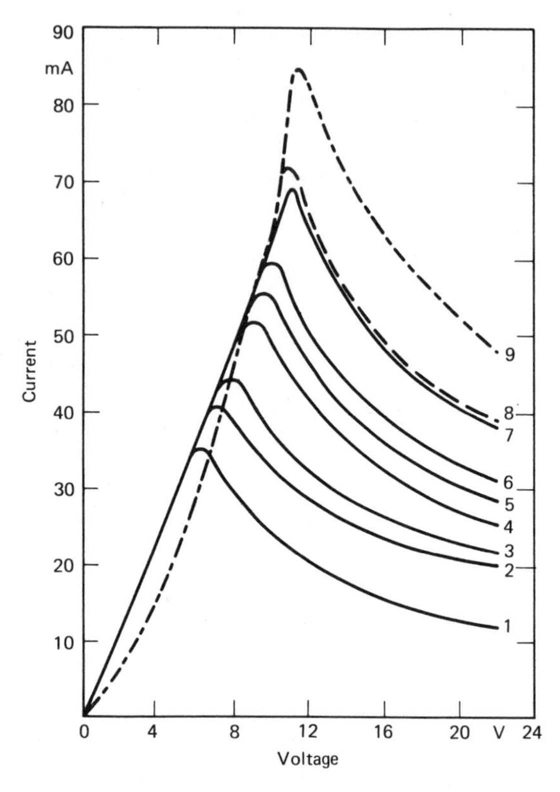

FIGURE 86. Influence of the dissipation constant on the PTC characteristic. Curve 1: still air, 290 K; 2 and 3: mineral oil and gasoline ~ 340 K; 5 and 6: mineral oil and gasoline ~ 300 K; 8 and 9: mineral oil and gasoline ~ 300 K with convection cooling; 4 and 7: Water ~ 340 K and ~300 K. Results of Hanke and Löbl.[709]

power input and translating the observed resistance increments ΔR into temperature increments ΔT:

$$\Delta T = \frac{100\Delta R}{R\alpha} \qquad (\alpha \text{ in } \% \text{ K}^{-1})$$

For ΔT values less than 2 K, D is approximately constant since α is approximately constant and the dissipation process is nearly the same. This does not apply to boiling liquids, especially in the cryogenic range. With increasing power input several phenomena can occur at the interface between thermistor and liquid. These may have opposing influences on the measured dissipation constant (as well as the time constant). At very low power levels,

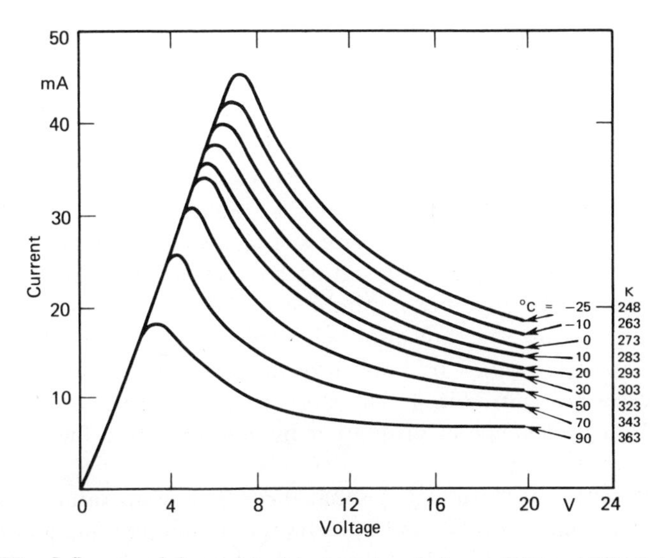

FIGURE 87. Influence of the ambient temperature between 248 and 363 K (−25 and +90°C) on the PTC characteristic. Results of Hanke and Löbl.[709]

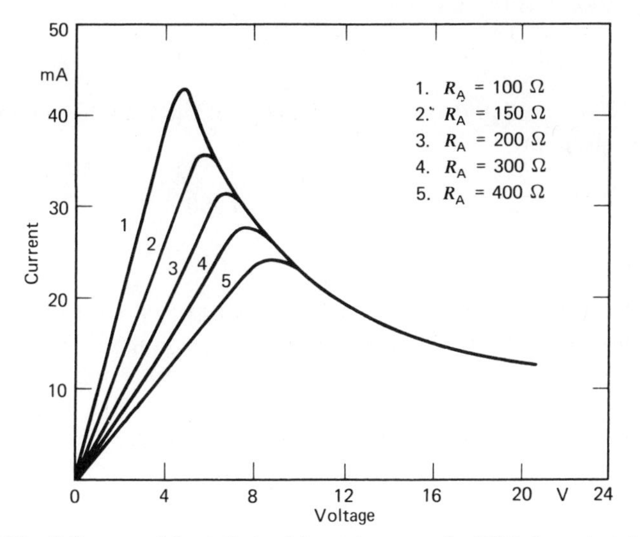

FIGURE 88. Influence of the initial cold resistance on the PTC characteristic. Results of Hanke and Löbl.[709]

heat conduction and convection supported by nuclear bubbling tend to increase the heat loss of the thermistor and therefore also the dissipation constant D. However, for each thermistor type a power level can be expected, at which a part of it is permanently insulated from the liquid by a vapor buffer zone, which reduces the heat exchange and therefore also D.

Several factors determine how large a fraction of the total immersed surface comprises the buffer zone: size and shape of the units, the capillary activity of its surface, and the resulting contact angle between vapor and liquid. This can be modified by bubble retarding or promoting coatings. Not only the bubble size, but also their tendency to adhere influence D. If they are rapidly removed by buoyancy or vibrations, an additional factor that increases D is introduced. Therefore many applications of thermistors are based on the change of dissipation by environmental factors.

TIME CONSTANT. It is well known that the rate of temperature change of a body with a temperature T_t and a medium of a different ambient temperature T_a is proportional to the temperature difference $T_t - T_a$, assuming that no first-order phase transitions in the body occur. The differential equation

$$\frac{dT}{dt} = (T_t - T_a)\frac{D}{Cp}$$

after integration has the solution

$$T - T_a = (T_0 - T_a)e^{-t/\tau}$$

with $T_0 = T$ at $t = 0$ and $\tau = Cp/D$. For $t = \tau$, the body has changed to a temperature fraction

$$\frac{T - T_a}{T_0 - T_a} = \frac{1}{e} = 0.368$$

of this initial temperature difference $T_0 - T_a$.

In other words, the thermal time constant represents the time required for a thermistor to change 63.2% of the total temperature difference between its initial and final temperature when subjected to a temperature change under zero-power condition. This time is evidently proportional to the heat capacity and inversely proportional to D, which is identical in dimension and numerical value with the dissipation constant D:

$$T = \frac{C}{D}\,\frac{\text{W sec K}}{\text{KW}} = \text{sec}$$

These relations are the basis of numerous applications.

The trend to miniaturization of sensors increases the risk of self-heating, although only response to ambient temperature is desired. The main incentive for applying smaller sensors is the reduction of the time constant. Since heat transfer processes that determine the dissipation constant are not only proportional to the mass of the units but are also influenced by the conditions of the surrounding medium, the rapidly decreasing heat capacity at lower temperature must be met by more sensitive test equipment. This led to the development of a number of high-resolution bridges for dc and ac operation. A four-terminal bridge operating with 1 mV sensor voltage has a resolution and stability of 0.003% of the resistance over several hours.[449,450]

The preceding considerations have dealt only with temperature decay after power input has been cut off or the ambient temperature had changed. Another interesting aspect is the thermal inertia of a thermistor with rapidly varying power input. This problem will be treated in some detail on pp. 300–303. It is quite obvious that the negative resistance of a thermistor operating beyond its voltage peak can only exist at dc or low frequencies. For high frequencies its temperature cannot follow the current fluctuations any more and remains constant. Therefore its resistance becomes positive. There must be critical frequency for each thermistor, determined by its thermal material and dissipation constant, at which its ac resistance is zero. Critical frequencies between 0.1 and 10 kHz have been observed for well-coated beads or oxide films with good heat sink.

6.4. STABILITY

6.4.1. Facts

CONVENTIONAL THERMISTORS. As for every electrical component, the stability and reproducibility of thermistors have always been primary concerns after electrical properties attracted the interest of users. There was a time when instability and unreliability of thermistors were taken for granted, at least in comparison with other temperature-sensing devices such as thermocouples and metallic resistance thermometers. Large-scale thermistor applications are a tangible proof that engineers have generally ignored this assumption. Systematic work by producers and users has helped dispel this myth, which often was based on the misconception in defining stability. A sensor with a temperature coefficient 10 times larger than others might change its value 10 times more and still be equal in stability with another one of lower sensitivity. Therefore, producers of thermistors always have been asking for a just figure of merit that compares the observed resistance changes to sensitivity at each temperature, in other words, spells out in-

TABLE 42

Sensor	Long-Term Drift D (K yr^{-1})	Sensitivity S (K or V K^{-1}) Reciprocal of Amplifier Gain Required for 1 V K^{-1} Monitor Output	$M = S/D$
Thermocouples			
(a) Pt-Pt/Rh 90/10	0.1	8.3×10^{-6}	8.3×10^{-5}
(b) (Chromel-Alumel)	0.1	3.5×10^{-5}	3.5×10^{-4}
Quartz-Crystal[a]	0.01	3.5×10^{-5}	
Silicon-bulk resistor	0.1	2×10^{-3}	2×10^{-2}
Platinum resistance sensor	10^{-3}	10^{-4}	0.1
Thermistor	5×10^{-2}	3×10^{-2}	0.6

[a] Given by the temperature coefficient of oscillation.

stability in calibration drifts. This viewpoint has been strongly endorsed at the "5th Symposium on Temperature, June 1971" by Swartzlander[451] and also in discussions. Swartzlander defines a figure of merit by the ratio

$$M = \frac{\text{Sensitivity per K}}{\text{Calibration drift per year}}$$

for which he gives data (Table 42) based on interchangeable thermistors available from a number of producers.

The reason for the relatively low rating of platinum resistance thermometers is of course their low sensitivity. Their use requires either a high-gain amplifier or a high voltage across the Pt sensor to obtain sufficient signal amplitude for automatic monitoring. The intrinsic high stability of the Pt sensor can be lost by drifting of the amplifier. Nonlinearities can also be added to the intrinsic nonlinearity of the sensor ($\sim \pm 0.25$ K) over 100 K range.

To implement the suggestion for compensating nonlinearity by a programmed reference voltage would complicate the circuitry.[452] This work was done to compare temperature sensors for the Apollo and other space instrumentation where a very critical analysis was mandatory and where the best commercially available thermistors were used.

For normal applications, experimental studies with ordinary thermistors are important. Research workers at Bell Telephone Laboratories had observed for uncoated thermistors a calibration change of 50 mK yr^{-1} after preaging at 378 K. Geological, meteorological, and oceanographic investigations, where long-range stability is very important and recalibration not as easy as in the laboratory, have initiated the collection of more data that, since measured under variable field conditions, not always were as favorable as in laboratory tests.

It was found[453] that the stability of bead thermistors in geological bore holes was not very good, with a drift rate of −120 to −140 mK per month, observed over a period of 17 months, which resulted in total calibration errors of up to several degrees.* In this case, the operating conditions and the environment (humidity in the soil) may have done some harm.

Droms[454] has presented stability data for eight different groups of thermistors, bare, glass-coated or sealed, with resistance values 2000–2400 Ω at 398 K, held at 373 K or 473 K for 91 days. He confirms that glass-sealed units are the most stable and reports that the calibration error after this period was 6–22 mK at 373 K and 55–330 mK at 473 K.

Glass-coated beads were found to be less satisfactory than glass-sealed units with calibration errors up to 6 K, possibly due to interaction between metal contact and glass coating. Bare units showed erratic behavior and errors exceeding 1 K. In this study the influence of drifting on the resistance at 273 and 323 K was measured. For each investigation the possible temperature fluctuations in the test environment cannot be ignored. Liquid baths are the most reliable medium and can be regulated within ±20 mK or better. But even vigorous stirring may not eliminate all spatial temperature differences. Furthermore, the effect of heat leaks from the clamps outside of the bath and possible interaction of the bath liquid with bare units may interfere with precise measurements.

Useful bath liquids for uncoated units should have the following properties:

1. Boiling point at least 50 K above required bath temperature.
2. Low viscosity.
3. Chemical inertness.
4. Good solubility in solvents of low boiling points.

Fluidized beds of sand or alundum are more convenient and can be used up to 1300 K.

Measuring thermistors in air has always been a problem. Not only their much stronger tendency for self-heating, but also the smaller heat transfer constant from the gas to the solid, can cause larger errors than found in a liquid bath. Pippin[455] has studied the influence of environmental factors such as daily room-temperature fluctuations on the overall reliability of thermistor resistance measurements for 84 days. The temperature of ovens operating at 293 and 303 K varied from 0.044 to 0.012 K, respectively, per

* The apparent temperature of the medium is lower, not the true one.

1 K temperature change in the environment. The daily variations were 0.043 and 0.014 K. For temperature shift of Fenwal GB 3111 (1000 Ω) thermistors he found 55 mK per month when self-heating during measurement was held under 20–27 mK at 293 or 303 K, respectively.

Cryogenic Sensors

Ge Resistance Thermometers (GRT). Ge resistance thermometers have been always considered as very stable and reproducible. The true temperature deviation ΔT_G after 50 cycles between liquid He and a room-temperature alcohol bath was found to be $\pm 70\ \mu K$.[456] It is defined by the equation:

$$\Delta T_G = \Delta T_T - (\Delta T_S + \Delta T_t)$$

in which ΔT_t is the apparent temperature deviation of the test unit, ΔT_S is the systematic error, and ΔT_t is the deviation due to a bath temperature change, which was found to be zero within the time of 50 cycles.

ΔT_s and ΔT_t were determined by keeping a second GRT permanently immersed into the same depth of liquid as control He during the entire cycling time. The resistance measurement were made with a current of 10 μA permitting a precision of 2 mΩ. The absence of a slope when plotting 5 consecutive temperature readings calculated from the corresponding resistance data of the control unit 2 indicated that the He bath was constant ($\Delta T_t = 0$). The systematic error is given by the average deviation from the mean

$$\Delta T_S = \frac{\left(\sum_{n=1}^{50} \Delta T_n^{\,2} \right)^{1/2}}{n}$$

This result is in line with earlier stability investigations by Blakemore, Schultz, and Myers,[210] who estimated $\Delta T_S = 0.1$ mK deviation by thermal cycling between 4.2 and 300 K. Even after 2500 cycles between 77 and 300 K the low-temperature characteristics remained unchanged for one GRT. Here the general rule applies that intrinsically stable units have good chance to have a long-term stability.

Groups of Ge thermometers that had been calibrated at the National Bureau of Standards, USA, the Physicotechnical and Radiotechnical Measurements Institutes USSR (PRMI), Iowa State University, and by the National Standards Laboratory, Australia NSL (between He and Ne boiling points) were used as secondary thermometers and their resistance temperature characteristics were compared with other scales (the acoustic scale of National Bureau of Standards, the gas thermometer of PRMI and NSL, and the magnetic scales of ISU and the Kammerlingh Onnes Laboratory, Leiden). It was found that the overall spread between these scales is 30 mK. Two-thirds of these effects are caused by systematic deviations of the scales. For the remaining 12 mK, intrinsic instabilities of the Ge thermometers could be responsible together with technical difficulties of attaining equal heat sinking of the units. For one Ge thermometer (Cryocal type) relative calibration shifts of 1 mK and 2.5 mK at 25 K, and 3.6 mK at 30K are listed. Details of the calibration technique were given and supplemented by a special study to minimize errors in expressing the resistance temperature characteristics by polynomial interpolation. Calculated temperatures can differ up to 5 mK depending on the interpolation procedure. The maximal order of the polynomial required to reduce the interpolation error to 2–3 mK differs from one thermometer type to the other. Values between 6 and 12 were proposed in the range 1–30 mK.[214] The European complement to this work is a study by van Rijn and co-workers.[457] They used five of the same Ge thermometers that the Australian group (Kemp et al.) had for their comparison of the low-temperature scales, and added two from the Compagnie Generale d'Electricité (France). As to their reproducibility between 1.5 and 30 K after cycling and repeated cycling up to 300 K the investigated 21 thermometers could be broken down into the following groups:

Calibration change at 20 K:	40	30	25	2–5	$\pm$ 0.3 mK
Number of units	1	1	1	3	15

This clearly points out that highly reproducible Ge thermometers can only be found by pretesting and selection with a possible rejection ratio of about 25%. It is, however, disquieting, as the authors found, that even carefully selected thermometers may change a few mK as found for the five units used in the international comparison project. Still more disturbing is the fact that one unit, previously constant for several years, changed 30 mK and thereafter remained very constant again. The paper does not specify the national origin of these mavericks. However, verbal communication on similar experience seems to indicate that it could happen anywhere.

The Dutch team was able to obtain optimal calibration for different temperature ranges by using different degrees of polynomical such as:

4 between 1.5 and 5 K within 0.4 mK
12 between 1.5 and 30 K within 0.2 mK
3 between 13.81 and 23 K within 0.5 mK

Carbon Resistors. Composition carbon resistors used in cryogenic measurements differ in stability with the brand and batch of resistors and their sensitivity. Anderson[458] surveyed the existing data and condensed them into these statements: Allen-Bradley and Speer resistors retain their calibration over 2 yr within 1% equivalent to an temperature error of 30–60 mK at 4.2 K. Over short periods stability can increase by a factor of 10, but differences between Allen-Bradley and Speer (1002) become more pronounced, the latter being less stable. Cochran and co-workers[459] found for Allen-Bradley resistors an initial rate of 1 mK wk^{-1} at 4.2 K. Observations by the author indicate the approximate validity of logarithmic time–deviation relation for Coldite resistors with the deviation increasing slightly faster than the log of time. In this case the units were stored at room temperature and occasionally measured in LH_2 or He.

Anderson dispels the myth that the long-term drift is caused by penetration of He into the resistor. It is intrinsic for the resistor composition, which as a complex structure of conductive carbon particles and insulating regions is bound to be "alive" under thermal cycling stresses.

Weinstock and Parpic[233] have observed a resistance increase of 2% in a recent production batch of 1000 Ω, $\frac{1}{8}$ W. Allen-Bradley resistors between the first and second thermal cycle from 4.2 to 300 K. This clearly demonstrates such internal changes, which can be expected to slow down with increasing numbers of cycles. This has been confirmed by a precise investigation with eight Allen-Bradley resistors. Some of them were reproducible even on cycling up to 300 K after the second cooling to ~4 K.[676]

Thermistors. Short-term reproducibility of 0.5 mK of oxide-type cryogenic thermistors for the He range has been reported by Schlosser and Munnings.[202] Using a hydrostatic head correction for immersion depth of the test unit and a precision manometer with an accuracy of ±0.03 torr, temperatures thus determined were related to the L 0904-100 kΩ thermistor resistance at constant filling level of the Dewar. Over a period of 4 days no change of resistance beyond the experimental uncertainty of 2×10^{-5} was observed. Repeated lifting of the thermistor from the bottom to the top of the liquid in the 25 liter Dewar increased its resistance linearly from 103.210 to 103.580 kΩ with a day-to-day reproducibility of 0.007 kΩ for each position. Uncertainties in barometer pressure could be responsible for this small deviation of $\sim 10^{-4}$ %. Using the temperature gradient by the head effect, the relative resistance sensitivity $dR/RdT = 1.41\ K^{-1}$ is calculated. The barometric pressure sensitivity at a constant immersion depth was 265 Ω $torr^{-1}$.

These data are in agreement with results found by the author while he was connected with the development and production of these and similar cryogenic sensors. His earlier statement[198] that the reproducibility was only about 30 mK was based at that time not only on experimental limitations but also dictated by the conservative attitude that a true figure for reproducibility should be found by each user under his own experimental conditions of bath temperature uniformity.

Schlosser, Munnings, and others have been correct in scrutinizing this modest claim of 30 mK by their own investigation.

Other L 0904 types such as those for the He, LN_2, and LOX range have the same short-term reproducibility.

A more challenging problem is always the long-term stability, which can be influenced by a number of factors difficult to control. Cryogenic applications imply in many cases quenching into liquified gases resulting in heat shocks. Their effect is especially severe since the brittleness of most materials increases with decreasing temperature. This is the reason why protective coatings, based on silicon-rubber compositions have failed in long-term life tests though they performed excellently at higher temperatures. Conventional epoxy and lacquer coating fared still worse, even in short-term tests. Vacuum evaporated SiO coatings on the other hand, though adhering well and nonpeeling in thermal quenching, are often porous and fail to protect sufficiently against undesirable atmospheric contaminations such as H_2O, NH_3, and so on. The detrimental effect of chemisorbed molecules on the electrical resistivity of grain boundaries has been reduced or eliminated by a special impregnation process.

Despite these precautions cryogenic oxide thermistors may drift because of the following factors:

1. Repetitive dunking into liquified gases may lead to microcracks with corresponding resistance increase. Practically never was loss of contact leads alone observed. Whenever one contact was lost, it always took also a small bit of oxide material away, indicating a region of a high stress in contact zone.
2. If units are removed from a liquified gas, they tend to frost in normal atmosphere containing some humidity. Upon further warming up, the frost melts and covers the entire thermistor with a layer of water which of course soaks the capillaries of the oxide body (despite impregnation). Repetitive dipping into the cold bath freezes this water and can lead to similar effects as the well-known frost cracking of stones.

It is unrealistic to make demands on calibration accuracy, reproducibility, and interchangeability that go beyond the limits outlined in the preceding chapter. Thermistor producers nevertheless strive for further improvements. However, progress in developing more stable units has been slow, since long-term tests are necessary to recognize the influence of variants in composition, processing, and contacting. The logarithmic time dependence of resistance changes helps to project future trends. It is, however, doubtful whether this can be applied to all degrading effects including contact deterioration. In the latter case the degrading process could be suddenly accelerated if thermal stresses resulted in a sudden resistance increase by partial loss of contact.

6.4.2. The Causes of Instability

General Remarks. Most of the data on stability of thermistors are only empirical, and no appreciable efforts have been made by the various reporters to relate them to physical or chemical changes. Furthermore, stability was only related to the fact that the primary electrical characteristics, resistivity and its temperature dependence, do not change. In many cases investigators were satisfied to measure the resistance change at a certain standard temperature after exposing the unit to different heat treatments, or simply to determine resistance change on the shelf. Consecutive measurements at several standard temperatures within the application range provided additional data on the stability of B.

Secondary electrical characteristics such as voltage dependence or electrical noise (often related to each other) are seldom measured in the initial state, and mostly neglected in life tests. It is known and accepted as fact that in most cases thermistors increase in resistance, except in highly reactive or corrosive atmospheres, which can completely change the chemical composition of the thermistor material and are normally avoided by encapsulation.

In analyzing the physical and chemical influences on resistivity and resistance-temperature characteristics, one has to distinguish between bulk and contact effects. It is possible to separate them by relatively simple test methods, although they are more time consuming than naive two-terminal resistance measurements in a bridge, which always give the sum of bulk and contact resistance. Nearly all published stability data lump these two resistance components together since in practical applications the stability of the entire system bulk plus contact is of interest and no one cares what might be the root of any observed changes. The attitude of the thermistor manufacturer must be quite different. He, who strives for perfection, likes to know the physical or chemical reasons for any changes at elevated temperatures or over extended periods. Therefore an analysis of these factors is of interest.

Bulk Effects. The resistivity and its temperature dependence is a function of the concentration of current carriers in the crystallites of the oxidic thermistor material, which is dependent on its defect structure. For each temperature an equilibrium between the atmosphere (its partial pressure of oxygen) and the defect concentration in the oxide exist. The equilibrium at sinter temperature is different from that at the operating temperatures, which are normally 800–1000 K lower. The cooling rate after sintering is not small enough to establish equilibrium at each temperature. Therefore the thermistor material is in an intrinsically instable state when arriving at room temperature (just like quenched steel in its mechanical properties).

The rate for readjusting this instable state to the new equilibrium conditions at or slightly above room temperature is very slow. In first approximation the temperature coefficient of solid-state diffusion is a measure of this rate. For the transition oxides used in standard thermistor compositions the diffusion rate at 1000 K increases approximately by a factor of 2 for a temperature increase of 30 K (NiO) or 50 K (CoO).

In an earlier chapter the point was stressed that sintered oxide materials are more complex than just single crystals with holes. The individual grains can range in size from 1 to 40 μm, depending on sinter conditions. The grain boundaries can differ in composition and electrical properties from the interior of each grain (microcrystallite) especially if the limited diffusion rates during cooling from the sinter temperature do not allow homogeneous distribution of defects within the grains. Therefore any thermal drifts at application temperatures connected with changing defect concentration will first involve processes in the grain boundaries. This applies especially to the effect of atmospheric contaminants such as H_2O, CO_2, H_2S, NH_3, and organic vapors that by chemisorption can change the resistivity of internal or external surface layers, in some cases such as H_2S, irreversibly. The omnipresence of such contaminants in living areas can contribute to the drift of thermistor characteristics. Model experiments with well-defined partial pressures of such contaminants can help one to understand instabilities and to find remedies to reduce them. The ultimate solution for eliminating these influences remains encapsulation with the sacrifice in response and dissipation constant.

Trolander, Case, and Harruff[429] have tried to relate bulk stability to structure conditions in spinel-type thermistor composition containing Mn-Ni oxides. The structure of such oxide systems used in thermistor materials has been discussed in detail in Chapter 3. Larson, Arnott, and Wickham[119] conclude that any phase change after sintering, for instance by annealing or heating to fire the silver contacts, can reduce stability proportional to the time of exposure to the higher temperature. By minimizing the possibility of phase changes during processing by proper selection of composition and sinter conditions, optimal stability can be obtained. Phase changes often result in mechanical microstresses that could create microcracks in grains or at grain boundaries with corresponding increase of resistivity.

So far only phenomena have been considered that influence the defect concentration and concentration gradients in the grain, the chemistry of grain boundaries, and the electrical effects of chemisorption at grain boundaries. The possibility of cruder effects such as the formation of micro- or even macrocracks cannot be ignored and is indeed sometimes the cause of bulk instability. Already the molding process can produce microcracks (laminations) if poor flowing of the powders results in inhomogeneous density of the molded compact and the sudden release of pressure during ejection from the mold causes sudden expansion exceeding the coherence strength of the compact. Tapered dies have minimized this effect. However, the internal stress pattern in the compact persists even to the temperatures prevailing during the sinter process and often becomes visible in warped sinter bodies but more often remains hidden in internal microcracks.

Thermal cycling of such thermistors has the tendency to amplify the microcracks, in some cases to the point where a macrocrack appears and destroys the unit. The thermal drift of such units with internal mechanical faults is in most cases accompanied by an increase of their electrical noise figure. It seems practical to screen out mechanically weak units by noise testing, provided that the background from contact noise is small enough. This condition cannot always be met, especially if leads are applied to silver coating by soldering or welding.

Long-range stability, also at elevated temperatures, can be influenced by the sinter conditions, especially sinter temperature. In general, long-range stability is expected to improve with increasing sinter temperature, sometimes only for the trivial reason that porosity decreases and thus also chemisorption decreases during storage. However, in many cases the resistivity also increases, thus resulting in a conflict of interest. A typical example is given in the Bell Telephone Patent U.S. 2,694,050 for a system Mn-Ni-Fe-oxides. For a drift reduction of a factor 3–10 by increasing sinter temperature from 1270 to 1570 (2000–2360°F), a resistivity increase by a factor of 20–30 had to be accepted.

A generalization of these results to other systems is not directly possible. In each case the cooling rates can also influence stability, and last but not least, a separation of bulk and contact effects on stability is necessary. It can happen that the improved bulk stability achieved by higher sinter temperatures is accompanied by a smoother surface, which can be detrimental to a good contact.

Contact Effects. The combined experience of many workers in the thermistor field during the years has led to the conviction that degradation of the electrical contact is the most common cause of thermistor instability. This applies to all kind of contacts. Let us consider the factors and processes that are detrimental to sustain a good contact for three different contacting types.

Disk with Fired on Silver Coating (*see pp.* 164–166). The silver coating after firing in an oxidizing atmosphere at 770–1125 K represents a mixture of fine silver particles and glass. Even under optimal coating conditions such as silk screening (screen printing), nonuniformities of the silver-glaze layer can exist which make the contact resistance vulnerable to changes. It is obvious that materials of three different expansion coefficients, such as the bulk of the thermistor, silver, and glaze may develop stresses and fatigue microcracks when thermally cycled. The variety of silver paints developed for this purpose can be considered as tangible evidence of efforts to minimize detrimental effects of this kind.

A second problem arises if a metal lead, usually tinned Cu, is attached to the silver coating by soft soldering or welding. The major problem, already recognized in other electrical components such as silver-coated ceramic capacitors, is the alloying of the thin silver film with the tin containing solder, resulting in the formation of the intermetallic compound Ag_3Sn that has poor adherence. Special solder alloys containing a few percent Ag have been brought on the market to reduce this effect. However, aftereffects of slow diffusion between Ag and solder, even at moderate temperatures, are still possible over extended periods.

Through decades, the reliability of the soldered lead connections has presented various problems, especially when stress and fatigue caused loss of leads. Improved solder compositions have reduced these risks. A full analysis of the mechanical properties of 18 soft solders up to 10^5 cycles has recently been made.[460] Capping of rod-type units eliminates the solder problem, but requires smaller tolerance to secure a good elastic fit.

It is often overlooked, even by developers of thermistors, that mechanically satisfactory contacts are often more or less nonohmic and voltage dependent. In this case chemisorption of O_2, H_2O, and NH_3 can influence the contact resistance considerably, sometimes in an erratic manner. Furthermore at temperatures above 435 K thin silver coatings have the tendency to "fade" either by migration or at still higher temperatures by evaporation. Even glaze-free pure silver films show this effect, which can also be caused by micrograin growth ("granulation"). The surface tension of metal films drives them to reduce their surface by forming a granular layer with a much lower effective contact area. Even if the bulk resistivity of a thermistor body is unaffected by a certain heat treatment, this reduction of the effective contact area must increase the resistance for purely geometrical reasons.

In some cases considerable Ag-whisker formation has been observed during long-term life tests with silver contacted disk thermistors.

A less-known source of contact deterioration is silver migration under dc conditions. This effect is basically not of thermal nature, although it tends to increase with temperature. It was first observed with uranium oxide thermistors for voltage regulation (Siemens type of 1940), which derated rapidly when used to regulate dc voltage although they had excellent stability with ac. During the late 1940s silver migration on the surface of Cu oxides thermistors with dc voltages applied was observed. By successive reversals of the applied voltage the electrical nature of this phenomenon could be clearly demonstrated.

Ag film electrodes only applied to the opposite ends of the dark oxide body (CuO or Co- Ni- Mn spinel) spread toward the opposite electode, if dc was applied for long time periods.

Since metallic silver migrated in the electric field, ionization had to play a role in this effect. Therefore use of contact metals with higher ionization potential was suggested to suppress migration and attain higher stability under dc conditions.[461]

Electromigration, that is, the electrolytic transport of metal ions in liquid or solid metals at high current densities, has been systematically studied for decades.[462] With the rapid expansion of microelectronics, high current densities could occur in many circuits, especially if semiconductors are contacted with thin metal films. The destructive effect of electromigration is proportional to the square of the current density, the latter being a linear function of the metal film thickness. For disk thermistors, though contacted with metal films of less than 10^{-3} cm thickness, the current density remains small even for self-heating conditions, since the current flows normal to the metal films. However, any contact configuration involving a voltage gradient in the plane of the contacting film, and even excessive thickness variations in contact surface, can result in electromigration. Whenever bars or rods are contacted at opposite ends with the current flowing in axial direction, electromigration can be expected. Its detrimental effect on electrical properties (stability) is slow, but increases exponentially with time. Any weak spot in the contact film has to carry higher current density, and the destruction rate increases as

$$DR \sim AI^2 \cdot \exp - \left(\frac{E}{kT}\right)$$

where A is the constant depending on cross section of the film (perpendicular to current flow) and E is the activation energy, which depends on the structure of the metal film. E decreases with the grain size and thus reduces the destructive effect of electromigration.[463]

So much has crystallized out from a large number of investigations that the current-induced mass transport (electromigration) occurs mainly at grain boundaries and is connected with the condensation of vacancies into voids at the grain boundaries. Temperature gradients result in divergencies in the mass flux directed toward the anode and thus cause cracks or holes in the metallic film.[464]

The electromigration in thin Cu, Ag, and Au films is toward the cathode, just the opposite direction observed in bulk. This reversal has been explained by the fact that the migration is confined to grain boundaries in contrast to bulk material.[465]

Independent of this contact migration effect, dc current in self-heated thermistors can produce long-term electrochemical changes in bulk and contact resistance. Many semiconductors, considered to be purely electronic, have in fact a small ionic component of conductivity. It is in most cases difficult to decide whether this component connected with ionic transport is caused by small impurities or is intrinsic to the material.

In the practical case of thermistor materials where for reasons of economy not always the purest raw materials are used, the existence of impurity ionic conduction is possible. In most oxides the metallic ions are smaller than the oxygen ions. Therefore their mobility is predominant and will determine the changes in the bulk material by dc current flow. At the contact electrodes the following effects can occur:

1. At the cathode: metal ions transferred from the bulk by the current will accumulate and decrease the oxygen-to-metal ratio in the oxide adjacent to the cathode (electrolyte), which might result in a barrier layer of higher resistivity.
2. At the anode: oxygen atoms appear by discharging O^{2-} ions. They can either combine to molecular O_2, thus forming a gas film, or react with the contact metal to form an oxide, which of course can have a different resistivity. The electrochemical oxidation together with the disappearance of contact metal of course increases the total resistance of the oxide thermistor. That such detrimental effects of dc on oxide thermistors, especially their anodic contacts, really exist, has been observed by the author in several cases. Even Pt electrodes up to 50 μm thickness are eaten away by anodic corrosion if applied as anodes on transition metal oxide bodies.

After 72 hr flow of 2 A ($\sim$53,000 C) at 750 K in sintered CoO, strong silver migration was observed from the Ag anode resulting in the following Ag contents in percent:

Anodic Interface Zone	Anodic Zone 1 mm Thick	Center Zone	Cathodic Zone
10	3.15	0.97	0.21%

A blank test at the same temperature using Co electrodes increased the O-to-Co ratio from the initial value of 1.015 to 1.049 after 90 hr at 2 A

(65,000 C), thus increasing the cation vacancy concentration and resulting in a considerable resistivity decrease. Similar effects can be found with Mn_3O_4 or compositions of both oxides, although they become more complex with increasing number of constituents. It can be clearly spelled out that electrolysis is another limiting factor to the use of certain conventional thermistor grades at higher temperatures under dc conditions. If the thermal stability is good enough, applications could be restricted to ac. The other solution would be a new composition. With gold electrodes no migration was detectable, but the resistivity decreased if the Mn content was considerable.*

Even with ac applied mass transport can occur in the semiconductor and contact material, if a temperature gradient within the unit continuously exists. This thermomigration, also called the Soret effect, has been discussed by a number of authors.[466–468]

So far all physicochemical considerations have dealt with processes that would impair the stability of disk or rod thermistors, having in common external contacts. For bead thermistors with molded-in leads, the problem of silver evaporation, granulation, and migration are nonexistent. Only the possibility of electrolytical deterioration can persist. However, they are generally limited to lower temperatures where the mobility of ionic current carriers is very small. More critical are the mechanical problems caused by the different expansion coefficients of the lead wires (Pt or Pt alloys) and the oxide bead. Thermal cycling can loosen the bond between them, thus increasing the resistance of the bead—in the worst case—to interrupt the contact entirely.

Sapoff[469] has given some empirical data on the stability of bead thermistors with or without copper oxide in the thermistor composition in comparison with common thermistors (disks, wafer, and chip) having external contacts (Table 43). His data apply to zero power resistance over 5000–10,000 hr within the biomedical range 300–315 K.

In Table 43 glass-coated beads are distinguished from beads in glass envelope. As already shown by Doms, a glass coating in direct contact with the bead reduces the stability, apparently by the stresses induced with the various expansion coefficients of glass, sintered oxides, and lead wires.

STABILITY UNDER SELF-HEATING CONDITIONS. So far published stability data were generally given for near-zero power load at the thermistors, since here the best test conditions are best defined. Under self-heating a variety of modes are possible, and power dissipation also enters the picture. Self-heating can be stationary or pulsed. In the latter case the pulse frequency and

* Unpublished work by the author.

TABLE 43 Resistance change

Type	Copper Oxide Free (%)	Copper Oxide Doped (%)
External contact	0.5–3	3–10
Bare (bead)		
strain relieved	0.1–0.5	2–5
Hermetically sealed bead	0.05–0.28	0.2–1
		0.1–0.5

duration must be defined. If pulsed heating occurs in mediums of higher dissipation, determined by high values of heat capacity and heat conduction or convection, the test condition is more severe, because higher thermal stresses will develop leading to fatigue cracks. Finally, the load itself, in percent of the load rating, will also determine the stability during tests. Dc effects such as Ag migration will be accentuated in self-heated thermistors, not only because of their higher temperatures, but also because of higher fields. Finally a basic problem exists in self-heated thermistors: the hot spot. Any region with local decrease in resistivity, already existing in the cool state, carries a higher current, reducing resistance further by self-heating. With a falling V/I characteristic setting in, a high current, constricted into a narrow channel, will take over the entire conduction and shunt the remainder of the thermistor body, leaving it cold except for heat conduction leaks. As a result of this sequence, the hot channel becomes overheated and can change its stoichiometric composition or even burn out.

This basic problem deserves a thorough treatment.

In earlier years the demands on thermistor specifications have been often unrealistic and unreasonable. They ignored the fact that high power rating and extremely small time constants and dimensions are incompatible with each other. In many cases compromises could be found between these physically contradictory conditions by modification of the dissipation factor, the radiation loss, and finally by proper choice of a thermistor material. The limiting factors in this case are thermal stability including the contact, specific heat, and heat conductivity. Some stability factors are treated in the previous chapter. Data for specific heats of the transition metal oxides are listed in Section 3.4.

In some cases limitations are imposed not only by the physical and chemical properties of the thermistor and contact material, but also by the design of the units. One typical case is the disk thermistor when operating in self-heating condition. The transition from the positive to the negative resistance branch of the characteristics can encounter considerable difficulties especially for large and thin disk thermistors. Thermistors with a high ratio of cross section to thickness combine high load rating with low inertia and low resistance.

However, in earlier phases of thermistor research at Siemens a very annoying effect was observed when large thin disks were operated in the self-heating mode. At the rim of the thermistor hot spots appeared, overbridging the resistance proper of the thermistor until overload of this current channel led to its burn-out. Spenke recognized the analogy of this effect to the well-studied phenomenon of the thermal breakdown in insulators, which had led to the theory of Wagner[470] based on the intrinsic existence of a weak channel in the insulator. Since it has to carry the current in a small cross section, its temperature increases rapidly thus also broadening the weak channel, while the remainder of the insulator remains "cold" and nonconductive. While Wagner concentrated his attention on the rising characteristics up to the voltage maximum, and neglected the possibility of stabilizing the current by series resistance, Spenke[471] extended his analysis to the negative resistance branch of the current–voltage characteristics.[472] His model of the disk thermistor consists of n parallel elements. Even if completely isothermal conditions are assumed, small local variations of the material constant B within the disk can result in variations of the voltage peak. The element with the lowest voltage peak reverts first to a negative resistance characteristic and heats up quickly, resulting in a voltage drop for the other $n - 1$ elements, which subsequently cool down because of lower power input. There will be a transverse heat flow from the hot element to the n-1 other elements, which tends to stabilize the power input in the hot element and leads to a stationary transverse temperature gradient in the disk with quasistable current–voltage characteristics. It is possible that more than one of the n elements contribute current channels, especially if transverse heat transfer helps to reduce the peak voltage. These considerations also apply to the parallel arrangement of individual thermistors where thermal coupling is small or negligible. Whenever load requirements cannot be met by existing thermistor types with limited power rating, parallel arrangement should be avoided and the load divided into a series of thermistors adding up to the desired operational resistance value. But if such low resistance values are not available, parallel shunting should be done only with sufficient thermal coupling. Two practical approaches to accomplish this are possible: (a) arrangement in piles (Figure 89) and (b) separation of the individual thermistors by thin micafilms, which are excellent electrical insulators but have simultaneously a relatively low heat resistance.

Spenke started out with the simple model of two thermistors identical in initial resistance (load $\sim$ zero) and temperature coefficient, but variable

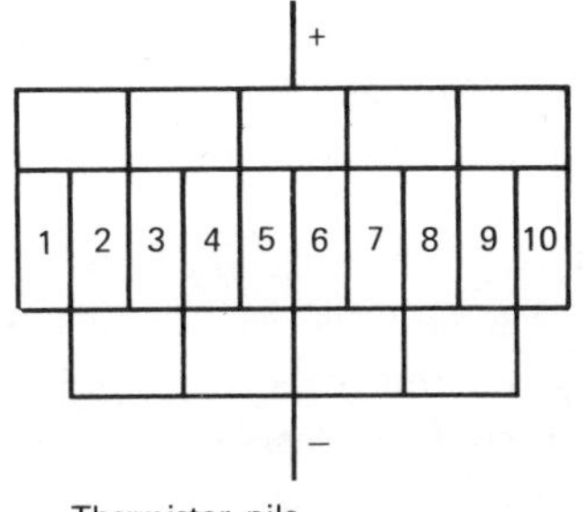

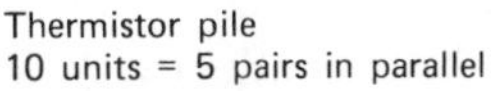

FIGURE 89. Thermistor pile.

thermal coupling. If the latter is smaller than infinity, the negative branch of the characteristics branches off to lower voltage values at increasingly lower currents as shown in Figure 90.

The temperature difference between the two thermistors becomes a maximum for coupling zero and disappears for perfect coupling, a result that seems to be self-evident. Advancing from this simple model, the stability conditions for a two-dimensional thermistor band, contacted at opposite sides of its length by metal sheets, were calculated. To simplify the theoretical treatment, thermal shunting of the transverse heat resistance in the thermistor strip by the metal electrode is excluded. While the latter is considered to be electrically isotropic, the assumption is made that the heat conduction exists only normal to the contact interface, but is zero parallel to it. This concepts brings Spenke nearer to the final problem of calculating the heat breakdown in the disk thermistor. Though the imposed condition of a highly anisotropic heat conductance of the metal contact seems to be physically absurd, it assumes a certain degree of reality, if one comes down to

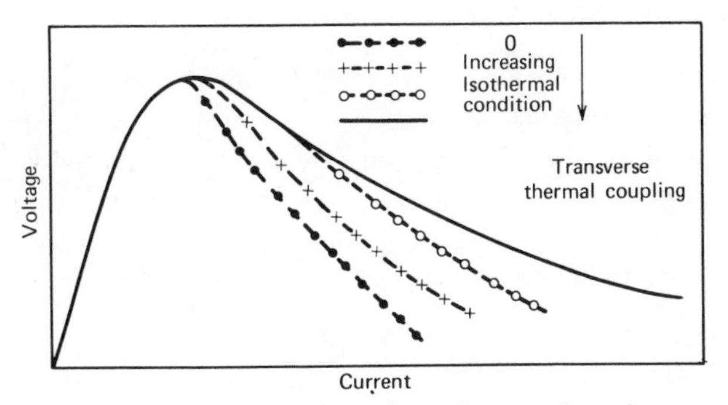

FIGURE 90. Influence of transverse thermal coupling on the voltage versus current characteristic (schematic).

the practical case of a disk thermistor with silver film electrodes of high transversal heat resistance and a current lead representing a heat sink.

Fired-on silver films usually have a thickness of $\approx 25\ \mu m$. Therefore a simple estimate for the heat resistance parallel and normal to the disk surface gives a ratio of $\geqq 10^5$. On the other hand the transverse heat resistance of the silver film is of the same order of magnitude as that of the thermistor body, assuming a ratio 10/1 of diameter thickness. For smaller ratios the thermal shunting effect of the silver and contacts decreases.

Some of Spenke's further calculations deal with disks contacted on both end faces with metallic heat sinks with isotropic heat conductivity. This case is irrelevant to commercial disk thermistors with attached wire leads and would only apply to disks held between spring-loaded metal sheets. The wire leads represent a heat sink of good, though limited, heat conductivity.

If attached in the center zone, their cooling effect produces a temperature gradient: Its transverse component stimulates the formation of hot channels near or at the periphery of the disk. Fluctuations in convection cooling can result in irregular motions of the hot spot at the rim, as observed in many cases.

There are occasionally more trivial reasons for hot spots to appear at the rim of disk thermistors. These include:

1. The Ag-film electrodes covering the disks have small irregular protrusions promoting a field concentration and corona effect. The resulting sparklets heat a tiny portion of the rim and create a region of lower resistivity.
2. Lower resistivity regions at rim have been formed during the sinter or the aging process, sometimes also by chemisorption in storage.

Whatever the reason for rim hot spots might be, they are annoying and certainly not beneficient to stability. There are several ways to avoid them:

1. Guard ring-type contacting, making the silver film coating somewhat smaller than the disk area.
2. Grinding of the rim to remove any areas of lower resistivity.
3. Increasing the thickness of the disks. This is least desirable since it involves selection of a lower resistivity material to obtain the same resistance value also tends to change the dissipation and time constant.

For rod-type NTC thermistors hot spot problems are practically nonexistent. In PTC rods they have the opposite effect: any local overheating in the initial stage of self-heating blocks further heating by forming a zone of high resistivity in series.

6.5. RADIATION DAMAGE

Systematic data of the radiation damage to commercial thermistors are scarce. The effects of electrons, neutrons, and α particles on germanium are better known by the work of Lark-Horovitz and co-workers.[473–476] Bombardment with these nuclear particles displaces atoms from normal lattice sites and creates vacancy-interstitial pairs (Frenkel defects). The results are different for n- and p-type Ge. Some of the radiation damage disappears at room temperature. Heating to 430–570 K leads to further recovery. Irradiation of n-type Ge with an integrated deuteron flux of 2.5×10^{15} neutrons cm^{-2} increases its resistivity by more than five orders of magnitude and converts it with still higher flux to p-type Ge with a resistivity only ten times the initial value. The threshold energy required to produce radiation effects at 78 K is 350 keV for electrons. Below 40 keV no subthreshold damage is detectable. n- and p-type Si can be made intrinsic with resistivities over 10^4 Ω cm by irradiation.

The radiation damage in oxide semiconductors is of special interest. Irradiation of Li-doped MnO with an integrated neutron flux of 1.5×10^{19} neutrons cm^{-2} led to partial compensation of the initially present Li^+ Mn^{3+} acceptors, resulting in increased resistivity.[479]

The resistivity ratio after to before irradiation is about 10 at 370 K but decreases with temperature by annealing out the damage.

From systematic studies of the annealing rate between 500 and 670 K, an activation energy of 1.8 eV for the recovery process was calculated with an annealing time of the order of a few hours at 670 K.

Only 25% of the radiation damage is produced by primary neutrons and 75% by α and tritium particles formed by fission of Li^6 present in the doped oxide. This is very important since it could also apply to Li-doped NiO and CoO. Nonstoichiometric sintered NiO with O excess of 2×10^{-5} atoms $mole^{-1}$ exposed to a fast neutron flux of 4×10^{17} nvt was shown to decrease in resistivity.[477]

The threshold energy for various radiation damage effects has often been blurred by subthreshold phenomena. This can be explained in part by the fact that changes in electrical properties, which can be complex in nature, have normally been used as criteria for assessing the damage. An independent method using EPR has shown that in n-type Si of 2 Ω cm resistivity and with 2×10^{17} cm^{-3} O concentration, the threshold energy can vary from 125 keV for 1 μA cm^{-2} to 280 KeV for 20 μA cm^{-2} electron current.[478]

Bombardment of silicon with 10^{14}–10^{17} ions per cm^2 of boron, phosphorous, or antimony with an energy of 200 keV produces an amorphous layer increasing with the temperature of the target from 100 to 300 K. At 830 K this layer recrystallizes.[480] A comprehensive analysis of radiation damage to other electronic components (Si-based devices) under space-flight conditions by a 1-MeV electron flux could give some estimate on the possible effects on Si bolometers. Choice of *p*-type Si and preexposure to high doses of 10^8 J/kg of low-energy electrons can reduce the radiation damage by subsequent irradiation with higher-energy electrons (radiation hardening).[481]

Thermistors selected for spacecraft application were irradiated with neutrons of over 2.9 MeV and a flux of $2.2 \times 10^{10} n$ cm^{-2} sec^{-1} together with a γ-radiation density of 42 J g^{-1} hr^{-1} for 91 hr producing only a transient decrease of their resistance.

Other commercial thermistors increased their resistance by a neutron bombardment of 10^{14} *nvt* and higher energy of 14 MeV, but recovered completely after 72 hr.[482]

Glass-coated oxide thermistors exposed to 10^{16} n_f cm^{-2} and 8.5×10^9 erg g^{-1} γ radiation apparently did not suffer changes in their temperature sensitivity. It can be expected that thermistors containing appreciable fractions of cobalt or uranium oxide are more damaged by nuclear radiation because of their susceptibility to neutron bombardment.[483] Irradiation of $BaTiO_3$ with fast neutrons from 10^{17} to $8 \times 10^{19} n$ cm^{-2} ($E > 1$ MeV) decreased the tetragonal c/a ratio of $BaTiO_3$ until a cubic phase appeared at an integrated flux of $nvt > 8 \times 10^{19} n$ cm^{-2}. Correspondingly, the dielectric constant as well as the Curie temperature decreased 23 K for $7 \times 10^{18} n$ cm^{-2}. Partial recovery occurred after annealing for 2 hr at 1070 K while radiation damage from β and γ irradiation disappeared at ~380 K. With an integrated neutron flux $>8 \times 10^{19} n$ cm^{-2} in $PbTiO_3$, metallic lead was formed by radiolysis. A sharp reduction of the ϵ peak from 6.2 to 3.8 ($\times 10^3$) at the Curie temperature has been observed when $BaTiO_3$ + 4 mole % $PbTiO_3$ was irradiated with a dosage of $nvt = 10^{18}$ fast neutrons.[484] In view of the connection between dielectric constant, Curie temperature, and PTC characteristics, appreciable radiation effects can be expected in PTC thermistors.[485] For a number of semiconductors mentioned in previous chapters or of potential use as temperature sensors the change of the carrier density by irradiation with a fast neutron flux ϕ is nearly the same and not very different for n-or-p-type semiconductivity. This has been shown for Ge, GaAs, GeSb, InP, and SiC with carrier densities between 10^{16} and 10^{18} cm^{-3}.[702]

7

Applications

Thermistors can be operated in three modes:

1. With nearly zero input responding to ambient temperature changes or impingent heat radiation.
2. Self-heated, responding to changes in dissipation conditions.
3. Indirectly heated and controlled by separate heating elements.

The applications are classified according to these modes. From the multitude of reported applications only a few have been selected; these represent the basics well enough that they can be used for other cases. It was therefore natural that a few older "classics" were treated more in detail.

7.1. MODE 1: OPERATION UNDER AMBIENT CONDITIONS

7.1.1. Temperature Measurements in General and High-Precision Measurements

If the resistance of the sensor is only to be a function of the ambient temperature, its dissipation constant has to be as high as possible and the applied power a minimum. This imposes limitations as to possible reduction in size and time constant. With dissipation constants normally in the range of 0.5–5 mW K^{-1}, the test currents for temperature measurements are pretty well fixed within a range of 10^{-5}–10^{-3} A for resistance values between 100 kΩ and 10 Ω (assumed temperature error by self-heating of ~10 mK). In many applications a compromise between self-heating and the required sensitivity of the circuit is necessary. The two-thermistor bridge with two identical, matched thermistors helps to compensate for the self-heating, while it also serves the purpose to compensate temperature fluctuations of the environment surrounding the object whose temperature is measured by the sensor. This part of compensation related to self-heating depends of course on the way the bridge is operated. For constant voltage at the sensor (bridge is nulled by changing the variable resistor in series with the sensor) the self-heating power in the sensor would increase proportionally to the resistance ratio of cold to hot. The resulting temperature error can only be precisely calculated if the influence of the temperature on the dissipation constant is known. If the unbalance of the bridge is used to measure temperature, and the detector has high impedance ($R_D > R_T$), which can be

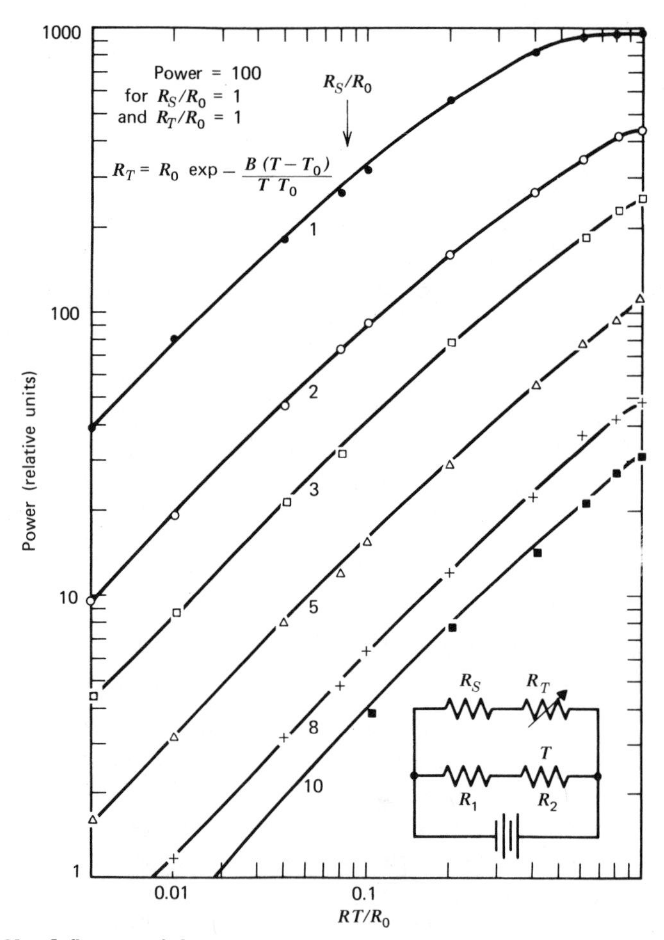

FIGURE 91. Influence of the series resistance and the resistance ratio hot/cold on the power dissipation in thermistors.

accomplished using a dc amplifier, the sensor would operate with nearly constant current ($\Delta i \sim +10\%$ for the condition $R_{series}/R_{thermistor} = 10$ and a sensor resistance $R_T = 10\%$ of its cold resistance). In this case self-heating power is only 12% of that for the matched reference sensor. For each resistance ratio hot to cold the circuit can be optimized to obtain reasonable self-heating compensation by controlling the impedance of the detector.

Figure 91 shows the influence of various series resistance values on the power dissipation in any sensor for hot/cold (R_T/R_0) resistance ratios 0.01–1.0 at constant bridge voltage. In cases where austerity is dictated by economical or technical reasons (miniaturization) half bridges are also used. It is possible to double the sensitivity of a complete bridge by arranging the sensors in diagonally opposite branches (Figure 92). In this case, as for the half bridge, the compensating effect of a reference thermistor is lost.

Goodwin[486] has made a comprehensive analysis of the optimal design of dc resistance thermometer bridges for wide-range temperature control.

Applying the simultaneous linear equations based on Kirchhoff's laws to bridge with metallic or semiconductive sensors, the following conclusions can be drawn:

For metallic sensors the resistance of the ratio arms should be greater, while for thermistors it should be less than the minimum sensor resistance. This minimizes for a given sensitivity of the bridge the required power in the sensor, which is important to reduce self-heating. The detector resistance should be greater than the maximum thermistor resistance. Since thermistor resistance increases rapidly at low temperatures, it becomes increasingly difficult to meet this condition. The effect of the detector impedance on the relative thermistor power relative to the minimum value necessary for matching impedance is shown in Figure 93 for three normal resistance value (assuming ratio arms of 100 Ω each).

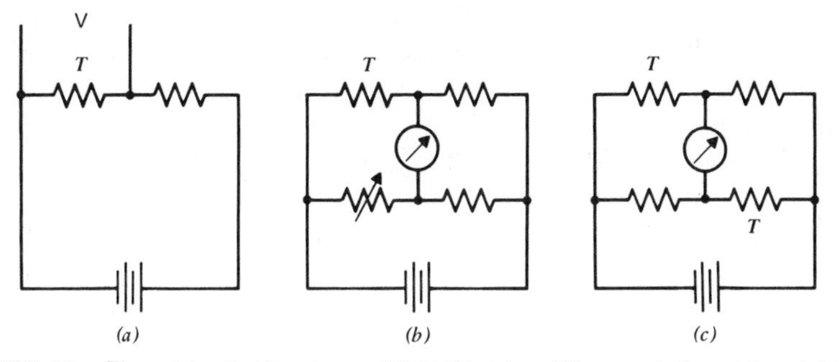

FIGURE 92. Thermistor bridge types. (*a*) Half-bridge, (*b*) normal thermistor, (*c*) two-thermistor bridge with increased sensitivity.

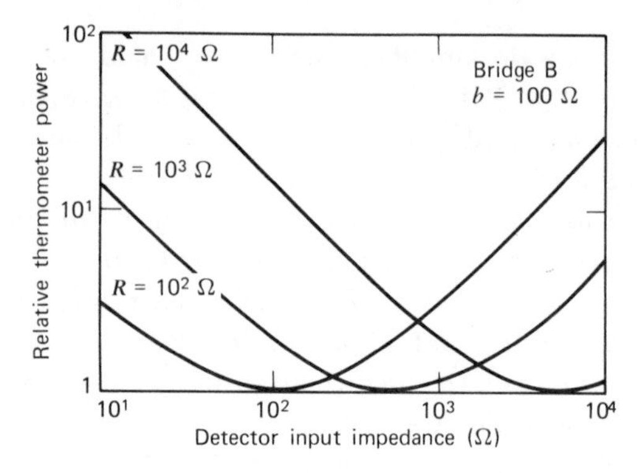

FIGURE 93. Relative thermistor power dissipation as function of the detector impedance. After Goodwin.[486]

A full mathematical analysis of the open-circuit output of a one-thermistor bridge as a function of the absolute temperature has been made by Boel and Erickson[487] and the influence of the bridge resistance ratio a and the material constant B of the thermistor has been calculated. The open-circuit output potential versus absolute temperature curve has an inflected shape that lends itself for linearization over an extended temperature range. The value of the output function at the inflection point is nearly independent on a change between ratios a from 0.001 to 1000. (T_i = temperature at which inversion occurs.)

If T_0 is defined at the temperature at which the bridge is balanced, the open-circuit bridge output is given by

$$\Delta e = \frac{a}{a+1}\left[\frac{\exp[B_T(1/T - 1/T_0] - 1}{a\exp[B_{T_1}(1/T - 1/T_0] + 1}\right]$$

By programming the function for Fortran IV the relative errors of a randomly chosen thermistor were calculated, but only for relatively small temperature intervals (308 and 348 K). Telethermometers using thermistors are commercially available.* These are thermometers made in different models varying the number of sensing channels from one to twelve.

The rapidly decreasing heat capacity of solids at very low temperatures ($c_p \propto T^3$) increases the risk of self-heating errors in cryogenic sensors. To reduce their power input, several highly sensitive bridges have been developed. Their resolution, however, is limited by the noise from the room-temperature reference resistor, the active elements of the ac bridge, and the detector. A low-noise ac bridge with an accuracy of $\leqq \pm 0.02\%$ for sensor values from 1–10^5 Ω has been described.[488] (See also reference 705.)

* Yellow Springs Instrument Co., Yellow Springs, Ohio 45387. (Table 44)

TABLE 44

Examples, Range: K	Maximum Error
233–423	±0.5 K(253–393 K)
	±1 K(243–253 K, 383–390 K)
	±1.5 K(243 and 403 K)
193–313	±1 K(213–313 K)
	±2 K(193–213 K)

HIGH-PRECISION TEMPERATURE MEASUREMENTS. In order to take full advantage of the intrinsic high-temperature sensitivity of thermistors, amplification of the bridge signal and suppression of parasitic side effects such as thermal and stray emfs is necessary. This can be accomplished by an ac bridge operating with a frequency that is not a subharmonic of the usual line frequency, for instance 27 or 33 Hz.[489]

The bridge signal amplifier has a gain of 120 dB, a bandwidth 0.3 Hz, and an input impedance of 10 kΩ. In the He range temperature changes of 4 μK can be detected, if the power dissipation in the sensor is held at 2×10^{-2} μW.[490] For room temperature this principle has been applied to a Fenwal GA 51P1 glass-bead thermistor with 100 kΩ nominal resistance and $B \approx 4100$.

The bridge operates with 25 Hz and its unbalance after amplification is detected by two phase-sensitive detectors phase shifted 90° against each other, thus balancing the resistive and reactive elements of the bridge. Temperature can be measured within ~14 μK. The total calculated noise from bridge and amplifier corresponds to ~5 μK, assuming that no extra contact noise comes from the thermistor.

TEMPERATURE RECORDING AND CONTROL. Starting in the late 1940s, thermistors were increasingly used as sensors for automatic temperature recording and control. Since these applications concentrate mainly on circuitry and equipment and do not add much new basic information, only references are given to pertinent publications on this topic.[491–493,285] The measurement of temperature profiles with traversing probes can be troublesome due to time delay and possible interference by the traversing operation. A multiprobe thermistor "comb" consisting of 68 bead thermistors in one stage and 36 in another stage (upstream) has been used to measure the heat transfer in the flash evaporator of a salt water desalination plant.[703]

POTENTIAL APPLICATIONS FOR HIGH-TEMPERATURE THERMISTORS. Measuring, monitoring, and controlling at high temperatures (<600 K) has been until recently the exclusive domain of thermocouples and the Pt resistance thermometer. However, low output in the first case and cost in the second, are handicaps for real mass applications, especially where no exact temperature measurements are necessary and calibration drifts of 10–20 K during long-range operations are tolerable. This is certainly true for monitoring pilot lights in gas ovens where only the on–off condition must be recognized by the thermal sensor.

The strong trend to reduce air pollution has stimulated the development of equipment that reduces unburnt fuel from automotive engines. The catalytic exhaust converter requires a certain temperature range for optimal performance and this could be a field for high-temperature thermistors, if the chemical interaction between the reducing exhaust gas atmosphere and thermistors can be avoided (encapsulation or selection of nonreducible material). High-temperature thermistor compositions based on ZrO_2, ThO_2, and CeO_2 doped with trivalent rare-earth oxides might have a good potential since there is a good chance that the contact problem can be solved in a reliable manner.

In search for a suitable substitute for the Pt catalyst, the less expensive $LaCoO_3$ has been suggested[494] and holds some promise.[495] It requires temperatures of about 700 K and would entail a high-temperature thermistor.

7.1.2. Thermistors in Meteorology, Oceanography and Geology

Temperature Measurement. Meteorological observations of temperature at high altitude, for instance, measurement of temperature, its gradient of air pressure, and velocity have for many years been used to make weather maps and are not yet made obsolete by satellite weather mapping. Contrary to earth-bound weather stations, all measured data must be transmitted to ground by radio waves. This requires transducers to produce an electrical signal for each atmospheric parameter. For many years thermistor rods were used in radio sondes of the United States weather services for temperature measurements. Badgley[496] has investigated the possible instrumental errors in radio sondes for temperatures. These are the lag and self-heating of the sensor, radiative heat exchange with the environment, condensation or evaporation of water, and frictional or compressional heating of air around the thermometer. The last two factors could cause an error of ~10 mK. More critical are wetting, icing, and evaporation due to the large amounts of latent heat involved together with the high heat transfer from solid or liquid surface films. On the other hand these play a minor role above 10 km altitude. Special attention was given to lag and radiative transfer by simulation in laboratory experiments in a bell jar that was ventilated by an air stream with 305 cm sec^{-1} velocity corresponding to the usual rate of rise of a radio sonde balloon, air pressures from 14 to 788 m torr and an ambient temperature of 296 K. Uncoated thermistor rods of 0.109 cm diameter and white-coated rods of 0.069 cm diameter were heated by ac current.

After switching it off, their resistance was measured as function of time in a dc bridge. The thermal lag coefficient is defined by the equation

$$\frac{dT}{dt} = \frac{1}{\lambda}(T_e - T)$$

where T_e is the ambient temperature, and T is the thermometer temperature; λ is normally considered to be constant.

The integration of the equation leads to

$$\log T = \frac{t}{\lambda}$$

and the plot of log T versus time should be a straight line with a slope λ^{-1}. The fact that in actual measurements the slope is not constant within one time characteristic is easily explained by the finite size and heat capacity of most sensors, resulting in an internal temperature gradient that is superimposed to the external gradient. A complete dimensional analysis of this problem requires taking into account the ten variables in Table 45. These can be condensed in six nondimensional factors as shown in Table 46.

A logarithmic plot of the product of two of these factors, π_1 and π_4, versus Reynolds number produces a straight line, using published data for density, specific heat, and the test data (Figure 94).

TABLE 45

Symbol	Physical Quantity	Dimensions*
λ	Time constant of the thermometer	T
V	The speed of ventilation	LT^{-1}
L	A length parameter of the thermometer (diameter was used)	L
ρ_a	Density of the air	ML^{-3}
ρ_s	Density of the thermometer	ML^{-3}
C_a	Specific heat capacity of air	$L^2T^{-2}\theta^{-1}$
C_s	Specific heat capacity of thermometer	$L^2T^{-2}\theta^{-1}$
k_a	Specific thermal conductivity of air	$MLT^{-3}\theta^{-1}$
k_s	Specific thermal conductivity of thermometer	$MLT^{-3}\theta^{-1}$
μ_a	Viscosity of air	$ML^{-1}T^{-1}$

* Badgley chose for temperature the dimensional symbol θ to discriminate against T (time).

TABLE 46

$\pi_1 = \lambda V/L$	$\pi_4 = k_s/k_a$
$\pi_2 = \rho_s/\rho_a$	$\pi_5 = \dfrac{C_a\mu_a}{k_a}$ (Prandtl number)
$\pi_3 = C_s/C_a$	$\pi_6 = \dfrac{VL\rho_a}{\mu_a}$ (Reynolds number)

Figure 95 shows explicitly the linear relationship between time constant and the atmospheric pressure for 293 and 333 K for white-coated thermistors, indicating an increase by a factor of 10 for a pressure decrease of a factor 100. This increasing sluggishness at high altitudes has to be taken in account in atmospheric temperature measurements. Finally the radiation equilibrium between sensor and air has to be considered. At higher altitudes the radiation unbalance between the blackbody emission of the sensor and its absorptivity for sun radiation can produce temperature errors of several degrees. The possible error for eight probable combinations of atmospheric temperature and pressure and different radiation conditions are estimated and range from +1.05 to −1.98 K from day to nighttime condition. Replacement of the usual rod thermistor by spherical (bead) type would reduce or eliminate directional variations with respect to the sun.

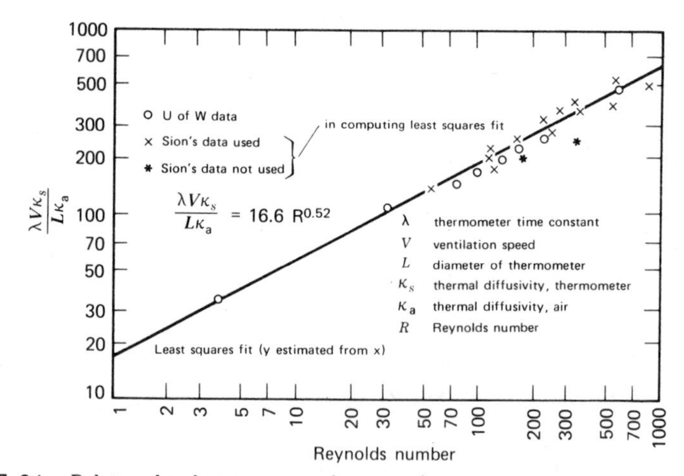

FIGURE 94. Relationship between two dimensionless parameters connected with the temperature response of ventilated cylindrical thermometers. After Badgley.[496]

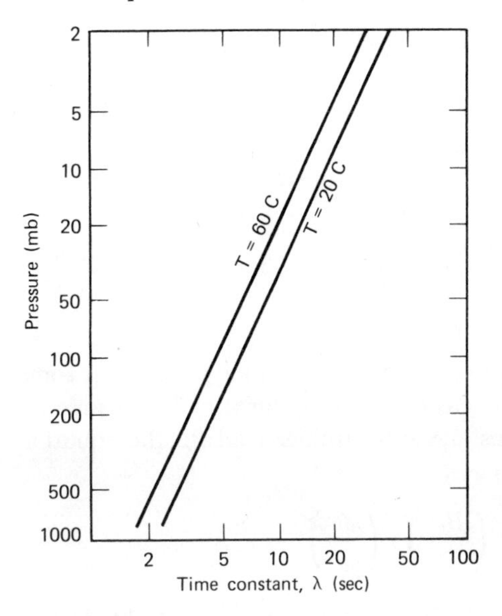

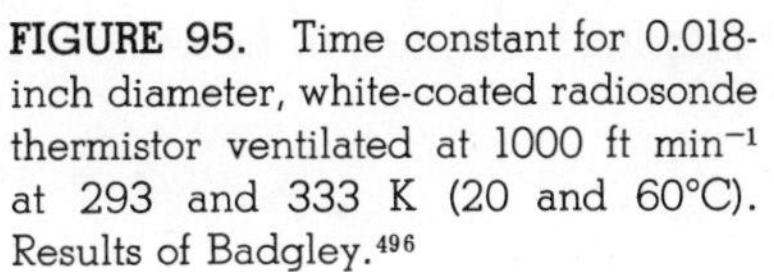
FIGURE 95. Time constant for 0.018-inch diameter, white-coated radiosonde thermistor ventilated at 1000 ft min^{-1} at 293 and 333 K (20 and 60°C). Results of Badgley.[496]

Application of 0.038-cm-diameter aluminized bead thermistors with self-heating power of ~8 μW permitted gas temperature measurements within ±0.35 K. This includes errors by infrared or solar radiation.[497]

Further progress was made by using smaller aluminized beads* of 0.023-cm diameter, which operated satisfactorily up to 39 km. Errors by self-heating were below 250 K, by simulated solar heating ~0.1 K. For higher altitudes up to 60 km, oxide thermistor films on 0.02-cm-diameter quartz fibers could be used.[498]

Staffanson[499] has developed a mathematical model for the film-mounted bead thermistor treating the convection, radiation, self-heating, and dynamic lag of the thermistor along with the convective, radiative, and conductive behavior of its supporting wires and film. For high altitudes, film sensors would optimize the response. Recently laser backscatter from the Raman rotational spectrum of N_2 has been suggested to measure temperature profiles up to 2 km. This could replace sensor methods.[500]

Atmospheric Pressure Measurements. Pressure transducers based on heat conductivity of gases lose their sensitivity above 10 torr (see Section 7.2.5) except for convection measurements. An ingenious way to overcome this handicap was found in 1860 by Whymper, who took a teakettle and a thermometer up to a highest peak in the Alps and determined the decrease

* Victory Engineering.

of the boiling point of pure liquids with the reduced atmospheric pressure at higher altitudes. This principle, called the hypsometer, opened the way to transduce pressure differences into temperature differences, which can easily be expressed by electrical signals as shown by Blackmore.[501]

The vapor pressure p of a liquid at a temperature T is in the first approximation given by

$$p = p_0 \, e^{-L/RT}$$

on the other hand,

$$r = r_0 \, e^{B/T}$$

where L is the molar heat of evaporation, p_0 is the vapor pressure at some reference temperature, and R is the gas constant (1.96 cal mole^{-1} degree^{-1}).

Differentiation of both equations and substitutions leads to the equation relating resistance to pressure changes:

$$dr = -\left(\frac{BR}{L}\right) r \left(\frac{dp}{p}\right)$$

The first factor can be considered as nearly constant. The second (r) is a temperature function and requires corrections based on the calibration curve of the thermistor. In this investigation VE 5 Ia units with a resistance of 5 kΩ and $B = 3500$ at 373 K were used, resulting in a sensitivity of 5 Ω for 1 μtorr pressure variation at 760 mm. Using boiling chips did not completely prevent temperature fluctuations in the vapor phase.

Though the peak-to-peak noise was ±5–10 μtorr for the glass-sealed sensors used, the average could be estimated to ±10% of the peak. These fluctuations set the limit to the usefulness of this method, and any further improvements must be directed to reducing them.

OCEANOGRAPHY. The thermal microstructure in the deep sea has been extensively studied with thermistor probes in 2000 m water depth. Velocity effects were eliminated by the circuitry and small power dissipation. The vertical temperature structure was determined with a resolution of 0.05 K and microstructures 8–75-cm thick were detected. The technique of lowering the probes and protecting them against sea water were described. The pressure effect was ~0.12 K per 1000 m depth. The bridge output was linearized with a maximal nonlinear deviation of ±4 mK over 10 mK.[502,503]

For thermistors used in geothermal investigations, see references 504–511; for measurement of heat flow over land see references 512–519. A thermistor controlled proportional thermostat eliminates noise in a broad band geophysical accelerator caused by the temperature dependence of the elastic properties of the used quartz fiber. It has a stability of 10 μK over a period of days despite fluctuations of the ambient.[704]

7.1.3. Biological and Medical Applications

The intrinsic high resolution of thermisors seemed to invite their application in life science and clinical practice. Besides their high sensitivity several other factors favor this trend:

1. The need for a small temperature range, only 268–323 K, and even smaller for clinical thermometers. Therefore problems of long-range stability and linearization are easier to solve.
2. The increasing need for continuous recording, remote measurement, and linking the readings with other electrical systems such as servo controls or computers.

There are, however, also detrimental factors that would apply to all sensors with electrical output. In life science and in clinical medicine, the need for a fail safe principle is more dominant than in industrial applications. A mistake here can mean a life and not merely material loss.

Even a remote possibility of failure or misreading has to be reduced, if not eliminated beyond any reasonable doubt. The more intricate an electrical sensor and its readout system is, the greater is the possibility of failure. Assuming absolute electrical and mechanical stability of any electrical sensor material, its contact and lead connections can also be sources of trouble. Still more critical can be the readout system. The fact that some of the electrical sensors, the thermocouple excepted, depend on an external power source, is in itself a handicap. The manufacturers of electrical sensors for medical use are cognizant of this danger and have a built-in redundancy in their instruments or a signal light that alerts against any possibility of wrong reading. A bipartisan round table discussion on clinical thermometry with a panel of physicians and manufacturers participating was held and taped during the 5th Symposium on Temperature in Washington 1971. Of course it also included application of thermistors, their merits and demerits. The fact that biological phenomena are often very temperature sensitive—the small temperature bandwidth of a few degrees between "good" and "bad" clinical condition of humans is given as example—makes it mandatory that self-heating of an electric sensor be reduced to a minimum compatible with sustaining sufficient sensitivity. Quartz resonator and thermocouple meet this condition best; they also permit size and mass reduction, especially thermocouples. However, the small sensitivity of the latter and the more complex readout instrumentation for the resonators have left enough room for thermistors in this field. To make them still more competitive, continuous efforts have been made to reduce their size (mass and heat capacity). The present practical limit of ≈ 0.01 cm diameter of the sensor excluding the leads is the result of two factors: first, the self-heating;

second, the limitations on homogeneity within a small mass of semiconductor and uniformity from piece to piece considering the grain size within sinter bodies. The problem of interchangeability (tolerance from unit to unit of the resistance-temperature characteristics) and long-range stability, so important for clinical measurements, have been solved by producing bead thermistors with an overall tolerance of ± 0.05 K between 270 and 313 K. This implies conformance to a standard within 1 mK and, if necessary, combination of two thermistors in one sensor to compensate for differences of the characteristics. When only small temperature differences are to be measured without concern about the accurate absolute values, thermistor sensors can be thrown away after application to patients with dangerous or highly infectious diseases, thus eliminating tedious sterilization procedures. Linearization methods for application are treated in detail in Chapter 6.

Special Examples of Clinical Applications.

1. Hypodermic needle for subcutaneous temperature measurement, 0.7 mm o.d. The main difficulty is the insulation within the needle.[520,521] The thermistor is inserted into a capillary drawn from Pyrex or Jena Glass. This is pushed into the needle from the top, which is threaded and screwed into a bakelite plug. One lead wire is soldered to the tip of the needle, the other to a brass tube at the plug (complete sterilization is possible).

2. During open heart surgery the body of the patient is brought into a hypothermic state by cooling peripheral areas. The body temperature must be monitored to ensure the required cooling. In this case a small, well-insulated thermistor is introduced deeply into the body through the rectum or the esophagus and its resistance changes are measured in a flashlight-powered bridge using the deflections of a rugged microamperemeter.[522]

3. Measuring of the venous blood flow in the human lower limbs.[523,524] The thermistor is introduced with a long, flexible, nylon catheter into the blood vessel to keep the sensor in axial position. Measurements are made in conscious and anaesthetized patients and the effects of breathing, posture and self-muscle exercise studied. The accuracy is $+5\%$ over the range 40–870 ml min^{-1}.

4. Monitoring ventricular contraction by intranasal thermometry.[525] Possibly applications in clinical cardiology and experimental physiology are suggested. The medical conclusions drawn from this investigation go beyond the scope of this book. For earlier studies to monitor respiration in space flight see Henry and Wheelwright.[526]

5. Patients who had an opening cut in the windpipe to facilitate breathing (tracheotomy) require constant supervision to ensure that the small inserted pipe through which the air flows into the windpipe does not become clogged, which would result in suffocation or irreversible brain damage. A thermistor

sensor acts as monitor for the air flowing through this pipe and can activate warning signals at a central nursing desk.[527]

6. Heat radiation of the human body is nonuniform. Certain areas, especially when inflamed, emit more heat than others with normal cellular or metabolic activity. This has stimulated thermal mapping of the human body for the purpose of detecting cancer. Several thermographic imaging systems have been used for this purpose. A survey on the status of thermography has emphasized the advantages of sensing the skin temperature without touching it with the sensor.[528]

The thermistor bolometer has found wide use for this application, finally, because of its high sensitivity (0.1 K) and its optical resolution of 1 angular mil. Five-thousand patients have been thermographed at the Albert Einstein Medical Center for detection of pregnancy phenomena, orthopedic and dermatologic problems, rheumatoid arthritis, and cancer. A large-scale screening of about 3500 patients for breast cancer (sponsored by the American Cancer Society) has confirmed the value of thermography as a diagnostic method.[529]

7. For remote temperature measurements a multichannel transmitter was developed using the temperature–frequency conversion principle (Section 7.1.8.) and applied to telemetering the body temperature and rate of respiration of birds[530–532] in flight besides other physiological data. Miniature biopotential telemetry systems have been described by Deboo and Fryer.[533]

Deep-body transmitters controlled by thermistors were used in the study of ovulation and oviposition in fowl.[535] The influence of stimuli (food intake and sexual activity) on the temperature in the cranium[534] and the cerebral[536] cortex was investigated with Fenwal CB 3552 and YSI 44005 thermistors.

By stimulation of certain receptor areas, temperature differences of ~1.4 K were detected within the brain.

8. Telemetry based on modulated electromagnetic radiation (230 MHz) fails in underwater studies. Therefore in this case where the loss of body heat is very crucial and has to be monitored, ultrasonic radiation has to be used as the signal carrier. The signals from the transmitter, modulated by the temperature value of the YSI sensors 44011, are received by an omnidirectional hydrophone. The finite propagation speed of sound in water has to be taken into account for a critical evaluation of the data. Under controlled experimental conditions an accuracy of ± 0.2 K was obtained between 270 and 315 K.[537]

7.1.4. Determination of Molecular Weights

The classical methods to determine molecular weights are based on melting point depression (cryoscopy), boiling point increase (ebullioscopy), or vapor pressure reductions of a solvent by a solute whose molecular weight is to be determined. Another method is based on the molecular flow of a gas through a circular opening (effusiometry).* The latter is restricted to gases and vapors, but adaptable to very small quantities of test material. In all these cases small temperature differences are to be measured with great precision and semiconducting sensors are the best choice.

CRYOSCOPY. Zeffert and Hormats[538] reported a precision of 10 mK in cryoscopic determinations of molecular weight. The cryoscopic constant of a solvent, that is, its freezing point depression in K produced by 1 mole/1000 g solvent of a nondissociating or associating solute is given by the equation

$$E = \frac{RT_m^2}{1000\ H_F}$$

where R is constant, T_m is the melting point, H_F is the enthalpy change for melting 1 g of solvent.

E- values range from 1.86 K for water up to 40 K for organic solvents. With an temperature error of 10 mK the precision of a molecular weight determination is better than 0.1%.

For high molecular weights the observed depression becomes rather small and a precision within 1 mK or better is desirable. As an example, molecular weights of humic acids have been determined using thermistors.[539]

According to Kulkarni[540] the upper limit of molecular weights, which can be determined with an accuracy of 0.25% by using commercial thermistors, is 2000 in water and 5000 in benzene (molar depression 5.065 K).

After the basic studies had shown the usefulness of thermistors for cryoscopy, direct-reading meters were developed for field applications in pharmacy and biology.[541]

Cryoscopic measurements with a known concentration of solute free from association or dissociation tendencies can also serve to determine thermodynamic properties.[542] The fusion heat of pyridine has been found to be 1800 ± 8 cal mole^{-1} using a thermistor method with a precision 1 mK.

EBULLIOSCOPY. When measuring the boiling point increment of solvents to determine the molecular weight of a solute, the (mostly organic) solvent has the potential to change the stoichiometry of an oxide sensor over extended periods of time. The high sensitivity of the resistance characteristic against such changes make it mandatory to use glass-coated or otherwise encapsulated sensors.

* Described in Section 7.2.5

THERMOMETRIC DETERMINATION OF MOLECULAR WEIGHTS BY ISOTHERMAL DISTILLATION.

Principle. If a solvent and solution of a material with unknown molecular weight are in isothermal equilibrium with the saturated solvent vapor, a temperature difference between solvent and solution is observed. Since the solute depresses the vapor pressure proportional to its molar concentration, a pressure gradient exists between solution and solvent leading to solvent vapor condensation on the surface of the solution. The resulting temperature difference between solvent and solution permits calculation of the molar concentration. This method goes back to Hill,[543] and it is remarkably suitable for the microdetermination of molecular weights and is popular for polymers with molecular weights up to 10,000 or more. This method requires up to 30 min to establish equilibrium, and this is its main setback.[544]

Since the early 1950s much work has been done using this method in connection with thermistors.[545–549] A critical review of these earlier investigations together with his own work was made by Tomlinson[550] and led to an understanding of the optimal conditions for this method. He used a matched and aged pair of (F 1512/300) bead thermistors (Standard Teleph. and Cable Ltd., Footscray, Sidcup, England) in a bridge operating at 1000 Hz with a power dissipation of 4 μW for each thermistor.

Their high-temperature sensitivity of 3000 Ω K^{-1} required an exact control of the base temperature by radiation shields with Al foil and thermostating of the surrounding water bath within ± 1 mK. Optimal results were obtained with a test cell (Figure 96). At first solvent was introduced with a micropipette through the side arm of the glass cell with the thermistor pair and to the tips of the thermistors and the base of the cell. After reaching a steady-state condition in the resistance bridge, one of the thermistors received a drop of the test solution. The distortion of the cell equilibrium when replacing the solvent by solution drops had to be minimized. The temperature error of 5 mK by self-heating is identical for solvent and solution and is small compared to the observed ΔT *Values*. Small power thermistor dissipation is also desirable to minimize convection. The sensitivity ΔR for a given molar concentration changes by 3.4% K^{-1} for ambient temperature variations around 298 K.

The standard deviation of molecular weights in the range 100–700 was 1.9%. These considerations were extended to a number of solvent systems.[551] For routine tests a commercial vapor pressure osmometer (VPO Model 301 by Mechrolab Inc., Mountain View, Calif.) has been developed containing a matched thermistor pair with 6040 $\pm$ 24 Ω and a completely linear characteristic of log R versus $1/T$ within the operational range 309–311 K and a temperature coefficient of (4175 $\pm$ 0.015)% K^{-1}. The

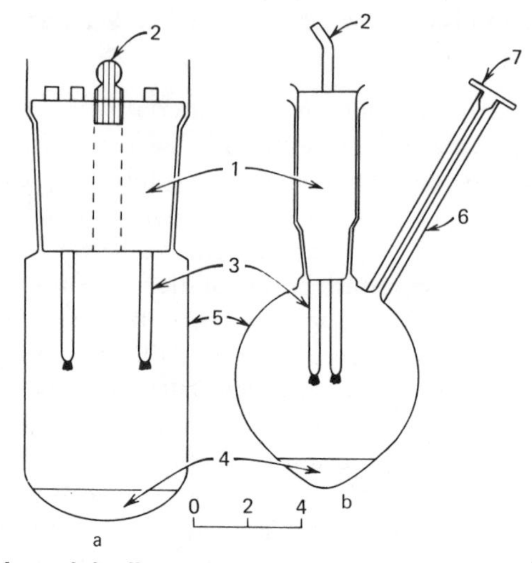

FIGURE 96. Isothermal distillation apparatus: (*a*) 1. P.T.F.E. stopper, 2. glass stopper, 3. thermistors, 4. solvent, 5. glass envelope. (*b*) 1. polythene stopper, 2. bridge leads, 3. thermistors, 4. solvent, 5. glass envelope covered with Al foil, 6. 1-mm-diameter capillary. After Tomlinson.[550]

instrument is calibrated with a pure solution of low molecular weight and the resulting cell constant determined by a number of factors such as solvent evaporation from the cell, thermal conduction, diffusion of solvent through the cell atmosphere, and solute diffusion through the drop. Calculations are made with an IBM 7094 Digital computer using Fortran language.

Adicoff and Murbach[552] have found that the cell constant a_1 for a series of solutes with molecular weights ranging from 135.16 to 1238 g mole^{-1}, did not vary more than $\pm 1\%$ for each solvent (in acetone, 454 $\pm$ 4; in toluene, 301 $\pm$ 2 Ω liter mole^{-1}). This agrees with the fact that the equation for the cell constant a_1 does not contain the molecular weight of the solute, but only data for the solvent. These data are its molecular weight M_1 and its heat of evaporation ΔH_v kcal mole and its density ρ_1, besides R (gas constant) and the material constant B of the thermistor and $r_{2.0}$, its resistance at pure solvent temperature:

$$a_1 = \frac{BRM_1\, r_{2.0}}{\Delta H_v \rho_1\, 10^3}$$

Since the resistance change is

$$\Delta r_2 = a_1 C, \qquad \Delta r_2 = \frac{BRM_1\, r_{2.0}}{\Delta H_v \rho_1\, 10^3} \cdot C$$

leading for maximal accuracy to optimal reading times ranging from 2 min for acetone and toluene to 6 min for 1,2-Dimethoxythane. The molecular weight of the solute is given by the known relation between its concentration and the depression of the vapor pressure of the solvent, calculated from the temperature difference between solution and pure solvent.

MELTING POINT DETERMINATIONS. Melting point determinations belong to the basic tools of organic chemists for identifying a substance; they supplement elementary analysis. For small samples they are done with the Kofler micro hot stage by visual observation of the melting under the microscope and simultaneous monitoring of the stage temperature in immediate vicinity to the mg sample. A thermistor of 100 kΩ at 298 K with B increasing with temperature from 4100 to 5500 and used in a bridge controlled by an audio signal permits the operator to concentrate on the visual observation only, and frees him from thermometer reading. For rapid calibration a plastic rotary type ohm–temperature converter can be used. The normal temperature range is 300–625 K. A refined version with built in thermoelectric cooling and direct thermistor temperature readout can be used between 195 and 315 K with an accuracy of ± 1 K down to 215 K and ± 2 K below.*

7.1.5. Calorimetry and Thermometric Titrations

In all applications where only temperature differences are to be measured with minor consideration for absolute values, thermistors are superior to most other temperature sensors due to their high sensitivity. One of these fields is calorimetry, where the effects of long-range shift are also unimportant or can be corrected by recalibration. On the other hand, fast response is necessary to avoid tedious corrections of postreaction periods. In this respect, the small thermistor (not necessarily bead type) is certainly better than glass thermometers and most resistance thermometers.

What are the limits of temperature resolution with commercial thermistors? If a 1000-Ω unit with a temperature coefficient of 4% K^{-1} is used, one might safely assume that its resistance can be measured precisely to 0.1 Ω in a thermistor bridge with a reasonably good null detector. This would correspond to a resolution of 2.5 mK and has been indeed sufficient for most earlier applications.

During the last 10 years more ambitious calorimetric investigations have been made with thermistors. In most of these cases the temperature resolution was increased to the order of 10^{-5} K by detecting bridge unbalances of one part in 10^7. Since various methods can be applied to improve the detection limits on bridges, their detailed treatment goes beyond the scope

* Supplier: Arthur H. Thomas, Philadelphia, Pa.

of this book. Only a few examples are given, each of them original in its approach.

1. Deviating from the conventional high-pressure combustion chamber ("bomb"), Wagner[553] followed the idea of Magnus and Becker[554] to use a metal block calorimeter that does not require corrections of heat capacity changes by evaporation of water from the water calorimeter or for heat produced by water stirring. The Cu calorimeter with a stainless cladded combustion chamber block is suspended in an evacuated chamber surrounded by a water-filled thermostat that was controlled within 1 mK. Three thermistors connected in series to average temperature difference were placed at a distance of 2.9, 5.0, and 7.1 cm from the bottom face of the metal block into 0.9-cm-deep holes at three equidistant positions around the perimeter of the metal cylinder providing nine thermistor sites (Figure 97). The used

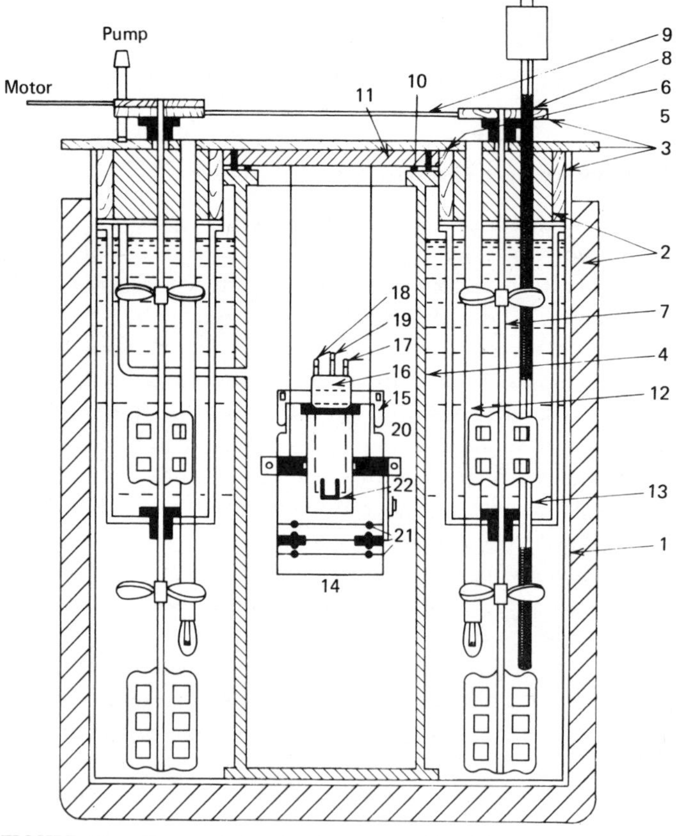

FIGURE 97. Thermostat and calorimeter. After Wagner.[553]

TABLE 47

Type	Resistance at 298 K	α(% K^{-1})	After 15 Months: Change of α (%)	After 15 Months: Calibration error (mK)
Siemens K 13200–4.0	200	−4.0	0	−32
VEB Hermsdorf HLS	300	−2.0	+0.03	−110
Valvo B 832007 P/150 E	150	−2.5	+0.04	−350

thermistor types are shown in Table 47, which also gives in the last two columns the change of α and the resulting calibration error found in long-range stability tests at 300 K. These results were found in 1960 and should not be generalized as more recent stability facts.

With a null detector sensitivity of 5.1×10^{-10} A per division a temperature sensitivity of 2×10^{-5} K was attained.

A calibration error without change of α is of course most desirable, since recalibration provides the same differential sensitivity in ohms per degree Kelvin and only one resistance value at an arbitrary reference temperature has to be remeasured. Simultaneous changes of α require full recalibration over the used temperature range.

2. Every electrical temperature sensor has a greater heat leakage toward the environment than a glass thermometer due to the much lower heat resistance of lead wires. For any precision temperature measurement this effect must be minimized or correctly compensated for either by experimental means such as an auxiliary heat source bucking the leakage or by calculation. For precision calorimetry Russian authors have determined the influence of the immersion depth between 2.5 and 13 cm in water on the apparent temperature error of thermistors KMT-1.[683] It amounts to 93 mK and becomes negligible above 15 cm. Other Russian thermistor types MMT-1 and KMT-4 show the same characteristics. These experiments were made near room temperature. For higher temperatures the immersion errors increase. Another source of error is the nonlinearity of the thermistor, not negligible even for small temperature intervals if precision is required. By introduction of a variable (temperature dependent) B into the basic equation for KMT-1, least square fitting leads to

$$\log R = -1.856346 + \frac{1811.3252}{T}$$

a value for the heat capacity of the calorimeter of 3158.7 cal K^{-1} is found that is constant within $\pm 10^{-5}$. A formula for the correction of the heat exchange with the environment suitable for thermistors has been given by Gadzhiev and Sharifov.[684]

For the rate of temperature change $v = dT/dt$ the general expression $-(T^2/BR)\,dR$ is valid. Applied to dt the pre- and postreaction periods of a calorimetric measurement two corrections constants C_0 and C_n are defined as $-T_0^2/BR_0$ and $T_n^2/-BR_n$ which contain T_0 and T_n as average temperatures of the two heat-exchange observation periods, R_0 and R_n as average resistance values, and ΔR_0 and ΔR_n as resistance changes per time unit. The final heat exchange corrections can be written as

$$\sum v_0 = \frac{C_n\Delta R_n - C_0\Delta R_0}{R_n - R_0}\left[\frac{r_n + r_0}{2} + \sum_1^{n-1} r - nR_0\right] + nC_0\Delta R_0$$

$$\sum v_n = \frac{C_n\Delta R_n - C_0\Delta R_0}{R_n - R_0}\left[\frac{r_n + r_0}{2} - \sum_1^{n-1} r - nR_n\right] - nC_n\Delta R_n$$

r_0 and r_n are the final resistance values in the initial and the main period. Using these corrections, the temperature changes are accurate to 0.5 K.

3. The ideal calorimeter is, of course, adiabatic.[685] Lawrence described a calorimeter operating between 273 and 343 K with automatic adiabatic control provided by thermistors. It permits temperature measurements with a precision better than 0.1 mK and heat effects of less than 0.01 cal. The necessary stirring is accomplished by rocking the calorimeter through 180°.

An isothermal calorimeter for the study of the thermodynamic properties of biological systems with only 4 cc volume using a Peltier cooler, a 7.5 kΩ control, and a 100 kΩ monitor thermistor can be held within $\pm 2 \times 10^{-5}$ K.[679]

4. Whenever small caloric effects are to be measured, thermistor sensors are most suitable. A typical example is the determination of immersion and adsorption heats of fine powders.[555] Graphor, a partially graphitized carbon black with a specific surface area of 85 $m^2\,g^{-1}$, has an initial immersion heat in benzene of 2 cal g^{-1}. It was determined by using Western Electric 15-A thermistors in a calorimeter with a sensitivity of 7 mK cal^{-1}. The interesting result of this caloric study is that much of the adsorption of benzene vapor for Graphone and other fine graphite powders is due to capillary condensation between the particles rather than to multilayer condensation, and only a minor part of the surface area is exposed while the major part is in the pores.

5. Low-temperature calorimetry. The caloric effects of mixing gases are small. Therefore their calculation from differences of thermodynamic data of mixture and components is less accurate than a direct determination. The Kammerling Onnes Laboratory in Leiden[556] has made direct measurements of the excess enthalpy of H_2-N_2, H_2-A_1, N_2-Ar mixtures at various compositions and of the ternary mixture 2 H_2-N_2-A at pressures between 30 and

130 atm. The temperature in the isothermal flow-type calorimeter operating between 150 and 293 K were measured with Keystone cryogenic thermistors (Type L 0904) below ambient and with glass-sealed bead thermistors at room temperature.

In related investigations on the influence of magnetic fields on the transport properties of NO, N_2, CO, and CH_4 a thermal conductivity cell with a relative resolution of 4×10^{-4} was developed.

Heat-capacity measurements between 1 and 35 K on small samples (1–500 mg) of material with heat capacities down to 5×10^{-8} J K^{-1} have been made using the chip bolometer described on pp. 100–101. In this case the silicon chip serves as heater, temperature sensor, and sample holder simultaneously.[557] Several options of wire attachment, bonding the sample to the chip, and thermalizing the calorimeter to minimize heat leaks from ambient are described in detail, together with an ac resistance bridge operating between 27 and 400 Hz. The necessary corrections are discussed for several modes of operation. The applied principle, the thermal relaxation method, is novel and differs from the adiabatic and ac methods. It has been extended to temperatures of 60 mK using slice cuts from a $\frac{1}{2}$-W 470 Ω nominal Speer resistor.[693]

6. Semiquantitative calorimetry to monitor chemical reactions. Nearly all chemical reactions are connected with production or consumption of heat energy (exothermic or endothermic processes), which can be measured quantitatively in a calorimeter. Often it is sufficient to monitor these resulting temperature changes without knowing the exact heat capacity of the system. This is the case when the rate of the chemical reaction or its endpoint is to be determined. In this case small mass and correspondingly small thermal inertia of the thermometer are crucial. Here again the more sensitive thermistors are competing with thermocouples. Another possibility is the determination of the concentration of a reactive component by its heating effect in the system. A typical example would be the continuous monitoring of water vapor traces in gases.

Harris and Nash[558] have applied this method by reacting water traces in H_2, O_2, CO_2, *n*-butane and ethylene with Ca hydride, which reacts instantaneously with H_2O to form $Ca(OH)_2$ resulting in the release of 28 kcal $mole^{-1}$ H_2O. If this reaction could be concentrated in the near vicinity of a 20-mg thermistor with a heat capacity of $\approx 5.10^{-3}$ cal K^{-1}, a sensitivity of the order of micrograms would be possible. Practical considerations do not permit this ideal arrangement. First of all direct contact of the sensor with the corrosive $Ca(OH)_2$ is undesirable. With increasing distance from the reaction zone the fraction of the reaction heat reaching the sensor decreases. Furthermore, this method is more attractive for monitoring the water concentration "on stream." Therefore a major part of the heat is carried away by convection. Despite these limitations, sensitivities of millipercent per ohm resistance change could be attained. The necessary heat transfer from this reaction zone to the sensors results in a slow response to concentration change (5–10 min). The method was developed with relatively large (Western Electric 17 A) thermistor disks of 0.5-cm diameter and 0.1-mm thickness with at least 20.10^{-3} cal K^{-1} heat capacity and would certainly be improved by using bead types. Instead of measuring the concentration of gas component reacting with a solid, it is also possible to determine the activity of a catalyst (Cu_2O) in decomposing H_2O_2. The temperature difference of thermistors placed on the up- and downstream sides of the catalyst is a figure of merit of the activity.[559] The reaction rate in polymerization and similar organic processes is easily monitored by physical measurements rather than by chemical analysis in view of the similar composition of mono- and polymer. Dielectric and optical methods would be appropriate for this purpose. A much simpler approach is to measure the temperature increase caused by the polymerization process, which is directly proportional to the fractional conversion as long the reaction remains adiabatic (i.e., without heat exchange to the environment.)[560]

A pair of identical 800-Ω Ohizumi thermistors with a resistance ratio 2.1 between 293 and 313 K is applied in a bridge operated with 2 V at 700 Hz and the bridge unbalance produced by the temperature increase of one thermistor in the reaction vessel and amplified to obtain a sensitivity of 1 mK with a signal-to-noise ratio 10/1. Using the known heat capacity of the reaction vessel and the polymerization heat of 16.76 kcal $mole^{-1}$, a temperature increase of 3.82 K per 1% conversion is found. The time lag in response is ≈0.6 sec. These data are better by a factor of 2–3 compared to thermocouples. The author extended the same method to other organic reactions.[561]

The high toxicity of CO together with its lack of odor and its ubiquitous existence near combustion devices requires a sensitive and rapid method of detection. A pair of matched thermistors, one in an inactive medium such as pumice, the other surrounded by an reactive absorbent bed for CO (Ag-$KMnO_4$-ZnO), are arranged in an Wheatstone bridge.[562] The unbalance of the bridge caused by the reaction of CO with the reactive bed permits a sensitivity of a few ppm CO in an air stream of 4–6.1 liter min^{-1}. The reaction rate is nearly independent of the temperature above 283 K and of the relative humidity between 30 and 100%. Lower humidities retard the reaction and in this case artificial humidification is necessary. The reactive bed is prepared by reacting $KMnO_4$ with $AgNO_3$ in the presence of ZnO, thus precipitating $AgMnO_4$ on the substrate. The wet product is pressed at 10 tsi, cured for 28 days at room temperature, and then broken up into small granules of 0.07–0.24-cm diameter. For the optimal flow rates a bed depth of 1.5 cm in a cell with 2.5–3.5-cm diameter is recommended. A general review of exothermic reactions of CO in presence of a number of oxidic, metallic, or organic (hemoglobin) reagents has been given by Belcher and Goulden.[563]

7. For measuring the rate of rapid reactions of two reagents brought together in a mixing chamber, the continuous-flow method of Hartridge and Roughton[564] has been modified by using thermistors (Stantel F 2311/300) as indicators.[565] The following reactions were studied, which can be broken down in steps either measurable or instantaneous:

(1) $K_2Cr_2O_7 + 2NaOH = K_2CrO_4 + Na_2CrO_4 + H_2O$
(a) $Cr_2O_7^{2-} + H_2O = 2(HCrO_4^-)$ measurable
(b) $2HCrO_4^- = 2H^+ + CrO_4^{2-}$ instantaneous
(c) $2H^+ + 2OH^- = 2H_2O$ instantaneous
(2) $NaHCO_3 + HCl = NaCl + CO_2 + H_2O$
(a) $HCO_3^{1-} + H^+ = H_2CO_3$ instantaneous
(b) $H_2CO_3 = CO_2 + H_2O$ measurable

The thermostatically controlled reagent solutions were forced by a driving pressure of 700 torr, which could be boosted to 1300 torr by applying vacuum at the exit, into a mixing chamber, consisting of a three-way capillary *T* stopcock with a capillary stem of 0.161 mm i.d. Several thermistors were inserted through holes bored into this test capillary, and a few more behind two larger volumes of 0.8 and 5 ml, respectively. The entire flow system including the sensors was imbedded into a paraffin block for thermal insulation and protection against breakage. With flow velocities up to 220 cm sec^{-1} the rate constant of the first reaction $K = 9.4\ sec^{-1}$ at 298 K was measured using time intervals up to 0.12 sec. The results with the second reaction lacked reproducibility, obviously caused by the effect of CO_2 bubbles.

During recent years thermometric titrations have attracted an increasing interest.[566–568]

Where ever a suitable chemical indicator is lacking and electrical and optical indicators for the end point are inconvenient, the thermal method finds its niche. Since the requirements of sensitivity and fast response for thermal monitoring of chemical reaction were previously referred to, no special applications will be further discussed except one. This is the analysis of binary mixtures by titration calorimetry, distinguished from the thermometric titration.[569]

Only binary mixtures with significantly different enthalpy changes for their reactions with a common titrant can be analyzed. Test mixtures were chosen with only slightly different free energy changes but large enthalpy changes ΔH_i, for instance

(a)	Sodium acetate	$\Delta H_i \sim 0$	$pKa = 4.8$
	and pyridine	$\Delta H_i \sim 5$	$pKa = 5.3$
(b)	glycine	$\Delta H_i \sim 11$	$pKa = 9.8$
	and phenol	$\Delta H_i \sim 6$	$pKa = 10$

The mixtures react quantitatively with strong acids or bases and the following equations hold:

$$n_T = n_A + n_B \quad \text{(Relation of the number of reactant moles)} \tag{1}$$

$$Q_T = n_A \Delta H_A + n_B \Delta H_B \tag{2}$$

After elimination of n_A and n_B,

$$n_A = \frac{QT - n_T \Delta H_B}{\Delta H_A - \Delta H_B} \tag{3}$$

Error analysis of Equation (3) shows that the error in n_A is mainly dependent on the inverse of the difference of enthalpies, in other words decreases with this difference. Measurements were made with a Tronac* thermometric titration calorimeter Model 1000 A using a 100-ml Dewar with a heat leak of 0.005 K min^{-1}. Using a thermistor of 100 kΩ and a bridge voltage of 12.0 V a sensitivity of 139.5 mV K^{-1} was obtained. The cal mV^{-1} equivalent C_0 of the reaction vessel was determined by electrical calibration and the heat capacity of the system C_t at each point of the titration period calculated by the equation

$$C_t = C_0 + \frac{C_s \cdot V_s}{S}$$

where C_s is the heat capacity per milliliter of solution, V_s is the amount (ml) in the Dewar, and S is the bridge sensitivity.

A correction is made for the deviation of titrant temperature from that of the Dewar. The measured thermograms temperature versus time were then converted into enthalprograms by introducing the energy equivalents for the pre, main, and post period. The enthalpograms for the titration of pyridine-sodium acetate mixtures with 1.009 m $KClO_4$ and phenol-glycine mixtures with 0.7042 m NaOH are given in Figures 98 and 99. After correction for heat of dilution of the titrant the differences between the found and the theoretical values decrease from 20% for low to 1–2% for high concentrations of a component. Extension to other than acid–base reactions was suggested.

* Tronac calorimeters 450, 550, 1050, 1150 from Sanda, Inc. Bala Cynwyd, Pa. 19004.

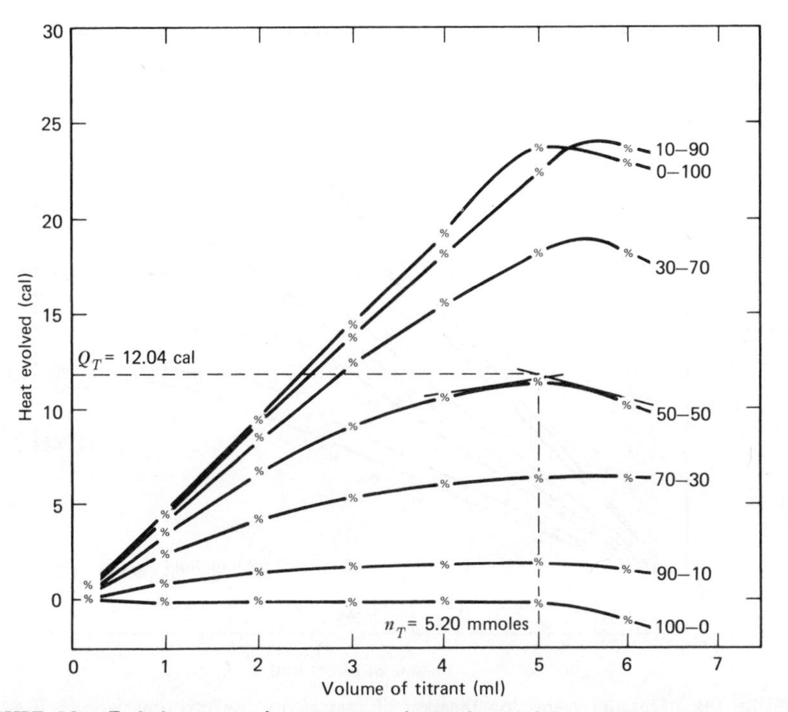

FIGURE 98. Enthalpograms for titration of pyridine-sodium acetate mixtures with 1.009 *M* perchloric acid. Curves are labeled with approximate mole % of acetate and pyridine, respectively. Sodium sulfate was added to enhance the end points. Titrant delivery rate was 0.1660 ml min^{-1}. Results of Hansen and Lewis.[569]

Thermometric titrations with thermistor sensors have found wide acceptance during the last few years. This is not only reflected in an ever increasing volume of literature[570–575] but also in the development of advanced instruments with sensitivities of 0.2 mK before and <0.01 mK after amplification, automatic equilibration of the reaction vessel, and electrical calibration. Other features are data printout on teletype and closed-loop precision temperature control in the environment both with less than 0.3 mK drift per week.

A thermistor bimetal thermocouple hemisphere electrode for simultaneous electrochemical and differential thermal analysis measurements can be standardized by using the known Peltier effect resulting from passing a charge across the thermocouple.[678]

7.1.6. Temperature Compensation and Control

One of the first examples of large-scale application of thermistors was the temperature compensation of aircraft electronic instruments in the late

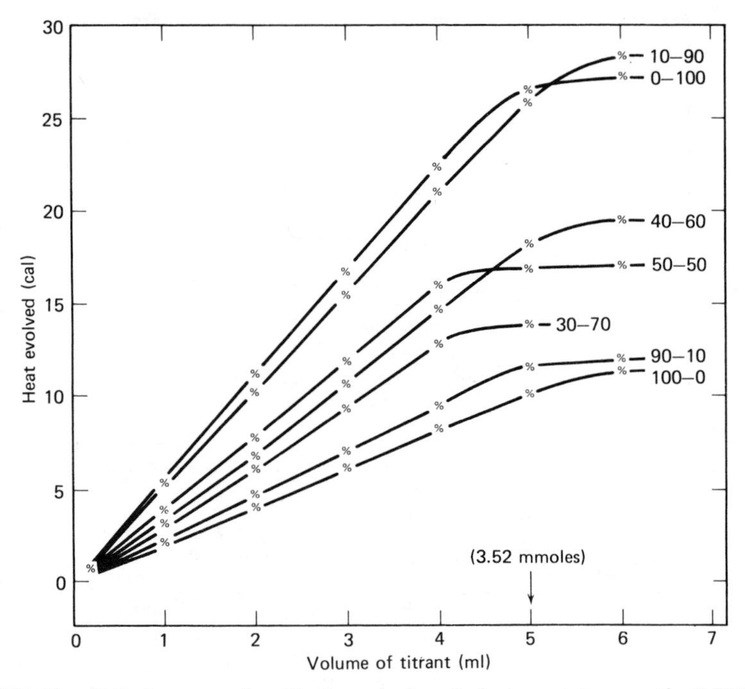

FIGURE 99. Enthalpograms for titration of phenol glycine mixtures with 0.7042 *M* sodium hydroxide. Curves are labeled with approximate mole % of glycine and phenol, respectively. Titrant delivery rate was 0.1660 ml min^{-1}. Results of Hansen and Lewis.[569]

thirties. In this, as in other applications mentioned in the classical publication of Bell Telephone Laboratories in 1946 and others to follow, the goal was the temperature compensation of Cu windings with a temperature coefficient of $+0.4\%\ K^{-1}$. It is quite obvious that a single series circuit consisting of a coil with the resistance R_c and a temperature coefficient of $+0.4\%$ K, and a thermistor resistance $R_T = 01.\ R_c$ and $\alpha = -4\%\ K^{-1}$ can only compensate over the temperature range for which α is approximately constant. Since $\alpha = -B/T^2$, it decreases for a thermistor with $B = 4000$ between 273, 323, and 373 K from 5.40, 390, or $2.95\%\ K^{-1}$.

A less naive network is necessary to accomplish better compensation (see Section 6.1.2). Some basic concepts on temperature compensation of instruments have been described by Pattee.[576] Even the temperature coefficient of magnetic core materials can be compensated by partial saturation of the inductance coil using a dc current that is controlled by one or several imbedded thermistors. With the advent of other semiconductive components such as diodes and transistors, temperature compensation found a new challenge since it had to deal with higher temperature coefficients than those of Cu coils. To avoid the thermal runaway destruction of transistors, their dc bias for the base or the emitter must be stabilized. This can be done by shunting the bias determining the voltage divider resistance with a thermistor that is in good thermal contact with the transistor If its temperature increases, the thermistor resistance and with it the dc base voltage decreases. This also reduces the collector current to a lower value, thus prohibiting runaway self-heating. Temperature compensation of transistors can also be accomplished by automatic limiting of the emitter bias, again using a thermistor-controlled bias circuit. This principle is, of course, useful in all transistor applications whether in amplification or oscillation circuits.

Wheeler[577] has given details on thermistor-compensated transistor amplifiers. For thermal drift compensation in oscillators see references 578 and 579.

A simple graphical method of selecting thermistors for the temperature compensation of Ge and Si transistors is based on matching the curves of the thermistor conductance and the collector current as a function of identical temperature abscissas.[580]

The nonlinear resistance-versus-temperature characteristic requires a linearization network as shown in Figure 100 for a class-B push-pull amplifier. For optimal compensation of an alloy-junction transistor a bias slope of -2.5 mV K^{-1} is necessary. The relations between resistance ratios of the thermistor branch of the compensating networks are

$$\frac{r_{323}}{r_{298}} = R_{323} \quad \text{and} \quad \frac{r_{273}}{r_{298}} = R_{273}$$

and the corresponding bias ratios B_{323} and B_{273} for the same intervals are given by

$$R_{273} = \frac{R_{323}(B_{323} - 1)}{(B_{273} - 1) + R_{323}(B_{323} - B_{273})}$$

Explicit calculations for two types of thermistors with 250 and 31.5 Ω at 298 K were given, using variable network parameters. The effect of compensation obtained in networks A, B, and C circuits are shown in Figure 101 for 2 N 109 transistors with a supply voltage of 6 V.

Compensative Stabilization of TV Deflection Circuits. With increasing temperature the current (and also the field) through the coils of the vertical deflection yoke decrease, thus changing the vertical linearity of the picture. If a thermistor is imbedded into one of the coils and in series with both, its resistance decreases, predetermined by the proper choice of its cold resistance; this can compensate the resistance increase of the coils and thus minimize the effect of temperature changes on the magnetic field (Figure 102).

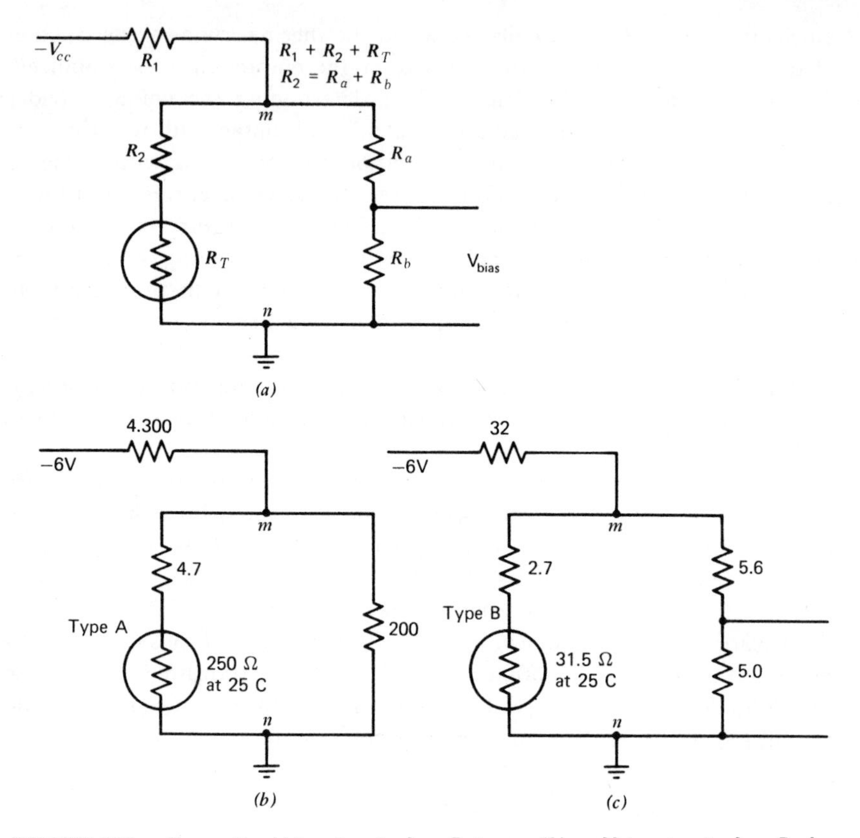

FIGURE 100. Generalized bias circuit when R_a is zero (*b*) and bias circuit when R_a does not equal zero. After Wheeler.[577]

The same effect can be attained by inserting a thermistor into the tube plate circuit. Its resistance decrease with temperature increases the dc plate voltage, and corresponds to the oscillator output feeding the grid of the vertical output tube. This also compensates for the higher resistance of the deflection coils.

A detailed description of the ambient temperature compensation of the gain in broadband carrier transmission systems has been given in U.S. Patent 2,806,200, September 10, 1957. Its transmission variation is held within ±0.1 dB between 240 and 345 K. Electronic humidity sensors based on sulfonated polystyrene have an exponential resistance range from 10^7 to 10^3 Ω between 10% and 100% relative humidity. They can be linearized by using the complimentary V versus I characteristic of a varistor ($I = KE^n$). The temperature dependence of n is compensated for by a thermistor in series with the sensor.[581]

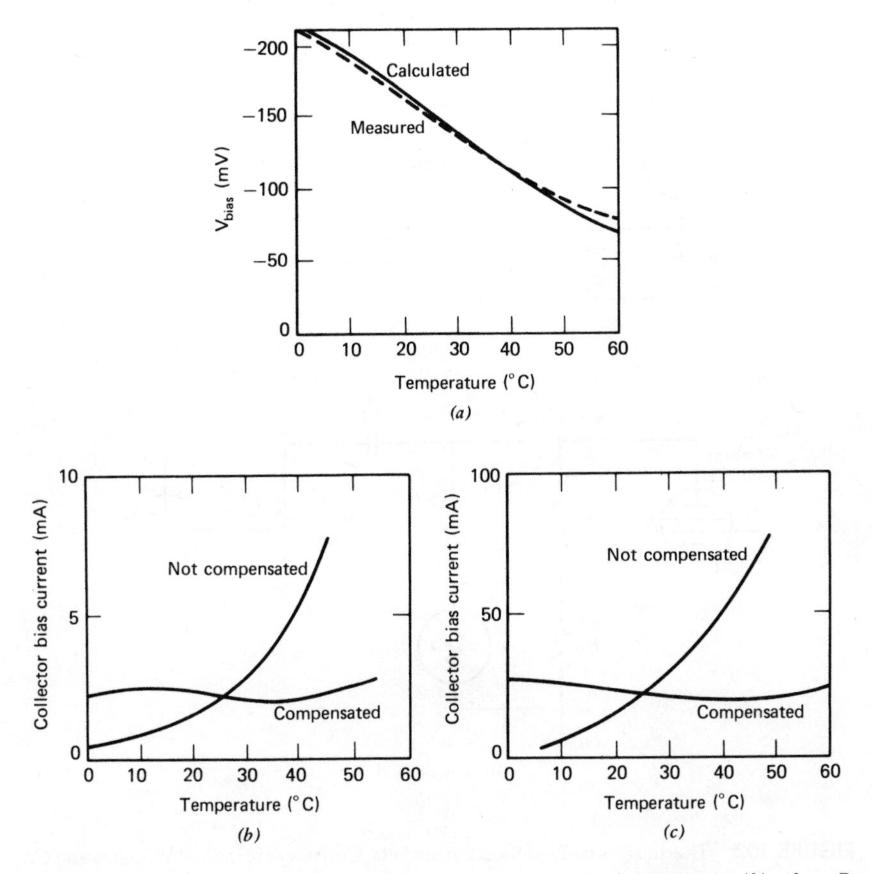

FIGURE 101. Bias (*a*) collector current versus temperature characteristic (*b*) when R_a is zero and characteristic (*c*) when R_a does not equal zero. After Wheeler.[577]

7.1.7. Radiation Sensing—Bolometer, Ultrasonic Sensor

Thermistors have found many applications in radiation (remote) thermometry. They have, contrary to many photoelectric devices, a flat, nearly uniform response over a broad wavelength range. On the other hand, thermal detectors tend to have a slower response, depending on their physical size and type of heat sinking. The classical form of the radiation sensor (bolometer) is the thermopile or the resistance thermometer. In order to provide sufficient sensitivity, the radiation-sensitive area has to be large enough to collect as much impinging energy as possible, making these areas intricate in design by using a plurality of thermocouples connected in series within a small space or a fine densely wound grid of temperature-sensitive resistance wire. Thermistors seem to be a logical solution, not only to

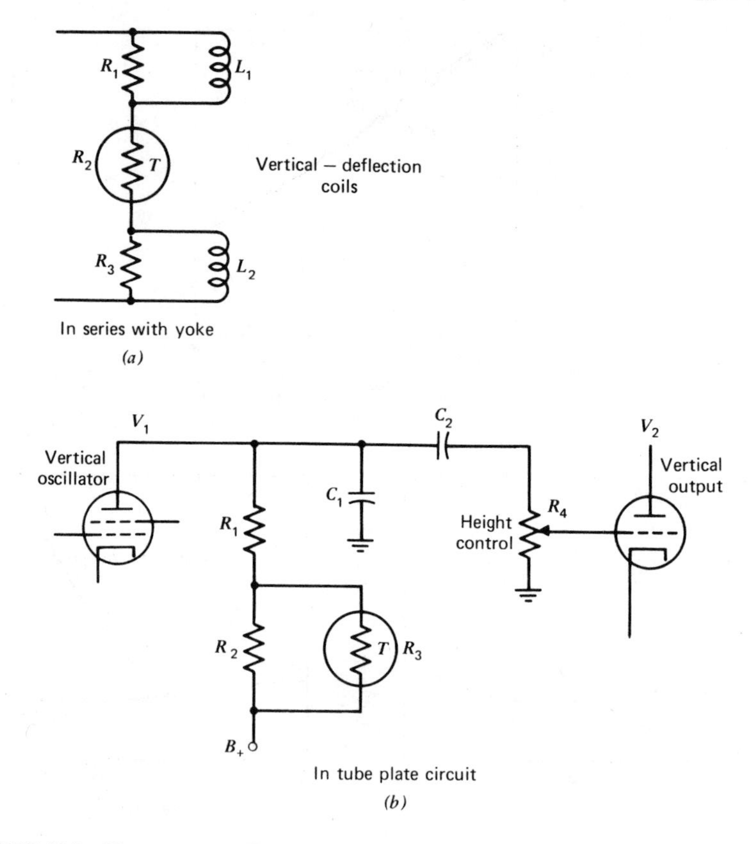

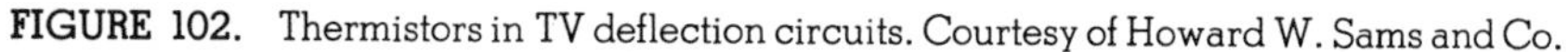
FIGURE 102. Thermistors in TV deflection circuits. Courtesy of Howard W. Sams and Co.

simplify the design but also to attain higher sensitivity. It is therefore not surprising that the first publication by Bell Telephone Laboratories on thermistors mentioned their use as area radiation sensors (bolometers). Further work was done by Brattain and Becker[582] and a final report on the development and operating theory of thermistor bolometers has been published by Becker.[583]

The increased interest in infrared for military applications ushered in a strong trend to develop infrared detectors with a good sensitivity up to 20 μm, and with smallest possible selectivity for certain wave bands (in contrast to photocells). Wormser[584] gave a review of the state of the art for thermistor infrared detectors. He pointed out the following three advantages:

1. Wide range of resistance values makes them adaptable to any amplifier tube or other electronic device such as photomultipliers.

2. Their sensitive area can be made very small, starting with linear dimension of less than 0.1 cm.

3. They are rugged in design. According to the desired size, beads, disks, or flakes can be used. In order to obtain fast response and high sensitivity, flakes 10 μm thick or films have been preferentially used.

It is, however, wrong to assume that bolometers using thermistors as detectors are always much more sensitive than those using metal wires or films. Hansen[585] has pointed out that high resistance bolometers with apparently higher voltage output have either higher heat capacity or electrical noise, resulting in nearly the same theoretical threshold of detectable radiation energy:

$$Et_{\text{min}} = 3.10^{-12} \text{ J sec} \quad \text{for } 0.01 \text{ cm}^2$$

where t is the duration of the radiation pulse.

The time constant of a bolometer is determined by the thermal sink used to mount the units. With quartz or glass sinks, values between 3 and 5 or 5 and 8 msec are obtained.

Wormser's work was done with two different thermistor materials of 2500 and 250 Ωcm resistivity, based on Mn-Ni and Mn-Ni-Co-oxide compositions. They were used as rectangular flakes with active areas varying by a factor of 12 and increasing resistance values within the same scope. It has been found desirable to limit the resistance values to less than 5 MΩ. The experimental procedure for studying the properties of thermistor detectors is shown in Figure 103.

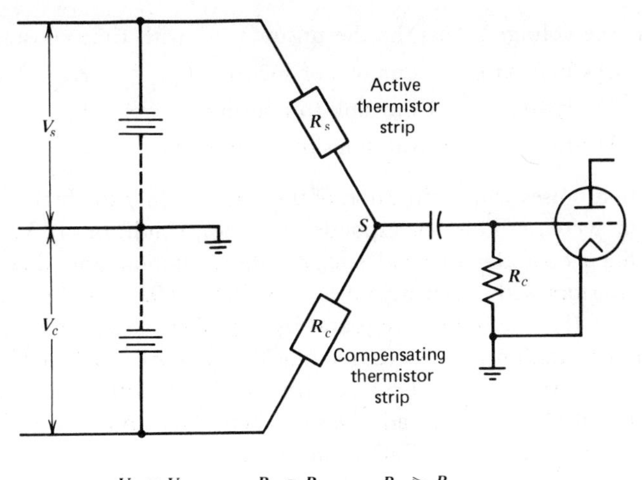

FIGURE 103. Thermistor bolometer bridge circuit. After Wormser.[584]

An active thermistor is exposed to radiation and connected to an identical thermistor to compensate for ambient temperature fluctuations. A positive-bias voltage is applied to the one thermistor, and negative of the same value to the other one, the connecting point S being at ground potential. With radiation input zero at R_B, no signal voltage appears at the input terminals of the amplifier tube. The bias voltage should be as large as possible to obtain a good signal-to-noise ratio, since for the high resistance values of these flakes the Johnson noise is predominant over the current noise. On the other hand the permissible bias voltage is limited not only by self-heating, but also by the steady-state voltage–current curve. As previously shown in the Section on voltage–current characteristics, thermistors have a negative characteristic after passing a certain peak voltage. It is desirable to operate bolometer flakes at bias voltages less than the peak voltage. For instance, at 60% of the peak at 298 K, the corresponding self-heating does not exceed 5–10 K. The absorption of the radiant energy and the corresponding resistance decrease are a complex effect influenced by the thermal resistance between flake and heat sink, and the heat conductivity of the flake itself.

The signal voltage produced by irradiation dW of the active thermistor is therefore a function of time and given by the following equation:

$$\frac{\Delta V}{\Delta W} = Fa\alpha V_B \left[\frac{1}{C_1} (1 - e^{-t/\tau_1}) + \frac{1}{C_2} (1 - e^{-t/\tau_2}) + \cdots \right]$$

where α is the fractional temperature coefficient of thermistor resistance, a is the absorptivity of the flake (absorption coefficient), F is the circuit constant given by the exact resistance of R_B and R_C. Further definitions are:

		with time constant
V_B,	the voltage across the thermistor	
C_1	Dissipation rate to quartz or metal	τ_1
C_2	Dissipation rate from sink to housing	τ_2
C_3	Dissipation rate from housing to ambient	τ_3

For radiation pulses with a duration of the order τ_1 only the first term has to be taken in account. This corresponds to the test conditions of Wormser.

The influence of the thermal sink on the radiation and detectivity of 3 MΩ thermistors with a sensitive area of $0.25 \times 0.02 = 0.005\ \text{cm}^2$ can be seen in Table 48. The data were found by exposing the active thermistor to the infrared radiation from a cavity blackbody source of 393 K that was chopped at a rate of 15 Hz by a square wave disk with a temperature of 300 K. 90% of the chopped radiation was between 3 and 20 μm with a peak at 7 μm. The impinging radiation on the thermistor was corrected for the emissivity of the blackbody and chopping disk. The applied bias varied with the heat sink since it was always held at 60% of the peak voltage

TABLE 48 Infrared detectors. Comparison of three types of thermistor detectors. Size of sensitive area: 2.5 × 0.2 = 0.5 mm². No. 2 thermistor material, resistance 3 MΩ at 298 K.

Thermal sink	Quartz	Glass	Air-Spaced Metal
Effective first time constant (sec)	0.002–0.005	0.005–0.008	0.020–0.040
Bias voltage (V)	212	130	81
Responsivity (V) rms/W av at 15 cps	705	585	1210
Reference band width (cps)	62.5	35.7	8.33
Johnson noise (μV)	1.7	1.3	0.62
ENI (W)	2.3×10^{-9}	2×10^{-9}	5×10^{-10}
Detectivity (W^{-1})	4.3×10^{8}	5×10^{8}	2×10^{9}

(which increases inversely with the thermal resistance to the sink). The better responsivity with poorer heat sink results in a much larger (10 times) time constant. Table 48 presents comparative data for these three sensors.

Responsivity (volt rms/watt at 15 Hz) does not require definition. Two other terms used in infrared detection are the equivalent noise input (ENI) and the detectivity. ENI is defined by the power in watts giving a signal equal to the Johnson noise in volts for a reference bandwidth sufficient to pass signal pulses from the detector. Detectivity, as the reciprocal of ENI, increases with better performance and is also used in Wormser's further investigation as a figure of merit. He found the following general rules for detectivity:

1. It increases with the first time constant, leaving all other conditions unchanged.

2. It increases linearily with the active area reaching a limit

$$D(\max) = \frac{10^{12}}{3(A/\tau)^{1/2}} \qquad \text{(Haven's limit)}$$

3. The responsivity measured by scanning in transverse and in the longitudinal direction of the sensor area with a radiation beam of 10^{-3}-cm diameter increases slightly near the edges of the active area, but otherwise is nearly constant in the center range.

4. The voltage output per watt remains constant up to radiation intensities 1000 times the Johnson noise level. The minimal radiation signal is of the order 10^{-9}–10^{-8} W.

5. The response to sinusoidal infrared radiation between $\frac{1}{2}$ and 300 Hz produced by chopping decreases with frequency. The different slopes of the

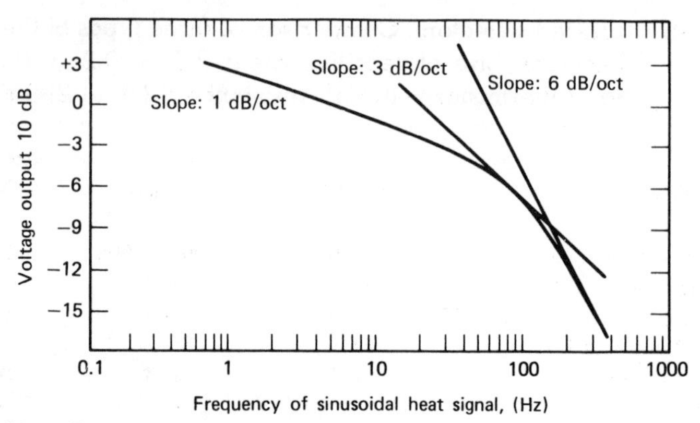

FIGURE 104. Frequency response curve for quartz backed thermistor detector. Results of Wormser.[584]

frequency characteristics correspond to the different time constants mentioned above (Figure 104).

6. Quartz-backed thermistors of 0.06 × 0.075-cm area blackened with Zapon lacquer were compared with fast vacuum thermocouples of the same receiving area blackened with gold through an infrared frequency range 2–19 μm using a blackbody source of 1183 K. The radiation was chopped with 10 Hz and dispersed with a KBr double monochromator giving uniform wavelength resolution. Table 49 compares the spectral responsivity of both sensors in relative units (thermocouple = 100). Figure 105 gives the comparison for entire wavelength range.

Some design data were given by Wormser for contacting, assembling and housing of his bolometers, which were manufactured by the Servo Corporation of America.

Hesse and Mendez[586] were able to use a commercial thermistor Siemens Type K15 2 kΩ with 0.77-cm diameter and 0.25-cm thickness, mass approximately 0.5 g, as bolometer for monochromatic light between 5000 and 6500 Å and obtained a sensitivity of $3.95 \pm 0.05 \times 10^{-4}$ W Ω^{-1} resistance

TABLE 49

Wavelength Range (μm)	Thermocouple	Thermistor
2–7	100	60–8
8–10.5	100	100
10.5–19	100	100–200

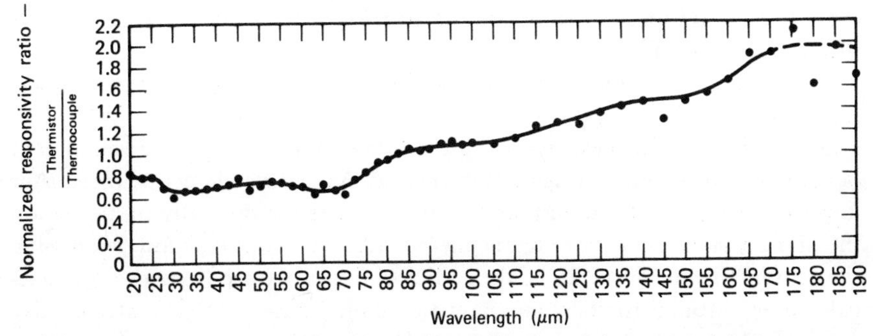

FIGURE 105. Relative spectral responsitivity of blackened thermistor detector as compared gold-blackened thermocouple. Results of Wormser.[584]

change. This required an integrating spherical cavity of 5-cm i.d. made of glass with a silver-coated inner surface and a small window hole for the radiation influx (Figure 106). The thermistor is mounted at the focal point of the reflecting sphere and covered with carbon black. With a temperature coefficient of -4.6% the observed resistance changes of 0.5–2 Ω correspond to temperature increments of only 5.5–22 mK. Ambient and self-heating must be reduced to a minimum to avoid serious interference with the measurements. The dissipation constant of K 15 is 8 mW K^{-1}. Therefore

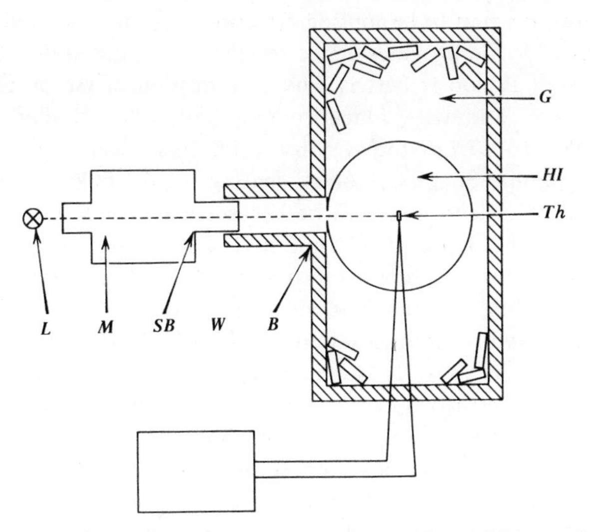

FIGURE 106. Schematic of test device: L, incandescent lamp; M, monochromator, SB, slit aperture; W, Wheatstone bridge; B aperture, G, housing with insulating rods; HI, hollow sphere integrator; Th, thermistor. After Hesse and Mendez.[586]

the test must be made with a limit of 10 mW, permitting not more 10^{-4} A through the thermistor.

The principle of integrating radiation input by focusing or optical immersion has found wide applications for horizon scanners with the advent of space technology. Barnes Engineering has been a leader in the development of commercial bolometers using this method. As sensing elements, thermistor flakes of 0.01×0.01 cm and 8.10^{-4} cm thickness are optically immersed on the plane surface of Ge or Si hemisphere to obtain an optical gain. Increasing the central thickness to $1.25 \times$ the radius of the curvature (hyperhemispherical dome) would further increase the gain considerably. Antireflective coatings are applied to these lenses. Detailed design data are given in U.S. Patent 3,109,097, October 29, 1963 (Barnes Engineering Company, Stamford, Conn.).

An example of a commercial bolometer using thermistor sensors is the YSI Kettering Model 65. It costs less than $600 and has seven different ranges of sensitivity between 0–0.25 mW cm^{-2} and 0–250 mW cm^{-2}. Frequency range is 0.28–2.6 μm and the voltage output proportional to incident radiation energy at all wavelengths (accuracy, $\pm 5\%$ full scale) is compensated for by a reference thermistor.

A comparative study of Ge and Si immersed thermistor sensors was made by de Waard and Weiner.[587] The earth atmospherical horizon provides two useful radiation sources of scanning. One is the carbon dioxide spectrum near 15 μm, the other the rotational water band of atmospheric humidity beyond 20 μm, the first to be considered more uniform thus providing a stable horizon. For CO_2, Ge provides the better detector in the range 14–16 μm and for water vapor silicon is better. The transmission of Ge or Si and of the thermistor flake, and with black coating had to be studied for optimal efficiency. For 16–17 μm the Ge absorption losses were low and the flake was nearly opaque. Successive black coatings of the flake added only little improvement. The lens radius in both cases was 0.0508 cm and maximal transmission at 16–17 μm for Ge and at 24 for Si. With thermistor flakes of 0.01×0.01 cm and a nominal resistance of 4 MΩ at room temperature, mounted at the base of the lens, and the resistance matched to another flake for temperature compensation as described by Wormser, the data in Tables 50 and 51 were measured using radiation pulses of 15 Hz.

The relative spectral response for Ge and Si immersed cells is shown in Figures 107 and 108. With increasing chopping frequency the response drops sharply, as shown for a Ge cell in Figure 109.

So far Ge and Si have only been used as optical immersion medium. Ge has also found application as a detector. For the very far infrared region up to 1000 μm, cryogenic bolometers with a high sensitivity and time constant of 10^{-7} sec have been developed by Low[588] using single crystal p-type Ge

TABLE 50 Typical performance of germanium immersed cell.

Property	Cell 5283–5	Cell 5284–5
Resistance	4.42	4.20 MΩ
Bias applied	45	45 V/flake
Time constant	2.5	2.3 msec
Responsivity (14–16 μm, 15 Hz)	3300	3600 V/W
Noise (15 Hz, 1 Hz)[a]	0.23	0.21 μV rms
NEP (14–16 μm, 15, 1)	7.0×10^{-11}	5.8×10^{-11} W
$D\lambda^*$ (14–16 μm, 15, 1)	1.4×10^{8}	1.7×10^{8} $W^{-1}(H_z)^{1/2}$

[a] Noise value for 1 Hz when chopped with 15 Hz.

TABLE 51 Typical performance of silicon immersed cell.

Property	Cell 5284–5	Cell 5286–5
Resistance	4.35	4.08 MΩ
Applied bias	45	45 V/flake
Time constant	2.3	2.4 msec
Responsivity (21 μm, 15 Hz)	3570	3670 V/W
Noise (15 Hz, 1 Hz)[a]	0.21	0.23 μV rms
NEP (21 μm, 15, 1)	5.9×10^{-11}	6.3×10^{-11} W
$D\lambda^*$ (21 μm, 15, 1)	1.7×10^{8}	1.6×10^{8} $W^{-1}(Hz)^{1/2}$

[a] Noise value for 1 Hz when chopped with 15 Hz.

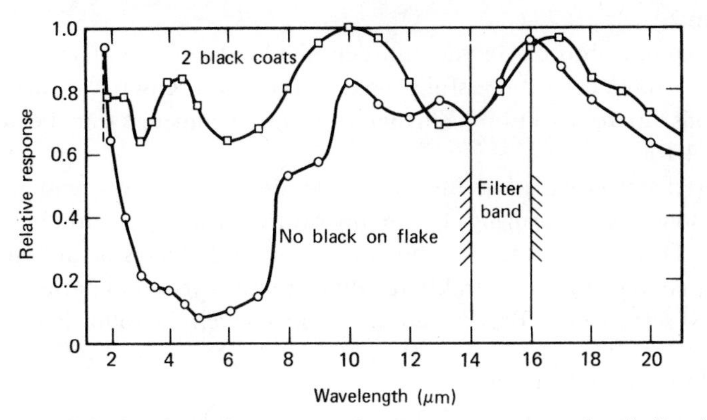

FIGURE 107. Relative spectral response of germanium immersed cell. Results of De Ward and Weiner.[587]

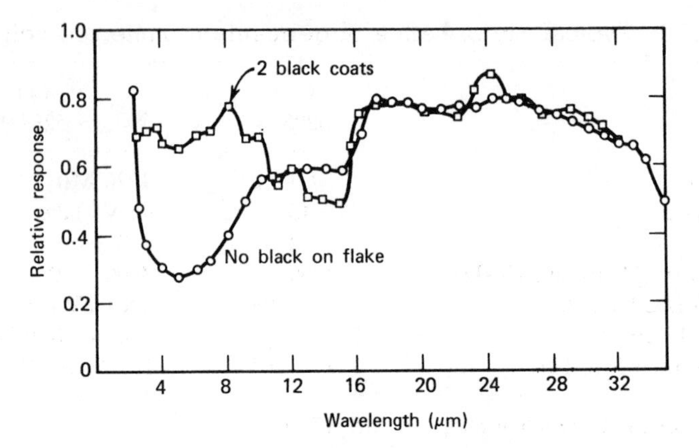

FIGURE 108. Relative spectral response of silicon immersed cell with and without black flake coating. Results of De Ward and Weiner.[587]

with 9×10^{15} cm^{-3} Ge acceptors compensated with 10^{15} cm^{-3} Sb donors. This material has a high absorption coefficient and a good temperature coefficient of resistivity. For optimal sensitivity, a light pipe and an integrating sphere were used in the design of this bolometer described in detail by Zwerdling, Smith, and Theriault.[589]

Cooling to He— temperatures and below not only reduces the thermal noise, but also the heat capacity of the Ge detector. At 1.55 K a NEP of 4×10^{-12} W Hz$^{-1/2}$ has been measured by Zwerdling, Theriault, and Reichard[590] at operating frequenies between 200 and 1000 Hz. This includes the noise of the preamplifier. A full analysis of the factors which determine the sensitivity of cryogenic Ge bolometers has been made by Low and Hoffman.[591]

0.30–0.6-μm films of Ge vacuum deposited onto mica with a rate of about 10 Å sec^{-1} have been successfully used as radiation sensors for rocket-borne operations up to 75 km.[592] Further data on bolometers can be found in Section 4.1.3.

Infrared detectors using thermistors as detectors have found numerous applications for measuring the temperature of bodies without physical contact. This is important for moving or rotating objects or in cases where physical contact is not desirable for other reasons (contamination or deformation of soft tissues.) Barnes and others have given the following examples:

1. Industrial applications.

   Hot axles in stationary machines or hot boxes in railroad cars.
   Hot spots in paper printing.
   Temperature of jet streams from rocket engines.[593]

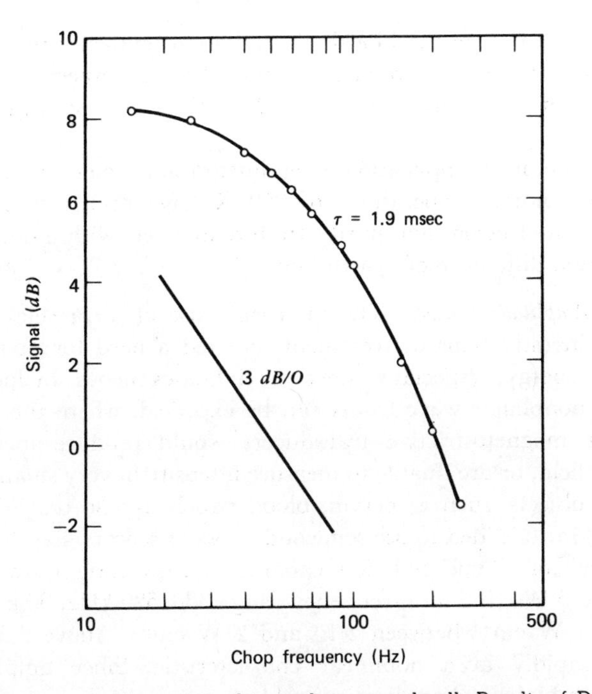

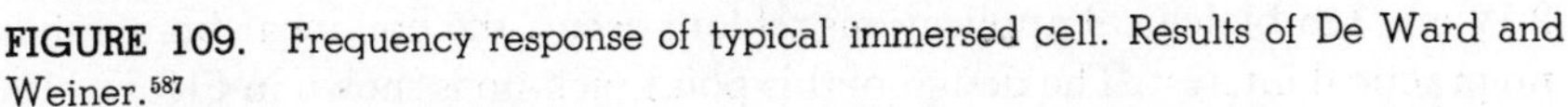
FIGURE 109. Frequency response of typical immersed cell. Results of De Ward and Weiner.[587]

2. Meteorological applications.
   Temperature of water or soil measured from airplanes, especially important because of the increasing awareness of thermal pollution downstream from power plants.
3. Medical applications (Section 7.1.3).
4. Nuclear applications.
   The detection of neutral particles in molecular beams has been greatly facilitated by using a low-temperature Ge bolometer detector with a responsivity of 10^5 V W^{-1}, a NEP of 10^{-13} W $Hz^{-1/2}$, and a thermal time constant of 10^{-2} sec. The detectable signal minimum is 10^7 molecules sec^{-1} for 0.01 cm^2 detector surface.[594]
5. Astrophysical applications.
   The detection of very weak radiation signals in radio astronomy for the 90–140 GHz range require drastic reduction of the noise temperature of the receiving equipment. In order to obtain a signal to noise ratio S/N of $\sim$10 the noise temperature should not exceed 3000 K based on the relation

$$\frac{S}{N} = \frac{T_S}{T_N} (\Delta f \Delta t)^{1/2}$$

where T_S and T_N are the blackbody radiation temperature of source and receiver noise for barrier diodes (1 and 3000 K, respectively), Δf is the band width of the receiving channel (250 HKz) and Δt is the integration time (1 hr).

For astronomical application long integration times are inconvenient. Much lower noise temperatures of 250 K have been attained by using He-cooled hot-electron high-purity InSb bolometers with a donor–acceptor concentration difference of $\sim 3.10^{13}$ cm^{-3}.[595]

Ultrasonic Amplitude Sensor. The increasing use of ultrasonics in medical-biological investigations or treatments created a need for point sensors of acoustical energy, especially since in inhomogeneous biological media extremely nonplanar wave fronts can be expected, where the usual piezoelectric or magnetostrictive transducers would produce incorrect data, distort the field, or are unable to measure intensity in very small areas of the irradiated objects, such as nerves, blood vessels and so on.[596] Labartkava, following Morita's idea to use semiconductors as sensors, has developed[597] a microthermistor (Type MT 54) capable of measuring ultrasonic energy levels up to 5 W cm^{-2} in a frequency range 300–520 kHz. The initial sensitivity is 0.1 W cm^{-1} between zero and 2 W cm^{-2}. Above 2.5 W cm^{-2} it increases rapidly in a nonlinear characteristic. Since amplitudes over 2 W cm^{-2} in biological applications seldom occur, the nonlinear branch has no practical interest. The design of this point pick-up is shown in Figure 110.

A hollow glass sphere of 0.5-mm diameter is filled with semiconducting material (nothing is said about the condition or density, whether sintered or only compacted, or whether the glass sphere was later shrunk to an existing semiconducting bead). The sphere containing the sensing elements with "several ten thousands ohm" resistance is mounted on a glass capillary that also protects the lead wires protruding from the sphere. The interesting

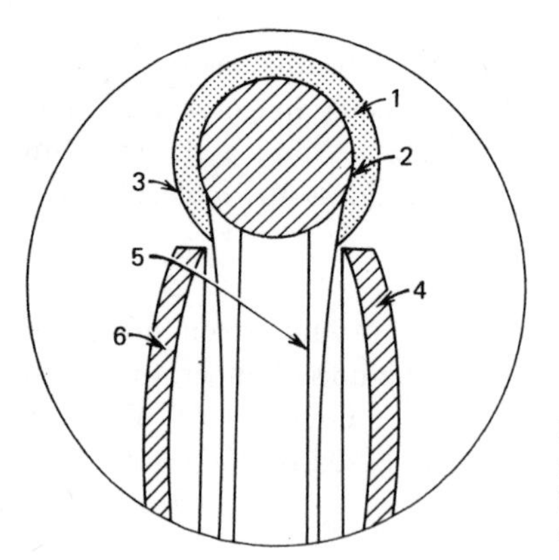

FIGURE 110. Ultrasonic pickup with thermistor sensor: 1. glass sphere, 2. thermistor bead, 3. Plexiglass layer, 4. glass capillary, 5. wire lead. After Labartkava.[596]

feature of this pick-up is that the glass sphere was covered by a layer of Plexiglass, which increased the thermal sensitivity of the unit by a factor 5–15 for 0.1 or 0.25 mm thickness, respectively. Other coating materials such as natural or synthetic rubber, applied 0.25 mm thick, produced only a sensitivity amplification between 4 and 6. The physical reason for this thermal sensitization is obviously strong ultrasonic absorption in the coating.

7.1.8. Temperature–Frequency Conversion

For remote temperature measurements the temperature–frequency conversion has created great interest. The produced frequency signal can be transmitted over long distance with a minimum interference by noise and digital presentation of values and is possible by counting the periods of the signal in time intervals of equal length. The temperature–frequency converter based on either piezoelectric, dielectric, and electroacoustic effects has been known for some time. They all have a relatively low sensitivity of 0.2% K^{-1} but a range of linearity of 100 K or more. Any increase in sensitivity has to be accounted for by a decrease of linearity range, for instance in a converter with 2.9% K^{-1} and range of 20 K consisting of a unijunction transistor (2 N 1671 A) relaxation oscillator with a thermistor as the resistive component (Figures 111 and 112). It has found an interesting application for temperature measurements in thermogravimetry.[598]

It has been always difficult to determine the exact temperature of a sample in a thermobalance, especially if in a solid-state reaction fast

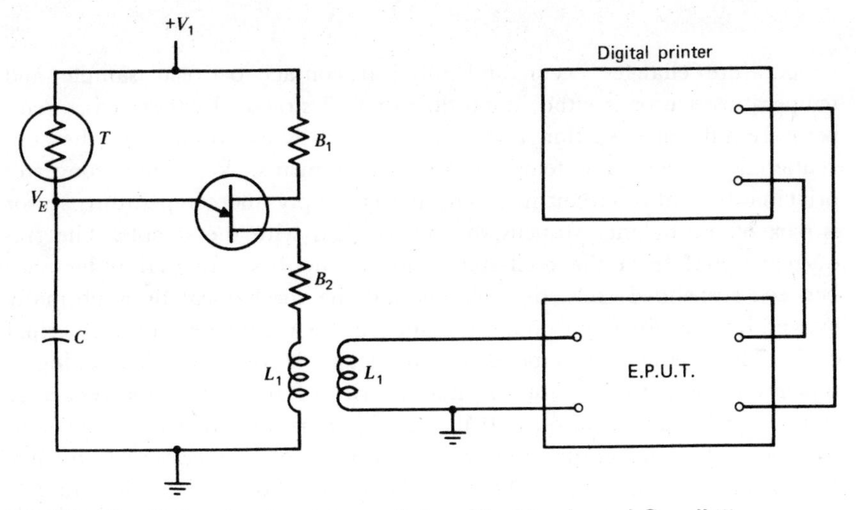

FIGURE 111. Unijunction transistor oscillator. After Manche and Carroll.[598]

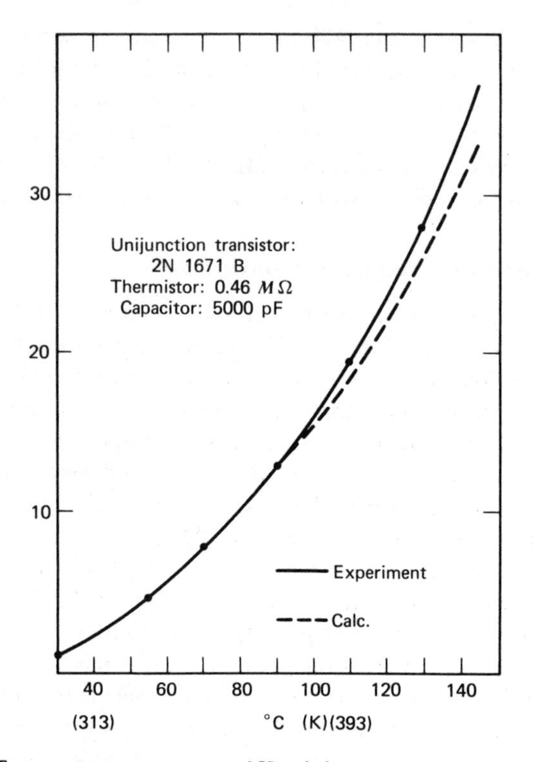

FIGURE 112. Temperature response in kHz of the unijunction transistor oscillator of Figure 111. After Manche and Carroll.[598]

temperature changes occur and physical contact between sample and temperature sensor is either impossible or undesirable. In these cases temperature differences in time and space up to 20 K can occur with the conventional methods for temperature measurements. With the converter principle the entire circuit including power supply and temperature sensor is part of the balance suspension and weighed with the sample. The frequency signal from the oscillator is picked up by a mutual inductance between suspended coils, thus eliminating the mechanical drag normally exerted by fine wire connections in amplitude measuring devices (or thermal emf). The frequency–temperature characteristic obtained by adding a subcircuit compensating for the nonlinearity of the thermistor resistance is shown in bead thermistors of 0.46 MΩ with $B = 4600$ K, which were used in this work. Further progress in linearization of temperature–frequency conversion was made in a systematic study by Lövberg[599] using an RC oscillator with a Philips E 205 CE/Pe 7K thermistor as the temperature

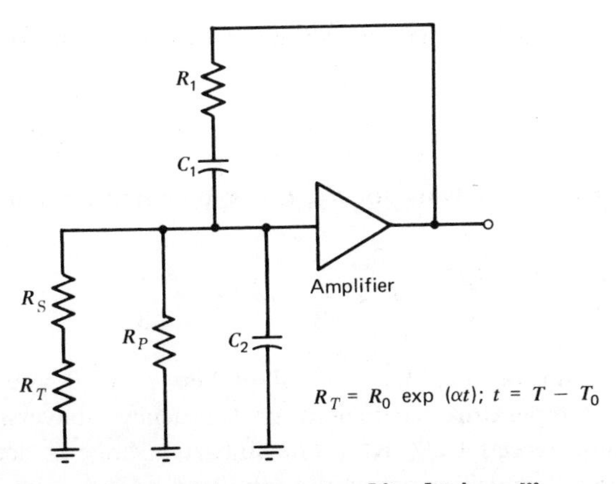

FIGURE 113. Block diagram of the converter. After Lövborg.[599]

sensor, which was prestabilized during manufacturing (Figure 113). Its frequency is given by (Figure 108):

$$f = \frac{1}{2\pi}\left\{\frac{R_p + R_s + R_T}{R_1 R_p (R_s + R_T)\, C_1 C_2}\right\}^{1/2}$$

The same equation applies to f_0 for a temperature T_0 with R_0 substituted for R_T. It is convenient to express the frequency f as function of the temperature difference $t = T - T_0$, using the basic equation $R_T = R_0 \exp (T - T_0)$. This leads to

$$f = f_0 \left\{\frac{(p + 1)\ (q + e^{\alpha t})}{(q + 1)\ (p + e^{\alpha t})}\right\}^{1/2}$$

with the dimensionless parameters

$$p = \frac{R_s}{R_0} \qquad q = \frac{R_s + R_p}{R_0}$$

The frequency temperature coefficient is

$$\beta = \frac{1}{f_0}\frac{df}{dT}$$

To make it independent of temperature requires that $d^2 f/dT^2 = 0$. This condition can be found by differentiating twice and taking $t = 0$; therefore

$$p = \frac{2 - K}{2 + K} \qquad q = \frac{2 + 3\,K}{2 - 3\,K}$$

with $K = 2\beta/\alpha$. Thus the necessary values of R_s and R_p can be calculated:

$$R_s = R_0 \frac{2 - K}{2 + K} \qquad R_p = R_0 \frac{16\,K}{(2 + K)\,(2 - 3\,K)}$$

In order to obtain a finite value for R_p, K has to be smaller than $\frac{2}{3}$. That means that

$$\frac{2\beta}{\alpha} = K < \frac{2}{3} \quad \text{or} \quad \beta < \frac{\alpha}{3}$$

Commercial thermistors can have α values between 3.0 and 4.5% K^{-1}. Therefore the temperature coefficient of frequency conversion for RC oscillators cannot exceed 1.5% K^{-1}. The linearity range is determined by two temperatures T and T_0 meeting the condition

$$\frac{f - f_0(1 + \beta t)}{\beta f_0} = \pm 0.1$$

The dependence of f on $K = 2\beta/\alpha$ can be derived from the restatement

$$f = f_0 \left[\frac{2 + 3K + (2 - 3K)\,e^{\alpha t}}{2 - K + (2 + K)\,e^{\alpha t}}\right]^{1/2}$$

Figure 115 shows that this range expands with decreasing α. Its temperature versus frequency characteristics in Figure 116. A circuit of a converter for 500 Hz at 298 K with $\beta = 1\%\ K^{-1}$ is shown in Figure 114. This converter maintains its calibration within 0.1 K for several weeks if the power dissipation in the prestabilized thermistor is kept below 25 μW.

The same principle was applied to a thermistor thermometer for linearized magnetic recording and telemetry. Roundtop pulses of 40 μsec duration and 3.5 V amplitude without a load or 200 mV with a load of 600 Ω terminated line are produced at a rate of 500 pps at a center value of the 245–335 K temperature range. The oscillator circuit temperature changes up to 333 K. Extreme load variations (including short circuit), and 10% change in supply voltage change the frequency only 1% corresponding to a temperature error of 0.02 K. The response is nearly linear (1.5% of full scale); the sensitivity $\approx$ 25 pps V^{-1}.[680]

7.1.9. Temperature Integration

In life sciences the influence of temperature is very important for the study of the relatively slow processes of growth and its variations. Evaluation of thermograph records can be rather cumbersome and impractical. A more

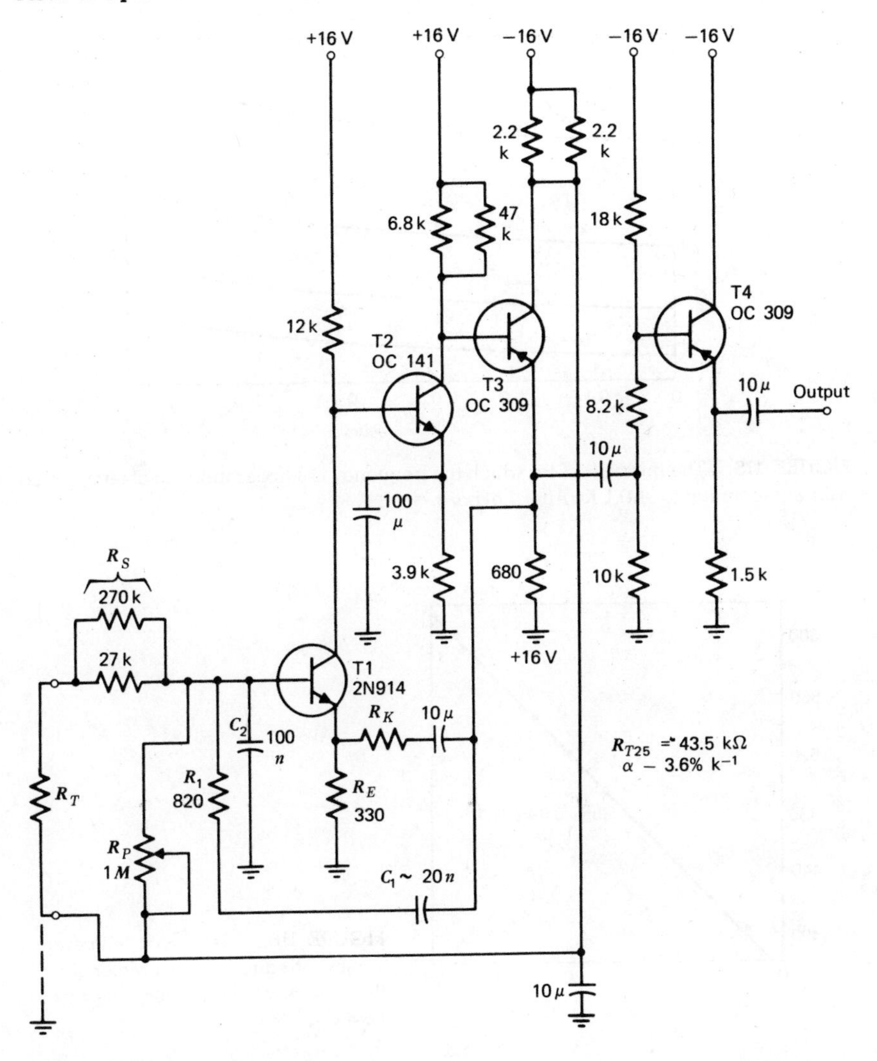

FIGURE 114. Diagram of a practical converter. R_T Philips type E 205 CE/P, 47 k Ω, R_K, Philips type B8 320 04/P4, 7 k Ω. After Lövborg.[599]

convenient and less expensive method uses thermistor beads at various locations to record glass house temperatures between 283 and 305 K over periods of 12 hr.

The frequency of a low-frequency oscillator using a silicon unijunction transistor 2N 2160 (or 2N 1671 A or B for closer tolerance) depends on the resistance of thermistor R_T in Figure 117.

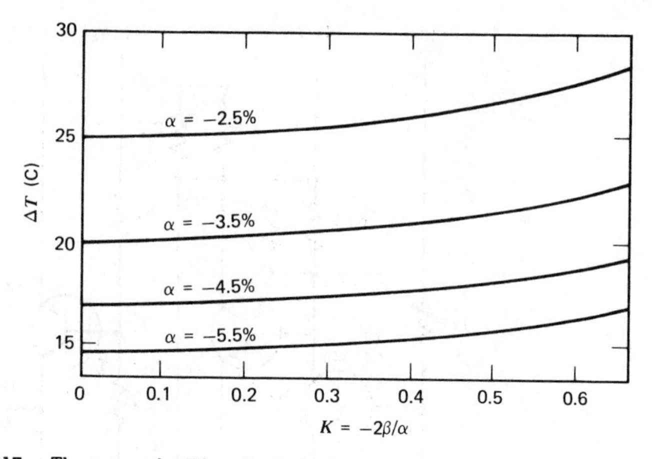

FIGURE 115. The interval ΔT for which the frequency is a linear function of temperature with an accuracy to ± 0.1 K. After Lövborg.[599]

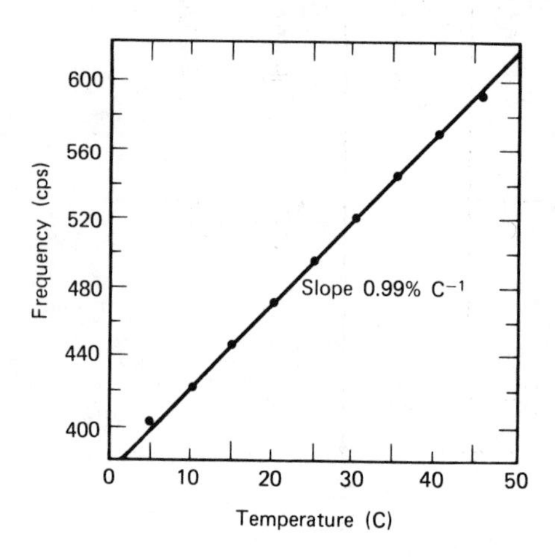

FIGURE 116. Measured temperature versus frequency characteristic of the practical converter. Results of Lövborg.[599]

The following voltages determine the period T of oscillation: V_I, supply; V_D, emitter diode; $V_{B_1-B_2}$, interbase; V_{EB_1}, emitter to base. Starting with the emitter at ground potential, V_{EB_1} rises exponentially while the capacitor C is charged through the thermistor R_t. When V_{EB_1} reaches a peak V_p the emitter becomes positively biased, thus dropping the dynamic resistance between emitter and base to a low value. The resulting discharge of C through the emitter stops at an emitter potential of about 2 V, because the

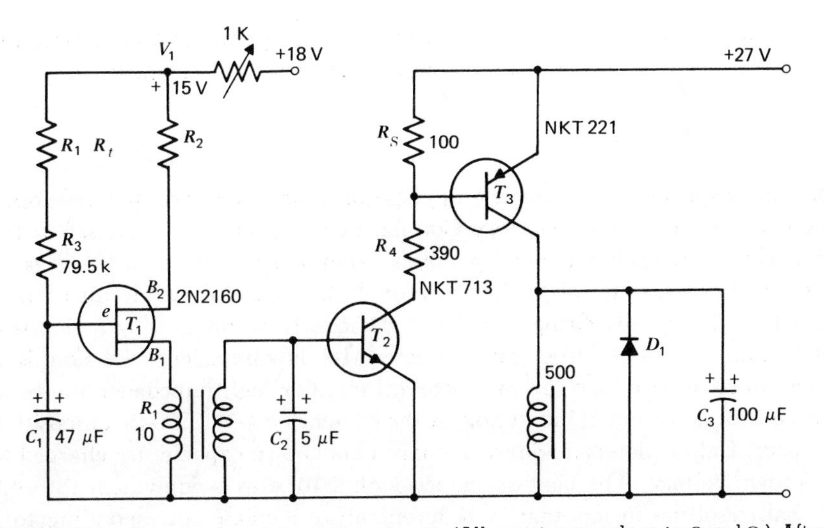

FIGURE 117. Temperature integrating circuit. (All resistance values in Ω or kΩ.) After Weaving.[600]

emitter starts to conduct and the cycle can repeat itself. The period is given by

$$T = R_t \ln \frac{V_I - V_{EB_1}}{V - V_{B_1B_2} - V_D}$$

or for small values of R_1 and R_2,

$$T = R_t \ln \frac{\eta}{1 - \eta} \quad \text{with} \quad \eta = \frac{V_p - V_D}{V_{B_2B_1}}$$

The resistor in series with the thermistor is chosen as two-thirds of the resistance value of the thermistor at midpoint of the temperature range. This is a practical compensation for the temperature dependence of α. With a bead thermistor YSI 44011 of 100 kΩ at 298 K in the circuit of Figure 117, counting rates of 7–14 per minute were obtained. Since the oscillation period, and so also the counting rate, is a linear function of the thermistor resistance, multiple sensing at different locations with thermistors deviating from a nominal value results in parallel shifts of the calibration curves. While the number of counts per minute increases between 283 and 305 K from 5 to 9.6, corresponding to an increase of 0.21 counts min.^{-1} K^{-1} or 4% K^{-1}, the influence of ambient temperature on the oscillator itself is less than 0.03% K^{-1}.[600]

A practical application of temperature–frequency conversion in a portable temperature integrator has been made for agricultural engineering.

7.1.10. Pulse Measuring

An unconventional calorimetric application is the thermergistor developed by Naval Ordnance Laboratory Corona, Fuze Department.[601] It serves the purpose of determining energy pulses ranging from 10 to 2.10^5 ergs in electrical circuits and of predicting reliably that an ignition circuit is energized for dependable firing detonating devices. In its integrating action it is the analog of a ballistic galvanometer. Its low-impedence version is a noninductive wire loop around a thermistor. For high impedence the bead thermistor is cemented into a hole at the center of a $\frac{1}{2}$-W 2500-Ω composition resistor. Calibration is made by discharging a known capacitance charged to a known voltage. The heat developed with 2.10^5 ergs is equivalent to $\sim$4.6 m cal, resulting in less than 3 K temperature increase corresponding to a resistance decrease of $<12\%$.

7.1.11. Miscellaneous

Temperature gradients are usually measured by a linear two- or three-dimensional setup of individual sensors or by corresponding motions of one or several single sensors. This would impose limitations in small volume elements. A novel solution could be the use of semiconducting InSb that has become anisotropic by inclusion of a second phase of parallel NiSb needles with a length of $\approx 5\ \mu$m and a thickness of 1 μm. Placed in a magnetic field perpendicular to a temperature gradient, the Ettinghausen–Nernst effect produces in a direction perpendicular to field and temperature gradient a field of $\approx 800\ \mu V\ cm^{-1}\ K^{-1}$.[602]

Semiconductors with high magnetoresistance are generally considered to be unsuitable for thermistors, since their resistance is a function of temperature and magnetic field. A remarkable exception is Zn-doped InSb, which has a magnetoresistance coefficient $\rho_F/\rho_0 \approx 10$ and $\rho_0 \approx 100\ \Omega$ cm at 300 K. However, this ratio remains nearly constant between 290 and 350 K or, in other words, the temperature coefficient of resistivity remains also constant at $\sim 2\%\ K^{-1}$ up to 10 kG.[603]

7.2. MODE 2. SELF-HEATED THERMISTORS (APPLICATIONS BASED ON VARIATIONS OF THE DISSIPATION CONSTANT)

The factors that determine the thermal conductivity of the medium surrounding the thermistor, and therefore its dissipation constant, also determine the thermistor resistance for a given self-heating power. The following applications are obvious and in extensive use.

Dissipation constant determined by	Application
1. Density (aggregate state, liquid or gas)	Level gauge
2. Specific heat of the surrounding medium	Level gauge
3. Gas or vapor pressure	Vacuum and pressure gauge
4. Flow rate	Flowmeter in gaseous or liquid state
5. Chemical composition	Gas analysis, chromatography

Since the observed resistance changes must be discriminated against ambient temperature fluctuations, these applications generally require a matched thermistor pair, one of them exposed to the variable heat conductivity condition changing the dissipation constant, but both held at the same ambient temperature. It is not so much the resistance itself but its temperature dependence, that should be identical for both units. Since in general only small temperature differences are involved, this condition is easier to meet, although precision tests can impose rather stringent demands. An analysis of the various factors involved in the operation of self-heated thermistors (physical dimensions, electrical characteristics, and environment) has been given by MacDonald.[604]

7.2.1. Surge Protection and Timing Devices

The time constant of a thermistor is normally determined by its heat capacity and dissipation constant.[605] This applies to ambient temperature changes. Under self-heating the power input plays an important role for the response time, which can thus be varied over a wide range, and opens many possibilities for time-delay circuits. In most cases the thermistor resistance is initially much larger than the series impedance Rs (for instance, in relay coils or filaments). Therefore the input voltage V_1 of the time-delay circuit is in the first approximation decisive for the heating power of the thermistor. As an example, if the latter drops by a factor of 20 to the value of the combined series resistance, the voltage across its terminals only drops to $V_{1/2}$ while the current increases by a factor about 10. Therefore in the first approximation the energy balance during the nonstationary heating process for each time element dt can be written

$$\text{thermistor input} = \text{heat dissipation} + \text{self-heating}$$

$$\frac{V_t^2}{R_T} = D(T - T_0) \qquad + \frac{C\,dT}{dt}$$

with

$$R_T = R_0 \exp\left[B\left(\frac{1}{T} - \frac{1}{T_0}\right)\right]$$

For all values $R_T \gg R_s$, a condition met in the major part of the self-heating regime, R_s can be neglected. Solving for the time between T_0 and a chosen R_T (e.g., $R_T = R_{s/2}$) presents considerable difficulties. Therefore the current as function of time is more easily determined by a set of experiments with variable input voltage and, if acceptable, series or parallel resistance values to the thermistor. Time delays between milliseconds and a few minutes can be obtained by selecting voltage and thermistor type. There is a possibility that for a given heat capacity, dissipation constant, and operational load (relay) resistance, the self-heating time is not long enough and reduction of the ampere input not permissible, since the final current would be smaller than required. In this case shunting of the thermistor with a variable fixed resistor permits one to prolong and adjust the time constant of the delay circuit within certain limits. These are given by the minimum voltage drop at the shunt to drive the thermistor characteristics over the peak and sensitivity to ambient temperature. For higher-power application this set-up has been successful.

Surge Protection. The use of surge protection for the filaments of electronic tubes has been one of the first large-scale applications of thermistors.

The low cold resistance of the tube filaments or of pilot lamps in series with filament results in a current surge that can shorten the life of the tube or lamp and eventually burn it out. In choosing the right series thermistor, a compromise must be found between maximal suppression of the current surge peak and the minimal residual (hot) resistance of the thermistor. In general it is desirable that the voltage drop at the hot thermistor does not exceed 5–7% of the line voltage.

For a chosen thermistor this voltage drop is a function of B (corresponding to the resistance ratio, and the dissipation constant D of the thermistor), in other words, its power rating. Practical experience has shown that limitations of the surge peak to 30% overload and a voltage drop of 6% at the hot thermistor in steady-state operation represent a good compromise. It has been suggested to eliminate the residual voltage drop completely by short-circuiting the hot thermistor with a relay in an auxiliary circuit. Surge control became obsolete when ruggedized tube filaments were used, although lamp protection still remains acute. The optimal conditions for surge protection in thermionic tubes have been discussed by Gano and Sandy.[606]

TIMING DEVICES. While in surge protection the circuit is energized normally in large time intervals, other timing devices, for instance, in telephone circuits, have a much higher repetitive rate. This does not give the thermistor enough time to cool sufficiently. Therefore its delay time becomes shorter with increasing rate of operation. To compensate for this undesirable effect, several solutions are possible.

1. The thermistor is short circuited by auxiliary relay contacts when the delayed relay switches and stays in this mode for sufficient time to cool the thermistor.
2. Instead of the self-heating time, the cooling time of a preheated thermistor is used for delay. When power is switched off, it cools and trips a relay with a time constant determined by C/D. The time necessary to restore the thermistor to its initial (heated) condition can be chosen freely by selecting the correct heating current. Even when no specific delay time is required, thermistors by their thermal inertia can prevent the tripping of relays by spurious surge signals.[607]

A critical factor that can influence the delay time is the ambient temperature. The final temperature of a thermistor in a delay circuit must be chosen high enough to minimize the effect of ambient temperature fluctuations. This, however, is detrimental to rapid recovery by cooling. A more practical solution would be to preheat the thermistor mildly by an auxiliary circuit to establish an artificial new ambient above the fluctuating ambient. Cost and space will always determine the compromise to be chosen.

It was previously mentioned that shunting with a fixed resistor can be used to make longer delays without imposing limitations on the voltage input. In this case the influence of the ambient temperature can become very critical if the voltage drop across the thermistor is only slightly above the voltage peak of its characteristic. A considerable drop of the ambient temperature could completely prohibit self-heating with the available voltage.

The dynamic behavior of 0.062-cm-thick plates of Ni-Mn-oxide disks with 65 at. % Ni and a diameter of 0.05 cm has been investigated with voltage pulses up to 1200 V and variable pulse length. For 10% resistance decrease the required pulse length dropped from 100 μsec at lower voltage to 100 nsec at 1200 V. The observed thermal response was in good agreement with theoretical calculations.[608]

7.2.2. Voltage Regulation

In 1937 Weise[609] had shown that with Urdox thermistors in a bridge circuit a voltage of 65 V could be held within 1% for a current variation

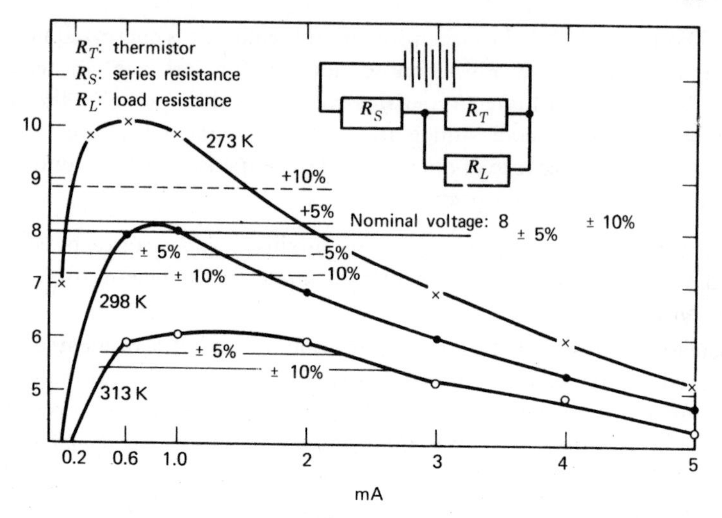

FIGURE 118. Voltage regulation at three ambient temperatures.

between −40% and +60% of the normal value if the current fluctuation rate was within the range for corrective action by the thermistors. In many cases the regulating thermistor is used in a simple voltage divider circuit as in Figure 118, although it can be shown that its stabilization effect is smaller than that of a bridge circuit.[610]

Communication technology stimulated the need for regulating smaller voltages at smaller power levels with faster response and led to the development in 1937 by Siemens of regulating thermistors meeting these requirements. They were made from uranium oxide sintered at ~1800 K in H_2 and had to be sealed in glass bulbs not only for mechanical protection but also to minimize the fluctuations of heat dissipation. It is quite obvious that units operating with the input of less than 20 mW at a maximal temperature of ~850 K would have good regulating performance only if protected against air convection. The technical data of these units together with those of present day regulating thermistors (Siemens) are shown in Table 52.

The influence of ambient temperature on the V-I characteristic (Figure 118) (see Section 6.2) has to be considered in specifying regulating tolerances for these thermistors. The high operating temperature of 850 K of the early types but also their lower dissipation constant made them relatively insensitive against ambient temperature fluctuations. The operational limits of a regulating thermistor can be recognized in Figure 119. O-*A*-*B*-*C* is its nominal V-I characteristic in series with a fixed resistor R_s and a fluctuating

TABLE 52

Type	Nominal Regulated Voltage (V)	Current (mA) Range for ±10% Voltage Tolerance	Upper Current Limit (mA)	Average Wattage Input (mW)
1937–1945				
HL2/0.5	2	0.2–1	2	1
HL2/2	2	0.5–5	8	4
HL6/2	6	0.5–5	8	12

		Voltages at			Upper Wattage limit
After 1955[a]		1.5 mA	5 mA	20 mA	mW
R51 R4/1/20	4	3.8–4.2 V	3.0–3.35 V	1.75–1.95 V	39
		0.6 mA	4 mA	10 mA	
R51 8/0.5/10	8	7.6–8.4 V	5.1–5.6 V	3.45–3.85 V	39

[a] Optimal values for ±5% voltage tolerance at the lowest current and 298 K.

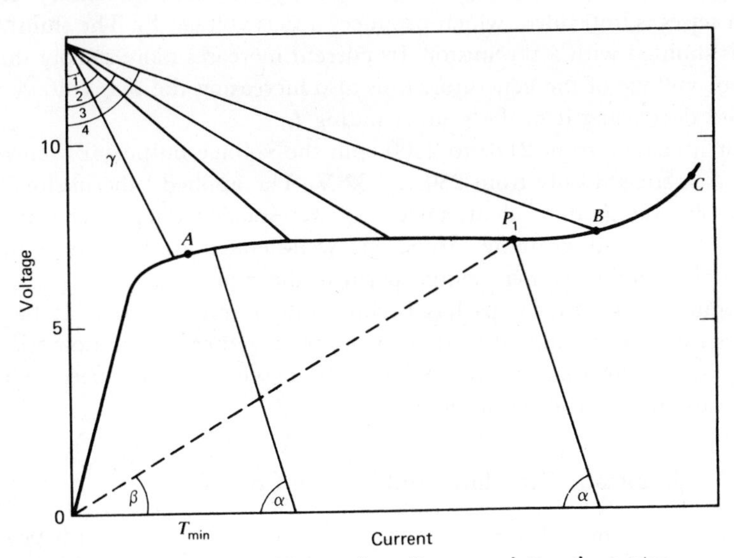

FIGURE 119. Operational limits of a voltage regulating thermistor.

line voltage V_2. The stabilizing limits are not only determined by this characteristic and its temperature dependence, but also by the load resistance R_2 parallel to the thermistor. Its load line intercepts the nearly horizontal part of the characteristic at a point P_1 that is determined by the angle $\alpha = \arctan R_L$. With decreasing R_L, α also becomes smaller until the load line intercepts the characteristic left at a point A, where no regulation is possible. The total resistance R_P of the thermistor shunted by the load resistance is given by the slope of a line from the origin to point B with $\beta = \arctan R_p$. On the other hand the load line of the series resistance intercepts the characteristic. Where this happens is determined by its slope angle γ, defined as $\gamma = \arctan R_v$ ($R_v = V/I =$ const). The point of intercept for a chosen load line of a series resistance R_s and a possible bandwidth of current fluctuations between I_1 and I_2 has to remain on the horizontal part of the characteristic left from P_1. With decreasing source voltage but constant R_s the operating point determined by $\alpha = \arctan R_2$ moves also to the left. This means that the current through the thermistor can only decrease to a value $\leqq I_{\min}$ where regulating stops. In general the bandwidth for maximal current ratio can be 5–20 depending on the allowed voltage tolerance. With increasing total voltage the current through the load remains constant, but increases through the thermistor.

rpm control of a generator has been accomplished by the use of semiconductors[611] Windmill-powered generators can vary in their rpm number by $\pm 25\%$. Into the shunt winding L of the generator, an auxiliary winding L_2 in series is imbedded, which produces a bias voltage V_2. The shunt winding is shunted with a thermistor. Its current increases more steeply than the output voltage of the generator, thus also increasing the bias voltage across L_2 and decreasing it in the shunt winding L_1.

For a change from 2170 to 2900 rpm the voltage output at a current of 31.3 A increased only from 230 to 232 V. The applied "thermistors" were probably voltage dependent, since they were made from a mechanical mixture of SiC and graphite. However, well-defined thermistors with only thermal resistance change should perform the same task.

Regulating of arc lamps has become rather obsolete now. It has been successfully accomplished with thermistors together with automatic guidance of sun mirrors by thermal sensing of geometrical aberration in a quadratic network of thermistors.[612]

7.2.3. Automatic Switching and Remote Control

The fact that the self-heating characteristic of a thermistor with peak and falling voltage branch simulates a bistable element offers many possibilities for sequential (or automatic) switching. In the first case a number of parallel circuits are switched on in a sequence determined by the choice of a thermistor in series, its time constant, its voltage peak, and the critical current to overcome it. Series of lamp bulbs or other elements in display and advertising units would be out of service if one single element fails. Therefore shunting each element with a thermistor of much large cold resistance has been successfully applied to keep the array going even if one or several units failed. Simultaneously the failing elements can easily be recognized, especially if they are lamp bulbs.

Sequential switching with thermistors can automatically replace defunct devices such amplifiers in remotely located circuits. It is necessary that the circuit after failing have so much higher impedance that the sequentially next thermistor receives enough voltage to overcome its peak (Figure 120). The following example shows how it would work. The line voltage is connected to the load resistance L_1 of 80 Ω by a permanent series resistor R_1 = 30 Ω and a second resistor R_2 = 10 Ω. Additional load resistances L_2, L_3, and L_4, all equal to L_1 are permanently connected to the line voltage and through R_1. However, in these parallel circuits R_2 is replaced by thermistors T_1, T_2, T_3, and T_4 with cold resistance values of 100, 130, 160, and 190 Ω. Since they have to substitute for R_2, their hot resistance has always to be 10 Ω. The necessary higher resistance ratios cold to hot can either be attained by using materials with increasingly higher B or by reducing the size and therefore the wattage rating of the units. For example, in the first case B has to be increased from ~3300 for T_1, ~3600 for T_2, ~3900 for T_3, and ~4200 for T_4. These figures are only approximations based on the simplifying assumption that the dissipation remains the same in all cases.

It has to be considered that each of the four auxiliary circuits has a leak current if permanently on line as stand-by. To avoid premature switching, the peak voltages of their thermistors must be higher by a good margin than

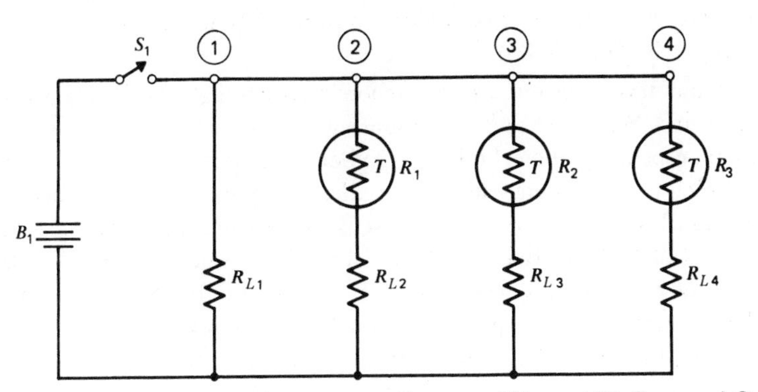

FIGURE 120. Sequential switching circuit. Courtesy of Howard W. Sams and Co.

their permanent voltage drop caused by the leak current. In the chosen examples this margin is 14–19 V, which could be considered as safe unless the ambient temperatures increases very much. After switching, each thermistor would dissipate ~10 W (~8% of total power consumption in each circuit or 12% of that in the device). This loss appears to be excessive; however, it is the price often to be paid for a fail-safe operation.

Mechanical switches can be eliminated if the remote selection of circuits is made by sending certain numbers of electrical pulses per second, which reduces the resistance of thermistors. For each circuit a thermistor with a different power sensitivity is chosen so that it "trips" with a different number of power pulses.

7.2.4. Liquid Level Sensing and Control

The dissipation constant of a sensor can increase by a factor of over 10 when it is immersed into a liquid. If it is operating with a constant voltage, cooling raises the resistance of the semiconducing sensor. Therefore its current drops from a value along the falling characteristic to a much lower one on the rising slope. The resulting current jump can be used to operate a warning device or a relay that controls the in- or outflow from a liquid tank. This principle has found wide use in automotive fuel tank level sensors and been suggested for hazardous liquids.[613] For cryogenic liquids it is especially attractive with thermistors for the following reasons:

Their large sensitivity, ranging from 10 to 25% K^{-1}, requires lower input power and reduces evaporation losses. The heat leak through the mechanical mounting and the lead connections is often sufficient for level measurement in liquid H_2 or He to obtain a resistance jump of a factor of 2–10 when passing through the liquid–vapor interface.

Operating with I approximately constant, the voltage drop across the sensor increases with the dissipation constant. Quantitative examples are given on p. 195 (Figure 85).

Cryogenic level sensing using carbon resistors had been suggested by Schwartz and Wilson[614] and has found wide use. However, their smaller sensitivity, especially above 20 K, has to be compensated for by higher power input, which often is undesirable in smaller cryogenic equipment because of evaporation losses. Immersion of the sensors into the liquid often produces cold shocks of >40 K that can be detrimental to their electrical stability. Even fatigue cracking has been observed. Last but not least, cryogenic thermistor level sensors are available with much smaller mass and heat capacity (0.09 mJ K^{-1} in He, 0.4 mJ in liquid N_2) than carbon resistors.

The factors determining the response of cryogenic thermistors for level sensors such as liquified He, H_2, N_2, and O_2 have been discussed by Sachse.[206]

7.2.5. Vacuum and Pressure Guages

The heat conductivity of gases in closed chambers starts to become dependent on pressure only if the mean free path of the gas molecules is not negligible compared with the distance between heat source and sink. This condition is increasingly met at pressures below 1 torr. At slightly higher pressure, convection can contribute to the increase of the dissipation, and vacuum meters with self-heated thermistors have indeed still a detectable though small sensitivity between 1 and 50 torr.[447]

For very low pressures the heat dissipation through the medium becomes small compared with the heat leakage through the leads. The conditions for optimal sensitivity have been investigated by Lortie.[615] She found a maximum sensitivity at 67 m torr. for a thermistor bead of 20 kΩ operating at 34 V. A more rigorous analysis of the factors determining the efficiency of a thermistor vacuum meter includes considerations of the optimal semiconductor surface and operating temperature (which were implicitly treated by Lortie in choosing the operating voltage).

The Knudsen formula for the thermal conductivity of gases in the region where the mean free path is not negligible compared with the dimension of the test bulb is written

$$Q/t = \lambda S p T^{-1/2} \, dT$$

A given heat transfer rate Q/t is proportional not only to the gas pressure P, but also to the surface S of the heat source, that is, the semiconductive sensor, the temperature difference between heat source and sink, and finally is inversely proportional to the square root of the ambient temperature. This led to the design of a thermistor manometer for the pressure range 10^{-6}–1 torr by attaching (soldering) two thermistor cylinders of 0.075-cm diameter and 0.022-cm thickness metallized with 50-μm-thick electrodes of manganin to 10-μm-thick round foils of 3-cm diameter.[616] Tin foil is most suitable because of its low specific heat, low emissivity, and its small tendency to corrode (see Figure 121). The thermistors, based on an iron oxides composition with addition of Mg, Mn, and Ti, had a rather high B value 5800 K and nominal resistance values of 35 kΩ at room temperature. The second contact for each of these two sensors was connected to the circuit shown in Figure 122. With the increase of the active surface, not only the sensitivity to pressure variations but also to fluctuations of the ambient temperature increases, thus partially offsetting the gain attained by this design. Compensation of ambient fluctuations has been accomplished by ingenious use of the voltage-versus-current characteristic.

Becker, Green, and Pearson had shown that the voltage-versus-current curve of semiconductors can have a maximum for the condition $B > 4T_0$. Since in this case $T_0 \sim 300$ K and $B = 5800$, the V-I curves must have a maximum. During the rising characteristic the sensor resistance is almost

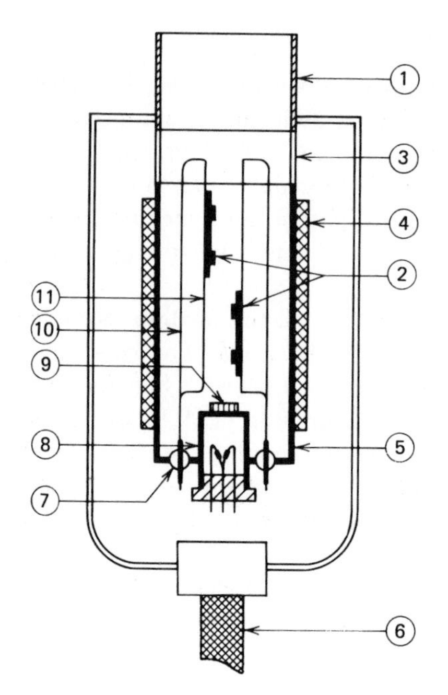

FIGURE 121. Cross section of the gauge head. 1. Vacuum joint, 2. thermistor system, 3. glass cylinder, 4. electric heater, 5. metal cylinder, 6. electric joint, 7. glass lead in, 8. thermistors for voltage control, 9. thermistor for temperature control, 10. supports of the system, 11. manganin wires (0.5-mm diameter). After Varicak and Saftic.[616]

entirely determined by the ambient temperature (no appreciable self-heating; practically independent of the gas pressure).

In Figure 123 the *V-I* characteristics of the thermistor are shown for two temperatures T_1 and T_2 at two pressures p_1 and p_2. A compensating thermistor is forced to operate in the rising branch by adding a larger series resistance than for the measuring thermistors. The bridge diagram with the entire circuit is shown in Figure 124. The currents I_c and I_m through the compensating and the measuring thermistor have to meet the following conditions:

$$V(I_c, p_1, T_1) = V(I_m, p_1, T_1)$$

$$V(I_c, p_1, T_2) = V(I_m, p_1, T_2)$$

and similarly for p_2.

Ideal compensation is restricted to a temperature interval of 10 K since the resistance versus temperature can only be approximated by a straight line within ± 5 K. Therefore an artificially raised ambient temperature of ~ 305 K is recommended. Precise measurements not only require compensation of fluctuations of the ambient temperature, but also automatic control of the bridge voltage, that is, the operating currents. This is accomplished by an auxiliary bridge parallel to the pressure-sensing bridge. It works on the same principle of identical thermistors, TV_c and TV_m at dual operat-

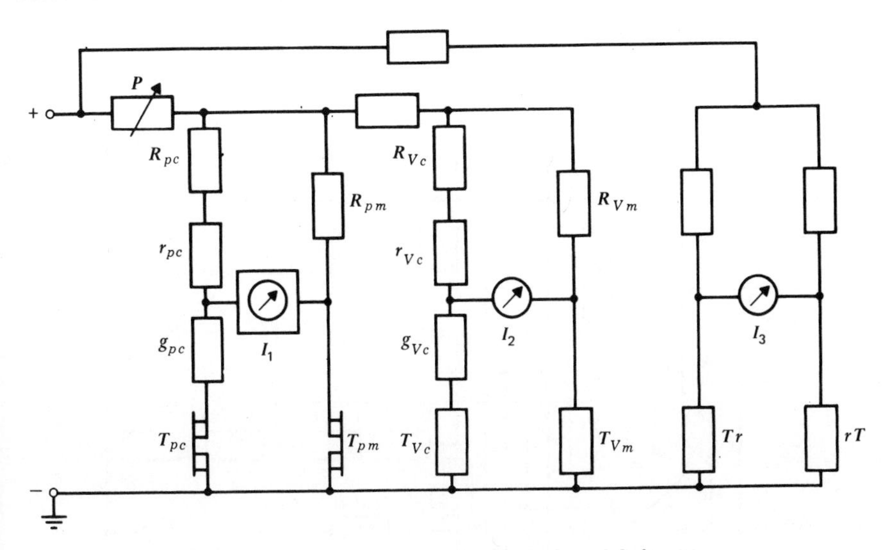

FIGURE 122. Schematic circuit diagram. After Varicak and Saftic.[616]

ing points. The first is only sensitive to temperature changes (similar to *Tpc*), the second to temperature and current fluctuations. Meter I_2 monitors voltage changes of 1 mV for 150 V power supply, which can be corrected by the resistance *P*. Finally the same principle is repeated to correct for temperature changes in the gauge head, which contains all measuring and compensating sensors (Figure 124). The bridge is initially balanced at a

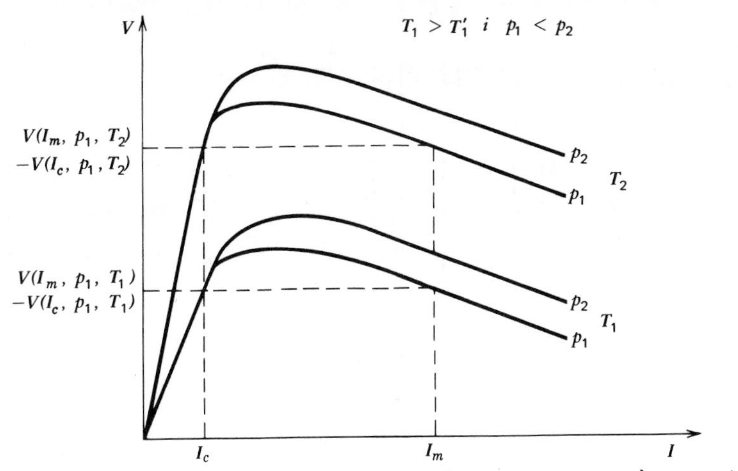

FIGURE 123. *V* versus *I* characteristics or two different pressures and temperatures. Results of Varicak and Saftic.[616]

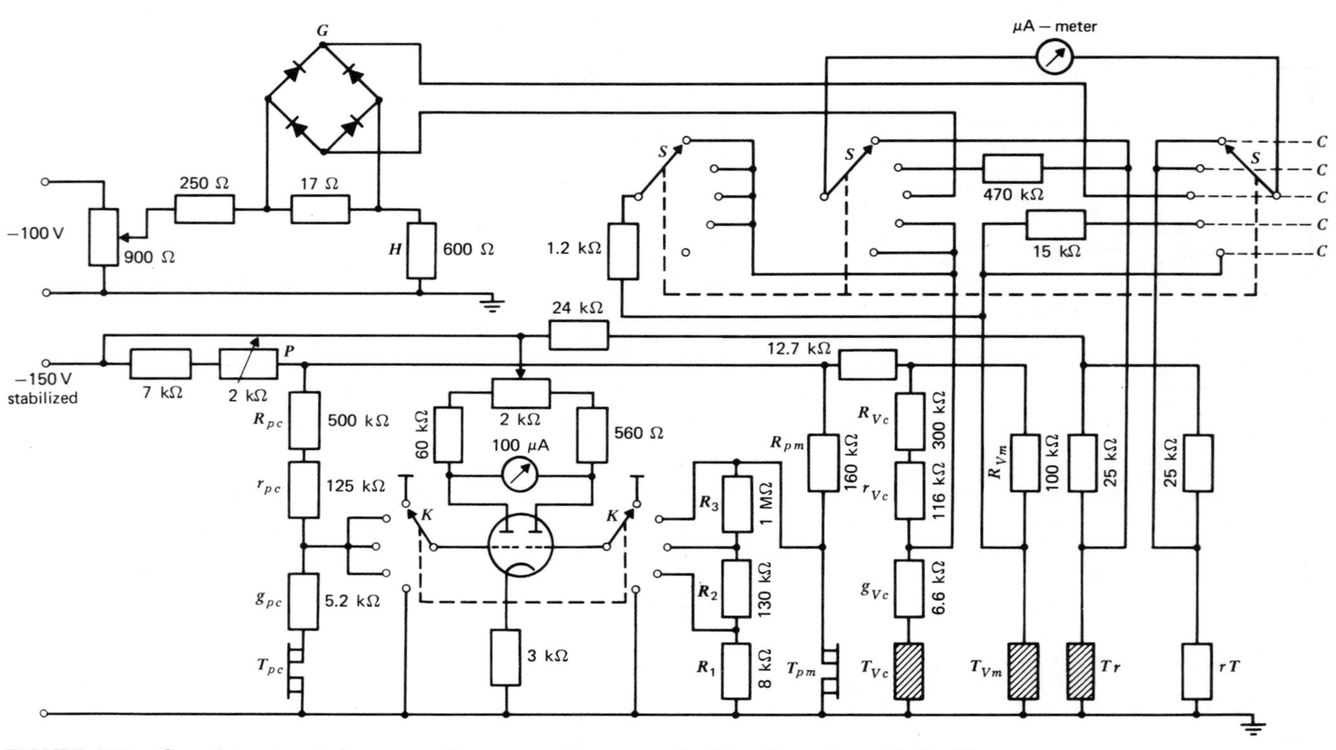

FIGURE 124. Complete circuit diagram with compensating network. After Varicak and Saftic.[616]

pressure of 1 μtorr at a current of 1.25 mA for the pressure sensing thermistor and 0.3 mA for its compensating counterpart. The voltage unbalance of the test bridge extends to 75 mV up to 60 μtorr, 630 mV up to 600 μtorr, and to 11.8 V up to 1 torr.

A fast response vacuum meter for the range 1–760 torr with stabilized operating conditions has been developed by Bulyga and Shashkov.[617] Operating thermistors at elevated temperature in high vacuum for extended time periods could result in minute changes of the stoichiometric balance and therefore the defect concentration. In oxides this generally produces drifts to higher resistance values, although hardly below 370 K. A study of Russian thermistors type KMT-1 and KMT-11 between 273 and 353 K at pressures between 760 and 0.1 torr failed to show any changes in resistance and B; apparent changes between 100 and 0.2 mtorr could be traced to temperature differences in the test equipment.[618]

A thermistor gauge operating at 373 K and between 0.1 and 1000 mtorr that can be baked out at 723 K without calibration changes could be most attractive for ultrahigh-vacuum systems.[619]

Pressure Measurements above 10 *torr.* As already mentioned, the heat conductivity method loses its sensitivity at higher pressure. Therefore another principle must be applied based on the boiling temperature of liquids (see Hypsometer on p. 227).

Effusiometry. Zemany,[620] following Dushman's[621] suggestion, described the application of thermistors (Western Electric D 176255) as a pressure gauge with linear response up to 200 mtorr when monitoring the molecular flow through a leak. This is made by punching a small hole in a thin Fernico cup and adjusting by etching with acid to permit a convenient flow rate. Figure 125 shows the design of the effusiometer. After it is filled to a pressure of about 200 mtorr, pumping through the leak starts and the decreasing pressure is continuously monitored by the thermistor pressure gauges and plotted logarithmically versus elapsed time. The resulting straight line permits reading out a time necessary for the pressure to drop to one-half its initial value. Since these half-times are proportional to the square root of the molecular weight, a full logarithmic plot of these quantities again results in a straight line with a slope of 1/2, which permits calibration of the device with one gas to avoid the more difficult absolute calibration by accurate hole size measurement (Figure 126). Furthermore, the effusion can be performed at any chosen temperature—as long as the useful temperature range of the thermistor is not exceeded or it is not subjected to strong temperature fluctuation that would result in erratic pressure readings. A small fraction of a cubic centimeter is sufficient for the measurement and valuable gases can be recovered.

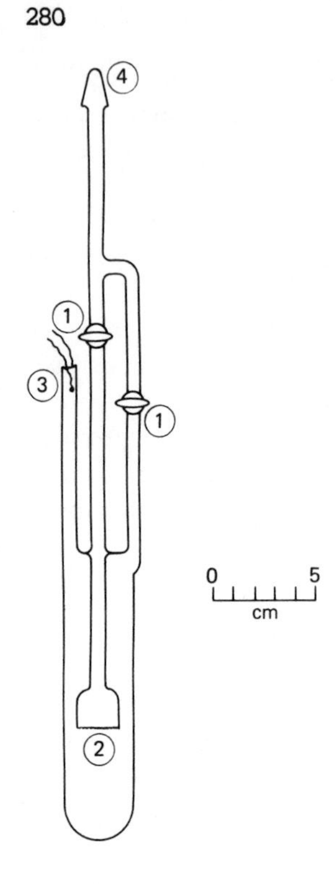

FIGURE 125. Effusiometer: 1. 4 mm bore stopcocks, 2. Fernico leak, 3, thermistor, 4. 12/30 ground joint. After Zemany.[620]

7.2.6. Flow Meters

Since the heat dissipation of a self-heated sensor is changed by variation in heat conduction or convection of the surrounding medium, the use of thermistors to measure the flow rate of gases and liquids was suggested rather early (Bell Telephone publication, 1946).[447] A thorough analysis of this application was made by Rasmussen[622] to study the relations between the time constant, sensitivity, dissipation factor, and flow rate in air and water in practical instrumentation. The general heat-transfer equation for a self-heated sensor is

$$\frac{C\,dT}{dt} = P - D(T - T_x)$$

where P is the power dissipated in the sensor; D is the dissipation factor, that is, the power to raise the temperature of the sensor in the steady state ($dT/dt = 0$) one degree above that of the surrounding medium; C is the heat capacity of the sensor; B is its material constant; T is the temperature of the

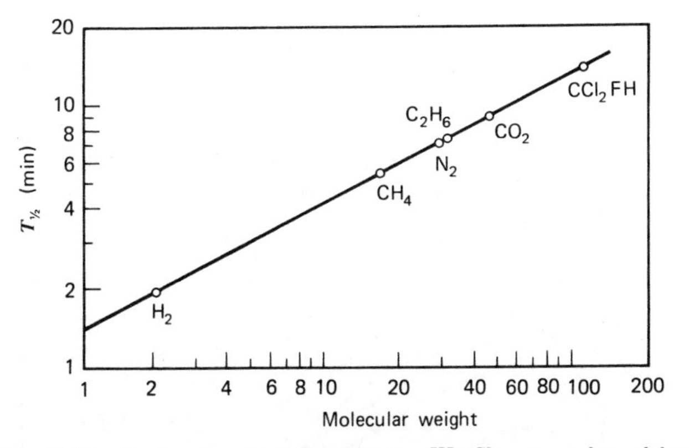

FIGURE 126. Calibration curve for effusiometer III. Showing plot of log molecular weight versus log of $T_{\frac{1}{2}}$ for several gases. Results of Zemany.[620]

sensor (T_0 in the initial state); and T_x is the temperature of the surrounding medium. The time constant τ for the nonsteady state is defined as C/D. It is reduced to a value τ' by power dissipation according to the equation

$$\frac{\tau'}{\tau} = \frac{1}{[1 + (BP_0/T_0{}^2D_v)]}$$

P_0 and T_0 are defined by the initial condition of the dissipation constant in the steady state,

$$D = \frac{P_0}{T_0 - T_{x0}}$$

The reduced time constant causes a decreased temperature sensitivity to the environment, but increases sensitivity to speed variations. For the velocity dependence of the dissipation factor the relation

$$D_v = \frac{P}{T_x}\left[\left(\frac{\alpha_x T_x}{\ln R/R_x}\right) - 1\right]$$

(α_x is the temperature coefficient at T_x) is derived. The zero-speed value D_0 for the thermistor used in water is 5 times its value in air, while its time constant is 25 times smaller. This equation can be verified experimentally by measuring V/I characteristics of a Veco 51A1 thermistor rotated with speeds between 0 and 75 cm sec^{-1}. Errors in the relative speed due to vortex motion were minimized or corrected. The results can be condensed in the following statements:

1. For the higher current values in the negative resistance region of the thermistor, the slope $-dV/dI$ is nearly independent of variations of speed. The total resistance V/I increases as expected with the velocity of motion (Figures 127 and 128).

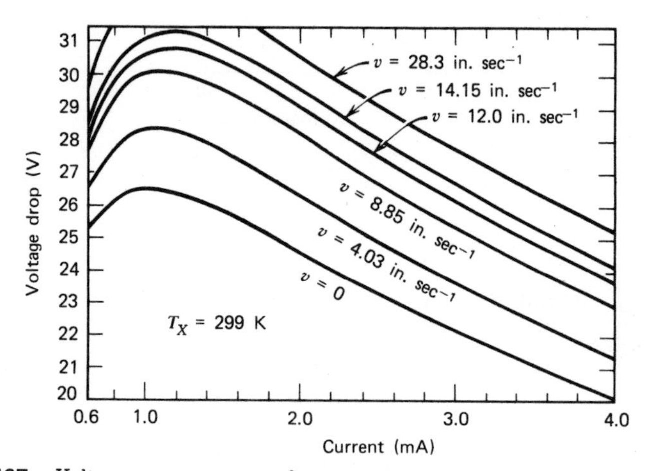

FIGURE 127. Voltage versus current characteristics at various speeds in air for a 51A1 thermistor. After Rasmussen.[622]

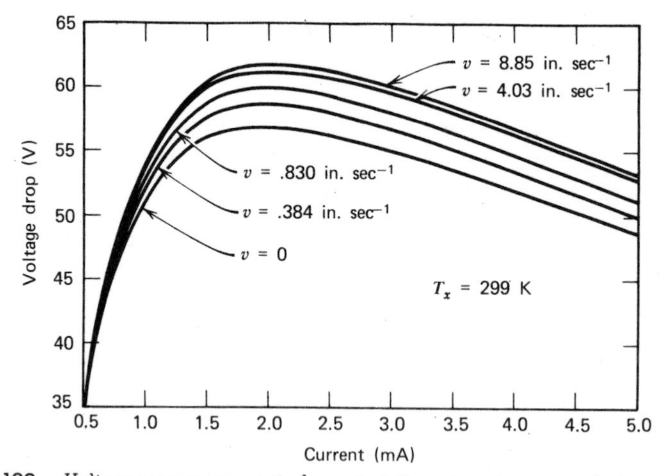

FIGURE 128. Voltage versus current characteristics at various speeds in water for a 51A1 thermistor. After Rasmussen.[622]

2. The increase of the dissipation constant as a function of the speed v is given by (Figures 129 and 130).

$$D_v = D_0 + a[1 - \exp(-bv)] \qquad a \text{ and } b \text{ are constants}$$

For a high value of v the exponential term approaches zero, the dissipation factor goes asymptotically toward $D_0 + a$, and

$$b = \left(\frac{1}{a}\right)\left(\frac{dD}{dv}\right)_{v=0}$$

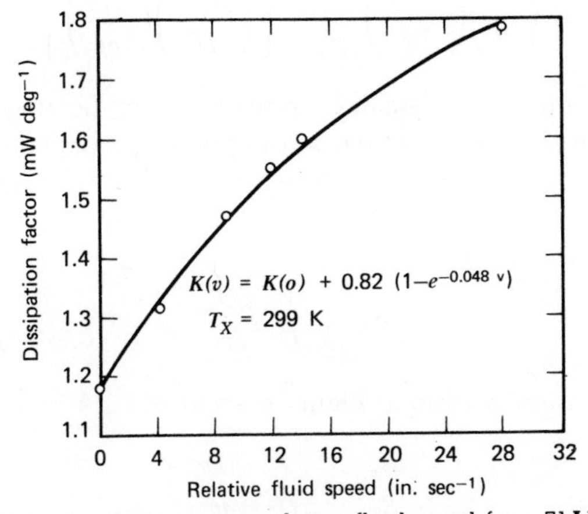

FIGURE 129. Dissipation factor versus relative fluid speed for a 51A1 thermistor in air. After Rasmussen.[622]

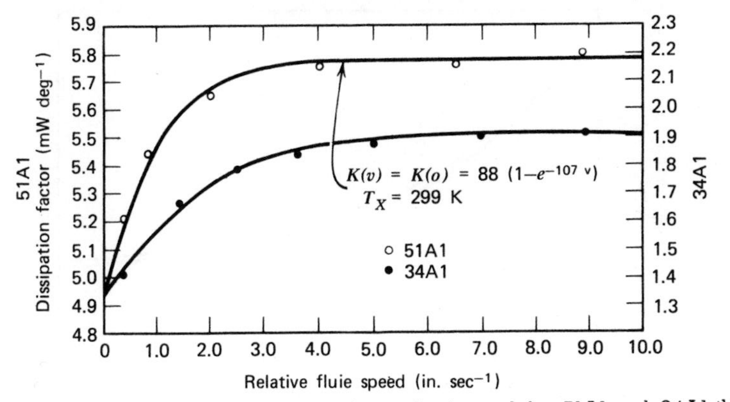

FIGURE 130. Dissipation factor versus relative fluid speed for 51A1 and 34A1 thermistors in water. After Rasmussen.[622]

In water the value $D_\infty + a$ is 18% higher than D_0 at zero speed. The environmental temperature and speed sensitivity at constant current is given by the equations

$$\frac{\partial T}{\partial T_x} = \left[\frac{1}{1 + (BP/DT^2)}\right] \quad \text{and} \quad \frac{\partial T}{\partial v} = \left[-\left(\frac{\partial T}{\partial T_x}\right)\left(\frac{P}{D^2}\right)\left(\frac{dD}{dv}\right)\right]$$

The corresponding voltage sensitivities are

$$S_T = \left(\frac{1}{V}\right)\left(\frac{\partial V}{\partial T_x}\right) = -\frac{B}{T^2}\left[\frac{1}{1 + (BP/DT^2)}\right]$$

$$S_v = \left(\frac{1}{V}\right)\left(\frac{\partial V}{\partial v}\right) = -\left[\left(\frac{S_T P}{D^2}\right)\left(\frac{dD}{dv}\right)\right]$$

where V is the voltage. For small currents (zero self-heating condition) the temperature sensitivity is large and sensitivity to speed small. With increasing self-heating current, temperature sensitivity decreases monotonically to a value $S_T \approx K/(P + DT_x)$ (Figure 131) while the speed sensitivity increases with current to (Figure 132)

$$S_v = \frac{1}{D}\frac{dD}{dv}$$

resulting in a sensitivity ratio at higher current of

$$\frac{S_v}{S_T} = \frac{P}{D^2}\frac{dD}{dv}$$

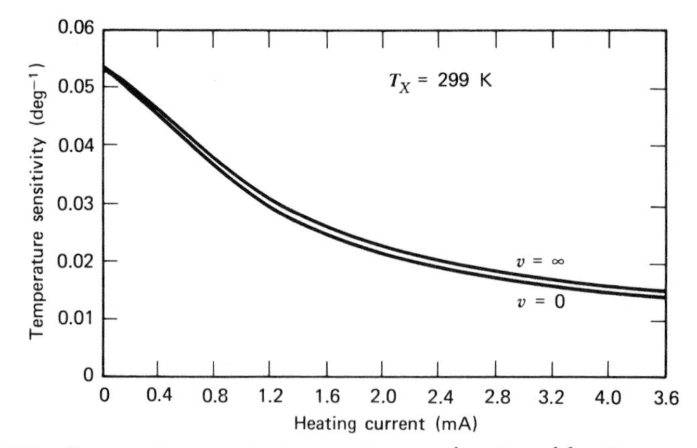

FIGURE 131. Temperature sensitivity in water as a function of heating current. After Rasmussen.[622]

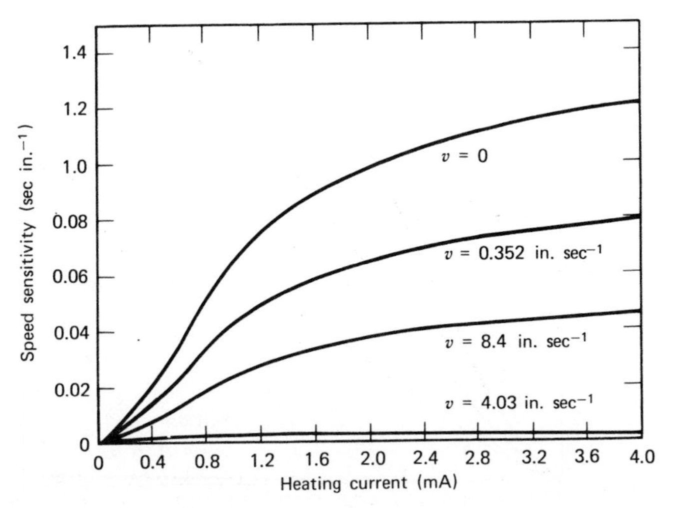

FIGURE 132. Speed sensitivity in water as a function of heating current. After Rasmussen.[622]

The temperature and speed sensitivity in water as functions of the heating current are shown in Figure 133.

Much shorter response times are possible when the sensors are operating not at constant current or voltage but at constant temperature. In this case the sensor current varies and is controlled by a servo system that determines the time constant of response.

The temperature and speed sensitivity at the operating current are then defined as

$$I_T = \left(\frac{1}{i}\right)\left(\frac{\partial i}{\partial T_x}\right) = \frac{1}{2(T - T_x)}$$

$$I_v = \left(\frac{1}{i}\right)\left(\frac{\partial i}{\partial v}\right) = \left(\frac{1}{2D}\right)\left(\frac{dD}{dv}\right)$$

The sensitivity for temperature and speed variations depends on the power dissipation. When D is small, constant temperature operation is more sensitive; when large, the constant current mode is to be preferred (see Figures 125 and 126).

Rasmussen applied his results to the calculation of the optimal operational parameters for temperature sensors dropped downward from the ocean surface with a speed of 1.5 m sec^{-1} to measure the vertical temperature profile. In order to obtain an approximation to 99% of the true temperature at each level, he calculated a required time constant in water of

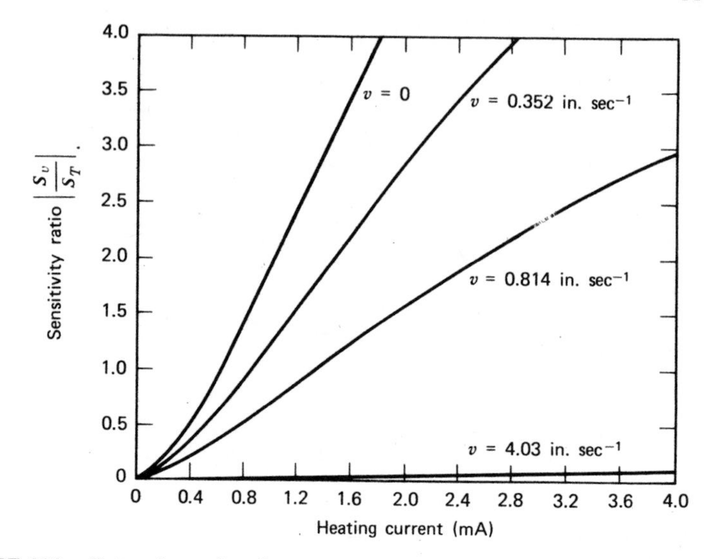

FIGURE 133. Ratio of speed and temperature sensitivities versus heating current. After Rasmussen.[622]

14 msec (corresponding to 350 msec in air). Since the fastest available bead thermistor with a time constant of 500 msec is still too slow, it requires heating to reduce its time constant from 500 to 350 msec. (or from 20 to 14 msec in water). Thus the conditions $D = 1.45$ mW K^{-1} in water, $B = 3600$ K, and $T = 290$ K lead to a heating power of 22.6 mW, which would heat the sensor 15.6 K above ambient. This reduces the temperature sensitivity by the factor $(290/305.6)^2 \tau'/\tau$. Since

$$\left(\frac{\partial V}{\partial T_x}\right)_i = iRS_T = \frac{PS_T}{i}$$

thermistor response will also be increased by a larger resistance value. If the thermistors are made with units of equal size but higher resistivity materials, a fringe benefit will arise from their normally higher B values.

Pigott and Strum,[623] fully aware of Rasmussen's basic studies, suggested and investigated a third mode of operation keeping neither current nor temperature constant. They measured the unbalance ϵ of a dc bridge with a bead thermistor VE Microbead Type No. TX 1647 in one leg as function of the flow velocity. The oval-shaped unencapsulated bead with a minimal and maximal dimension of 0.2 and 1 mm, respectively, and 3500 Ω at 273 K is placed into the axis of a Pyrex tube with 1.6 cm i.d. at a distance of 120 cm from the inlet and 23 cm from the outlet for a stream of methanol and carbon tetrachloride. After balancing the bridge at zero flow, the flow performed at 273 K was varied from zero to $\sim$9 cm sec^{-1} and the corresponding voltage unbalance measured with a constant bridge voltage producing a zero-flow power of 16.6 mW. The dissipation constant $D(v)$, as a function of the measured unbalance ϵ caused by a various flow velocities can be calculated from the equations

$$D(v) = \frac{P}{T_x}\left(\frac{-B/T_x}{\ln (R/R_x)} - 1\right)$$

$$P = V\left(\frac{-\epsilon}{r} + \frac{V}{r + R_0}\right) - r\left(\frac{-\epsilon}{r} + \frac{V}{r + R_0}\right)^2$$

$$R = \frac{V}{-\epsilon/r + V/(r + R_0)} - r$$

where V is the bridge voltage, r is the resistance of the fixed arms of the bridge, R_0 is the thermistor resistance at zero flow, B is the thermistor material constant, and R_x is the thermistor resistance at the temperature observed with dissipation by zero flow but with natural convection and is obtained by extrapolation of R_0 to zero thermistor power. If the zero-flow thermistor power is higher than 6 mW, Dv increases monotonically, as reported by Rasmussen, and is nearly independent of thermistor power. For power levels <3 mW, R approaches R_x with increasing velocity due to cooling. Therefore with ln R/R_x becoming smaller, $D(v)$ increases from 0.81 to 1.7 mW K^{-1} in methanol and from 0.56 to 1.0 mW K^{-1} in CCl_4 between flow velocities of zero and 8.9 cm sec^{-1}.

The reciprocals of the linear slopes of bridge voltage output versus velocity decrease in an approximately parabolic manner (Figures 134 and 135) with the thermistor power input and approach a nearly constant level above

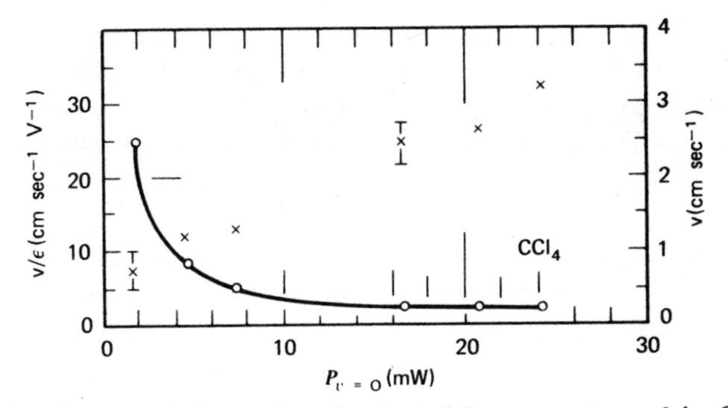

FIGURE 134. Reciprocal slopes ϵ/v as function of the power at v = 0 for CCl_4. Results of Pigott and Strum.[623]

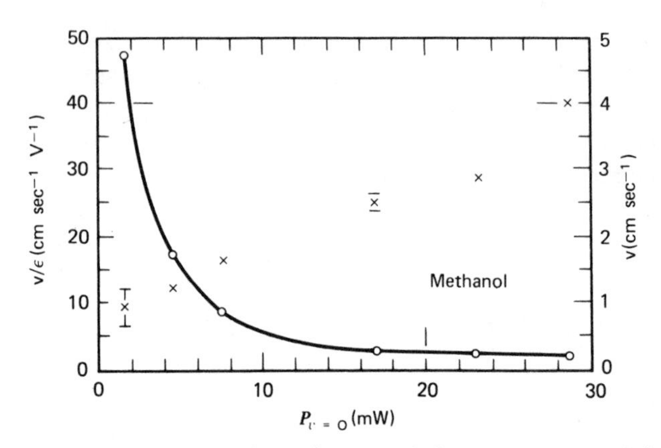

FIGURE 135. Reciprocal slopes ϵ/v as function of the power at $v = 0$ for methane. Results of Pigott and Strum.[623]

16 mW, indicating a constant ratio of bridge unbalance to flow velocity, nearly independent of the thermistor power. Measurements of intensity and direction of water currents in the ocean have been made with indirectly heated thermistors (see p. 228). The much larger sensitivity of PTC units has been used to measure air flow to pressures down to 10^{-6} torr with flow sensitivities from 10^{-3} to 10^5 cm sec^{-1}.[624]

For low flow rates of gaseous N_2 and H_2 (<4 liters min^{-1}) an indicating flowmeter using thermistors as sensors has been developed by the National Bureau of Standards. Its main advantages are low cost, ruggedness, high sensitivity, and relative small pressure error; its only drawback is its slow response to rapid fluctuations of flow.[625] Special applications are in fluid and aerodynamics (including acoustics). In this field ambient and self-heated operations (sometimes hybrid) are often converging and are therefore treated together. For the study of complex flow problems, thermistors anemometers have been suggested[626] and applied.[627] They are especially useful for detecting secondary flows and the spatial dependence of a wave form near a boundary for frequencies below 2 Hz.

In strong sound fields viscous heating at a boundary layer and micro-streaming occur in a region of a few tenths of a millimeter. The problem of point measurement of temperature with a sensor probe that does not protrude into the boundary layer and cause sound field distortion has been solved by Gould and Nyborg.[628] A Gulton type 343 uncoated spherical thermistor of 0.01-cm diameter is placed near the top of a shallow rectangular mold, its leads protruding near the bottom. The mold is filled with a liquid plastic, which is removed from the mold after hardening. Its top surface is milled off until the imbedded thermistor becomes visible and then hand lapped until a quarter of it has been removed. This ensures that the thermistor blends smoothly into the surface of the plastic holder, thus avoiding any distortion of the sound field. The sensor has a dual function. With negligible self-heating current, it detects temperatures in highly localized sound fields. Under self-heating it measures local microstreaming in the acoustic boundary layer. This principle has been applied by Walker and Adams[629] to detect and study the stream fields near a cylinder located in a low-frequency (120 Hz) sound field in air. A GE RO52 thermistor is imbedded into a wooden cylinder of 0.466-cm diameter in such a way that the exposed thermistor surface corresponded 10° cylinder arc. The cylinder (Figure 136) is mounted with its axis perpendicular to the sound field and can be rotated around its axis, thus exposing the thermistor at different angles to the sound field to determine whether the maximal streaming at this obstacle would occur at 45° as predicted by theory. The interesting result was that only for a high intensity of 117 dB was this prediction fulfilled. For lower intensities of 110 dB maximal cooling, that is, microstreaming, occurred at 90° (perpendicular to the acoustical field), caused only by the tangential component of the acoustical motion. Between these extremes a condition at 114 dB exists with equal cooling for all directions (Figure 137). This investigation demonstrates vividly the usefulness of thermistors for very delicate flow studies. The behavior of VE 32A8 thermistors under rapidly changing flow conditions and velocity turbulence at flow rates up to 72.1 cm. sec^{-1} in water test tunnels indicates a sufficient sensitivity at $v < 10.3$ cm sec^{-1} and a frequency response of ~0.5 Hz. The noise level from turbulence corresponds to temperature fluctuations of ~2 mK.[630]

7.2.7. Gas Analysis and Chromatography

Thermal conductivity has been used for gas analysis for nearly 40 years. Usually a hot wire filament was applied as a heat source. This could promote

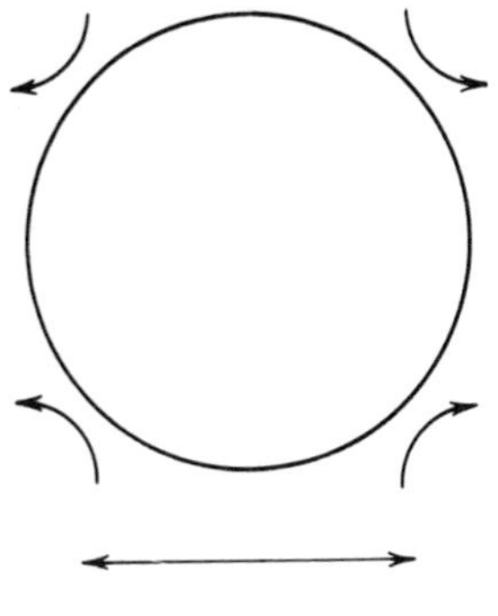

FIGURE 136. General nature of acoustic streaming (curved arrows) in neighborhood of cylinder placed in sound field. Acoustic particle velocity direction indicated by straight arrow. After Walker and Adams.[629]

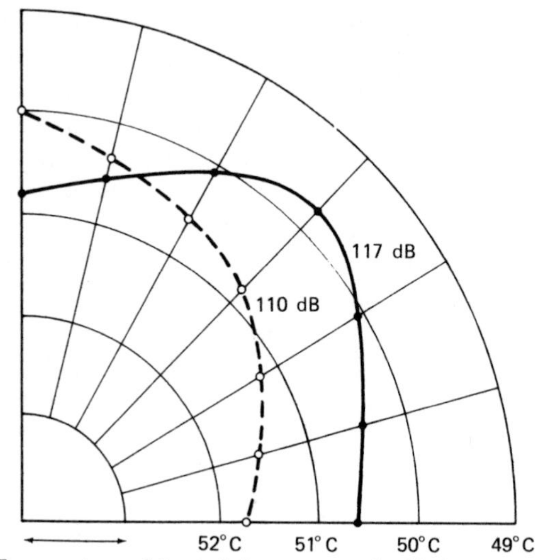

FIGURE 137. Temperature of heated thermistor imbedded in cylinder, as function of angular position with respect to sound field at various intensity levels. Acoustic particle velocity in direction of arrow. After Walker and Adams.[629]

reactions in a gas mixture, even only by catalytic action. The more temperature-sensitive thermistors can be operated at much lower temperatures. Therefore in the 1950s, serious efforts were made to use them as a controlled heat source for this purpose.[631–633]

A thorough theoretical analysis of factors that determine zero stability and sensitivity has been made by Walker and Westenberg,[634] who discarded the myth that stability of thermistors is not good enough for such applications. For a bridge circuit with a reference and a test cell, each containing a bead thermistor (Figure 138), which is balanced with reference gas of equal pressure in both cells, the voltage unbalance produced by a test gas mixture is

$$E = \frac{I(R_1 R_r - R_2 R_s)}{(R_1 + R_2 + R_r + R_s)}$$

If the spherical beads with radii a_r and a_s are centered in spherical cavity cells with the radii b_r and b_s, steady-state heat balance exists for the following condition:

$$I_s^2 R_s = 4\pi\lambda_s \frac{a_s b_s}{b_s - a_s} (T_s - T_c)$$

where R_r and R_s are the resistances of the reference and test thermistor, T_c and T_s are the temperature of wall and test thermistor, respectively, and

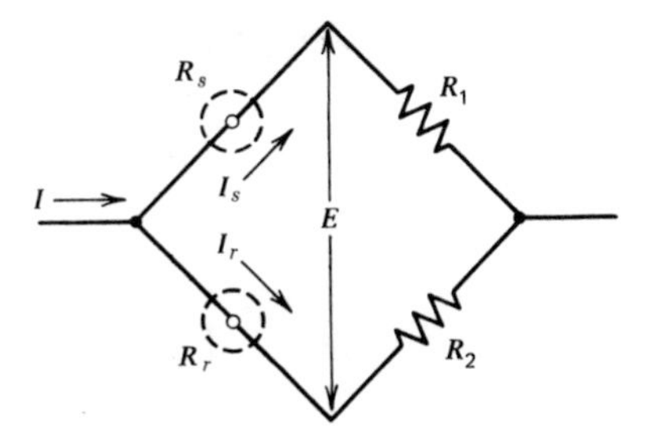

FIGURE 138. Basic bridge circuit for thermal conductivity gas analyzer. After Walker and Westenberg.[634]

λ_s is the thermal conductivity of the sample gas. A similar equation for the reference cell is obtained by substituting the index s by r. Introducing the temperature coefficient α,

$$R_s = R_{sc}[1 + \alpha_s(T_s - T_c)]$$

where R_{sc} is the test thermistor resistance at ambient temperature.

For thermistors with identical values of R_c and α placed into gas cells of identical geometry, the differential change of E is

$$\delta E = \frac{AR_cI^3}{2(A - I^2)\,(2A - I^2)}\left(\frac{\delta\lambda_s}{\lambda_s}\right)$$

with

$$A = 16\pi\lambda_s \frac{ab}{\alpha R_c(b - a)}$$

where $\delta\lambda_s$ is the differential change of heat conductivity and $\delta E/\delta \ln \lambda_s$ has a maximum at $I = (-4.38\ A)^{1/2}$. At this maximum the influence of current fluctuations is minimized and maximum sensitivity is attained.

N_2 with small additions of He or CO_2 has been analyzed with cell temperature of 315 and 306 K, respectively, and the calibrations are shown in Figure 139.

The precision for 0.1% addition was ±5% and smaller concentrations were still detectable, though with larger tolerances. Use of thermistors with higher resistance, or cooling the cell walls, are recommended for further increase of sensitivity. Resistance calibration was reproducible within ±1% over a period up to 6 wk. After that an occasional recalibration was necessary.

Chemical methods to detect eluted materials in a carrier gas stream have been increasingly replaced by physical methods; one uses the measurement of the heat conductivity of the gas mixtures. The replacement of the hot wire operating at constant current has not only offered higher sensitivity but also adaption to microanalysis because of their small size.

A stream of H_2 carrying separated fractions of light petroleum (bp 313–333 K) had effects of an opposite sign on a bead thermistor detector. When

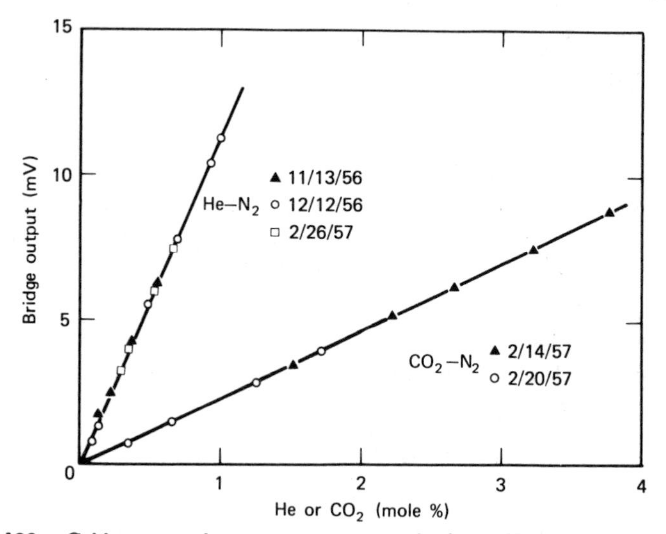

FIGURE 139. Calibrations taken over various periods of time. He-N_2 and CO_2-N_2 voltages actually have opposite polarity. Results of Walker and Westenberg.[634]

the thermistor was placed 0.3 cm from the column exit, cooling resulted, but at 15 cm heating was observed.[635] This reversal did not happen with N_2 as carrier gas. When eluting *n*-amylalcohol, the position effect appeared in both carrier gases. This led to the conclusion that the thermistor not only responded to the heat conductivity but also to the heat capacity of the surrounding medium (in the remote position).

Davis and Howard[637] developed an apparatus with a Standard Tel. and Cables Ltd. glass-coated Type 2361 thermistor as a sensor of 2.0 kΩ nominal resistance operating at 10 mA (~200 mW self-heating power) at temperatures from 290 to 440 K in N_2 as carrier gas. The lifetime of the sensor in N_2 or He seems to be unlimited, but hydrogen causes breakdown of the glass coating (probably reduction, if lead glass was used). Optimal sensitivity for a positive signal (resistance increase) was obtained within 2 cm from the exit of the column. Since for larger distance sign reversal occurs, the superiority of the pin-point thermistor sensor over the extended hot wire is obvious.

Sensitivities of 10–30 μg/ml N_2 for *n*-pentanol, saturated and unsaturated aliphatic hydrocarbons esters, alcohols, and ketones were found, with a small influence of the operating temperature and negligible noise level.

A basic study for understanding the various factors in the operation of thermistors in gas chromatography was made by using negative resistance equivalent circuits in several bridge configurations to calculate signal and noise output.[636] The signal output in chromatography depends in the first approximation on the current, temperature, and dissipation constant. The latter is more influenced by the composition of the gas stream to be analyzed. The relations between these quantities were delineated. The following basic equations are used:

$$\text{Voltage } IR = IR_0 \exp\left[B\left(\frac{1}{T_T} - \frac{1}{T_0}\right)\right] \tag{1}$$

$$VI = P = D(T_T - T_W) \tag{2}$$

where T_T, T_0 are the temperatures of thermistor bead in operation or before operation (R_0), T_w is the temperature of environment (carrier gas), P is the thermistor power (W), B is the thermistor material constant, and D is the dissipation constant of thermistor (W K^{-1}).

From these two equations the voltage–current characteristics can be calculated as

$$V = IR_0 \exp\left[-B\left(\frac{1}{T_0} - \frac{1}{P/D + T_w}\right)\right]$$

or

$$V = IR_0 \exp\left[-B\left(\frac{P/D + T_w - T_0}{T_0\,(P/D + T_w)}\right)\right]$$

Sensitivities in the ppm range have been attained in the chromatographic analysis of isoprophyl alcohol-benzene, benzene-toluene, cyclohexene-toluene, and methanol-water mixtures and for trace analysis of cyclohexane by using 2-kΩ thermistors VE A 111 in a bridge with the signal amplification using a Liston-Becker model 14 DC breaker amplifier in connection with an attenuator as shown in Figure 140.[638] This was necessary since linear amplification was only possible with signals below 300 μV. The noise level of the bridge was less than 0.2 μV and the drift 5–10 μV hr^{-1} due to temperature changes in the mounting blocks containing the thermistors.

The Leeds and Northrup amplifier 9835 A is another option with the additional benefit of a feedback circuit, eliminating the need for recalibrating the system for gain changes resulting from changing the attenuator. The optimal bridge voltage to obtain best sensitivities was investigated for 3 VE-thermistors A 111, AX 1039, and AX 1040 with 2, 8, and 80 kΩ resistance, respectively, at 298 K. It was found that the signals obtained by injection of *n*-hexane increased approximately as the square root of their resistance. Although the linear increase of noise with the signal has to be reckoned with, a higher resistance value for operating thermistors at higher temperatures remains useful. In many cases gas mixtures to be analyzed have reducing components, hydrocarbons, alcohols, or even H_2, especially if the latter is used as a carrier gas. In this case chemical reduction of the

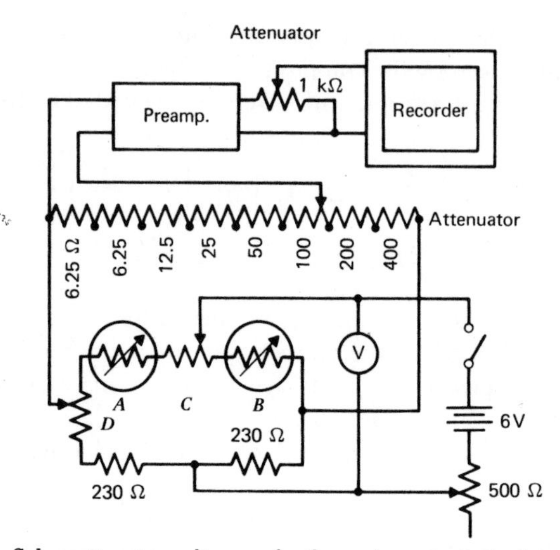

FIGURE 140. Schematic wiring diagram for thermal conductivity detector. A. Reference thermistor, B. Detector thermistor, C. Coarse bridge adjustment, 25-Ω Helipot, D. Fine bridge adjustment, 5-Ω Helipot. After Bennet et al.[638]

oxide thermistor must be taken in consideration. The effects of various glass coatings on the sensitivity, drift, and reliability of bead thermistors were investigated with H_2 and He as carrier gases. Double glass coating provided sufficient protection against reduction while sustaining reasonably fast response.[639] One basic problem in using sensitive detectors such as thermistors is their electrical noise, which can result from flow and ambient temperature variations and vibrations, adding to bridge and amplifier noise. The inherent electrical peak-to-peak noise of the thermistor itself normally does not exceed 0.3 μV, while the other noise sources can add up to 10–20 μV, thus making it impossible to take full advantage of the intrinsic sensitivity of thermistors. Diffusion- and convection-type detector cells reduce this noise, but at the same time reduce the speed of response by a factor of 10–20. Reducing the diffusion length to the detector by shortening the diffusion path has eliminated this disadvantage. This was accomplished by mounting the thermistor (Fenwal G 112, 8 kΩ) into the center of a cage made of perforated Pyramid 125 T.O. 035-inch nickel screen with a clearance of only 0.025 inch (0.6 mm) between screen and thermistor. This increases the signal to noise level by a factor of 50 permitting detection of $\sim 10^{-8}$ mole organic vapors per mole He.

The choice of the thermistor resistance is optional since both response and noise level increase with the square root of the resistance value, keeping the signal-to-noise ratio constant.[640]

More data on applications in chromatography can be found in references 641–644.

Mine safety requires monitoring of the methane concentration in air. A portable methanometer using a detection head with thermistors has an accuracy of 0.2% between 1–3% CH_4. An additional PTC thermistor provides temperature compensation.[644]

7.2.8. Hf Volume Control and hf–dc Conversion

The main disadvantage of normal volume controls with moving contacts are

1. limitations as to size reduction,
2. electrical noise from the moving contact, especially during dial setting,
3. nonnegligible phase angle,
4. difficulty of remote operation.

The early Urdox regulating resistors mentioned in Section 1 permitted a resistance decrease by a factor 1000 between 0 and 2.5 W input by indirect heating. For stationary equipment operated at line voltage, this high control input could be tolerated. For portable devices with miniaturized circuitry it was, however, not acceptable. Reduction of the physical dimensions was only half the answer to solving this problem. Internal heating of the thermistor by dc or low-frequency ac, both separated from the hf circuit by blocking LC networks, was more efficient for operating with a lower control input. Since the initial resistance is much higher than that within the regulating range, not only a sufficiently high controlling voltage must be available, but the self-heated thermistor must be protected against runaway by a suitable series resistor. As shown in Section 3.5, a sintered material in most cases represents a network of series and parallel microcapacitors, each of them formed by the interfaces between the microcrystalline grains. However, even if the effect of this distributed capacitance is negligible at higher frequencies, a "true" frequency dependence of the material can exist due to the relaxation spectrum of the atomic constituents of the thermistor material. This dielectric loss can cause an increment to the ohmic resistance measured at low frequencies.

For thermistors of 20 mg mass power, inputs of less than 1 W are sufficient to regulate their resistance by a factor 100. The limiting factor is given by the stability of material and contacts at elevated temperatures. Sealing into cans or bulbs eliminates fluctuations of the dissipation constant, but also increases the inertia of the system. With Co-Ni oxide thermistors, the influence of frequency between 0.5 and 200 MHz on the resistance is minimized and becomes smaller with increasing thermistor temperature.

RF-dc transducers based on the hot-wire thermocouple principle have been always vulnerable to burnout due to their quadratic response to current input. Replacement of the hot wire by a thermistor would moderate this effect, since its temperature increases much less than proportional to I^2. Solid-state diodes have made this application obsolete.

7.2.9. Volume Compression and Expansion

Volume compression to restrict the signal amplitude can use a thermistor in a similar circuit as for voltage regulation (Figure 141). The inversion of this principle for expansion requires a thermistor in series with the signal output and a shunting resistance as shown in (Figure 142). In the first case the output voltage increases with increasing input voltage slower (compression) and rises more quickly in the second case (expansion) due to the negative temperature characteristics of the parallel or series thermistor (and vice versa for decreasing input voltage). The circuit acts as an expander of the signal fluctuation, but not as amplifier since a part of the input power is lost in the network. Typical applications are: sensitizing relays, translators of linear into square law response for meters, pulse steeping, and increasing the

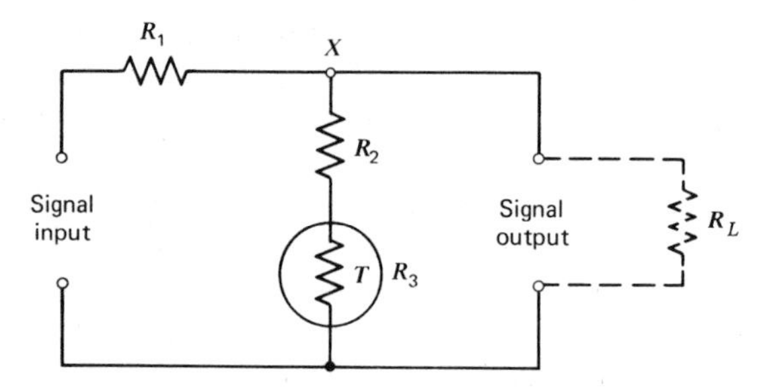

FIGURE 141. Audio compressor (limiter). Courtesy of Howard W. Sams and Co.

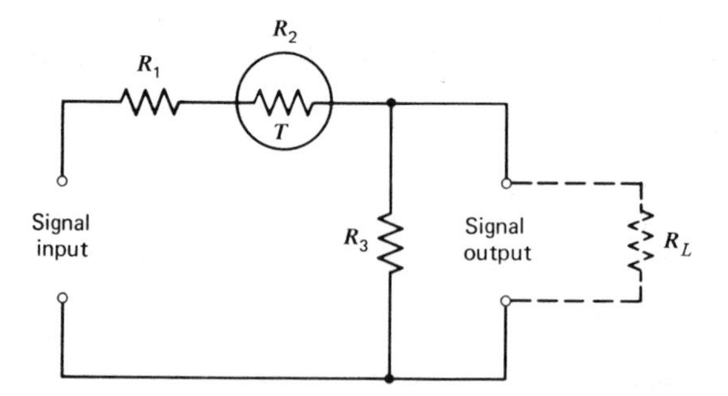

FIGURE 142. Audio expander. Courtesy of Howard W. Sams and Co.

negative feedback rate of an amplifier to maximize sensitivity of low signal levels. Another application is the reduction of cross-talk. Cross-talk in communication circuits caused by electromagnetic radiation from one circuit to another has always been a serious problem and has been successfully met by a volume limiter using a thermistor.[681] Figure 143 shows how it works:

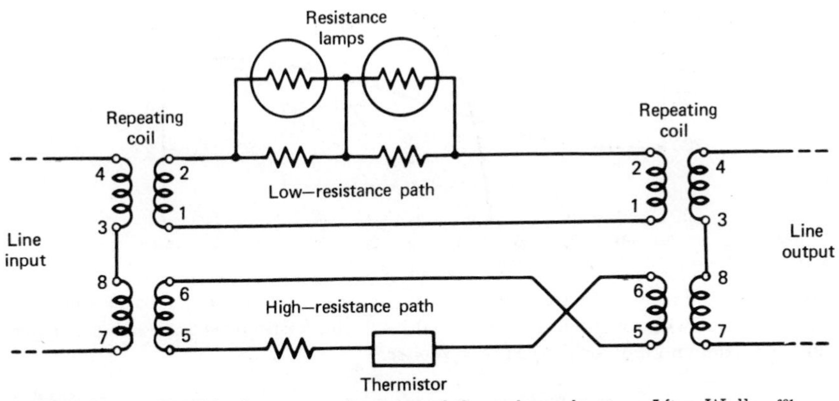

FIGURE 143. Simplified circuit schematic of the volume limiter. After Weller.[681]

The line input carrying the voice signal is fed into a transformer with two balanced output windings. One of them is connected to a low-resistance path consisting of two shunted lamps with positive resistance characteristic, the second to a high-resistance path with a thermistor and a fixed resistor in series. If one of the windings of the output is reversed, equal currents in both paths would result in cancellation of the output signal. Since for low input amplitudes the resistance of the two paths is very different, a signal appears at the line output. With increasing input the thermistor receives more power, is warming up, and its resistance would finally converge with the slightly increasing resistance of the lamp circuit (Figure 144). The signals at the output transformer increasingly buck each other, thus introducing a loss into the line that is controlled by the voice. The intrinsic inertia of thermistors does not respond to amplitude peaks, but prevents nonlinear distortion. The frequency response of the limiter between 200 and 8000 Hz is flat ($\sim$2 dB).

7.2.10. Automatic Gain Control

In communication network systems the output of each repeater element (amplifier) must be stabilized within a narrow margin. This can be accom-

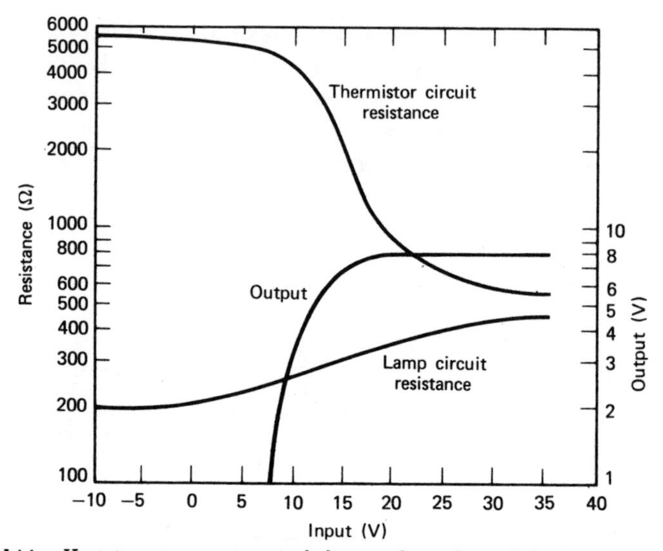

FIGURE 144. Variation in resistance of the two branches of the limiter as the input varies, together with an output in V given at the right. Resistances approach a common value at maximum input level. Results of Weller.[681]

plished by controlling the feedback with a self-heated or indirectly heated thermistor.[645–647] Figure 145 explains how it works:

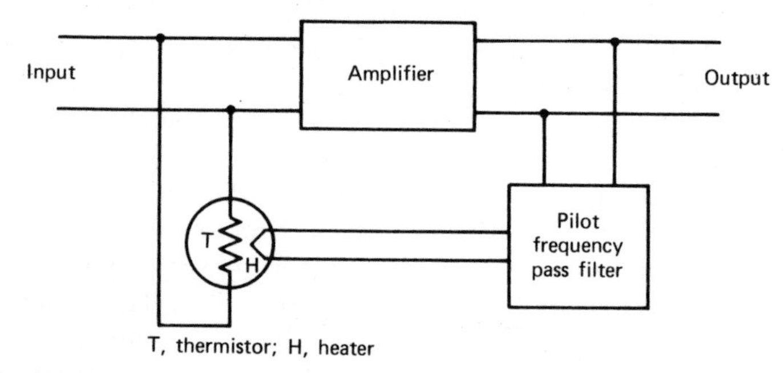

FIGURE 145. Automatic gain control by indirectly heated thermistor as negative feedback element. After Becker, Green, and Pearson.[447]

Increasing input at the amplifier would result in increasing output. A small fraction of the input energy is consumed in the thermistor shunting the input terminals, thus reducing the input power. More efficient is the use of an indirectly heated thermistor as a degenerative shunt. To be independent of the varying frequencies of the information signal, a pilot frequency signal is superimposed and amplified together with the information signal. Increasing the amplitude of the pilot signal, which is filtered out from the overall output by a pilot frequency pass filter and feeds the heater elements of the indirectly heated thermistor, reduces the shunt resistance at the input. This principle invites a number of modifications to regulate wider frequency bands than just the pilot frequency or to include a second thermistor for compensation of ambient temperature on the regulating thermistor element.

Automatic Oscillator Amplitude Control.[648–651] The sinusoidal wave output of high-quality signal generators such as the Wien-bridge oscillator can be stabilized by substituting a thermistor for one of the fixed bridge resistors (R_5) (Figure 146).

The regenerative feedback at the grid of the first tube from the plate of the second tube is automatically controlled by the thermistor in this feedback path, thus keeping the output of the oscillator-amplifier constant. The thermistor is heated by the oscillator output and is sustained at a value that keeps the RC bridge slightly unbalanced to produce a constant oscillator output with a minimal fraction of harmonics. The thermal inertia of the

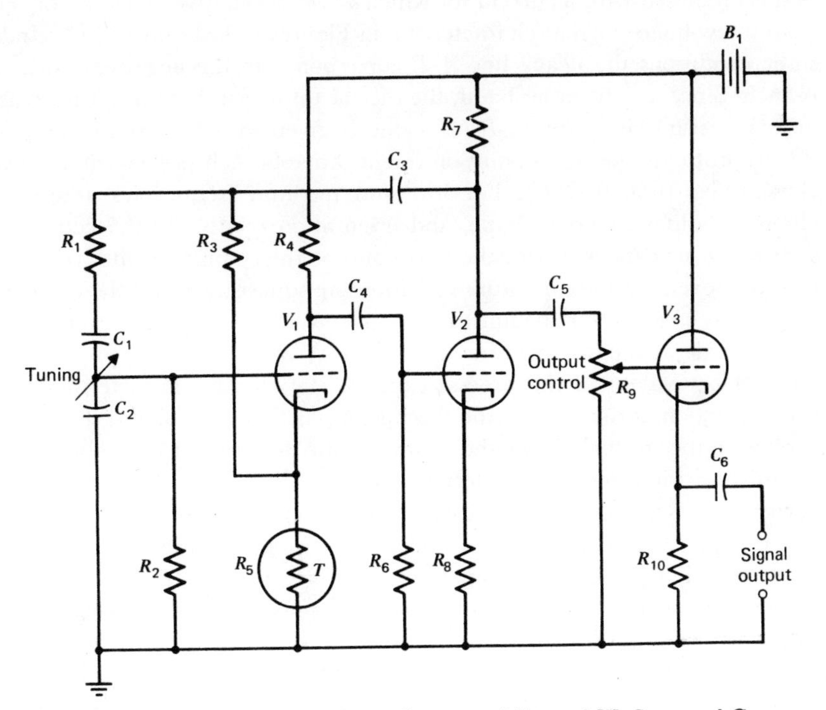

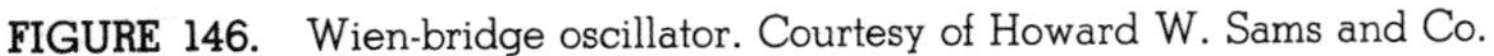
FIGURE 146. Wien-bridge oscillator. Courtesy of Howard W. Sams and Co.

thermistor must be large enough to prevent variations of its resistance with the oscillator frequency. To optimize precision, the stabilizing thermistor can be held at a thermistor-controlled ambient temperature. The same principle can be applied to the transistorized analog.

7.2.11. Filter and Oscillator for Low Frequencies

A self-heated thermistor with a finite time constant behaves like an apparent inductance, since its current has a time lag behind the applied voltage as any induction coil. This thermal lag is determined by the heat capacity of the thermistor and its dissipation constant. Since the time constant of thermistors can be made to the order of 10 sec, considerable apparent inductance values up to over 10^3 H can be simulated. If low-frequency ac is applied to a thermistor, its temperature and therefore its resistance changes are lagging versus the applied variations of voltage and power, and its apparent inductance will decrease with frequency. At sufficiently high frequency its thermal time constant is much larger than the oscillation period and it would behave like a normal resistor with negligible induction. Of special interest is the case when a low-frequency ac signal is superimposed on the thermistor that is prebiased with a current for which dV/dI is negative. This can be seen from the voltage–current characteristic in Figures 147(*a*) and 147(*b*). Under static conditions the heavy line *A–B* corresponds to the negative resistance branch. Since, on the other hand, the thermistor acts at high frequencies as a normal resistor, its positive dV/dI value is given by the straight line from the origin to an operating point on the negative branch predetermined by a chosen bias (line 0–*C*–*D*). For low and medium frequencies a series of ellipses of different area, shape, and orientation relative to the coordinate system can be expected with the bias point as the common center. For very low frequencies the temperature variations are quasistatic and the current is lagging by nearly 180° behind the voltage. With increasing frequency the phase angle decreases continuously until it practically disappears. The thermal equivalent of the process explains the elliptic characteristics. For slow reduction of the current the thermistor has time to cool and the voltage across the thermistor follows the characteristic determined by ambient temperature, dissipation, and material constant. However, for an alternating current this is not the case, and when the current has already dropped to a lower value, the temperature decrease is lagging, resulting in a dynamic characteristic with lower voltage in one half-cycle (decreasing current) and higher voltage in the second half-cycle (increasing current) compared to the static values. With increasing frequency the long axis of each elliptic dynamic characteristic tends to rotate counterclockwise around the center point on the static characteristic until the voltage is practically in phase with current (high frequency).[682]

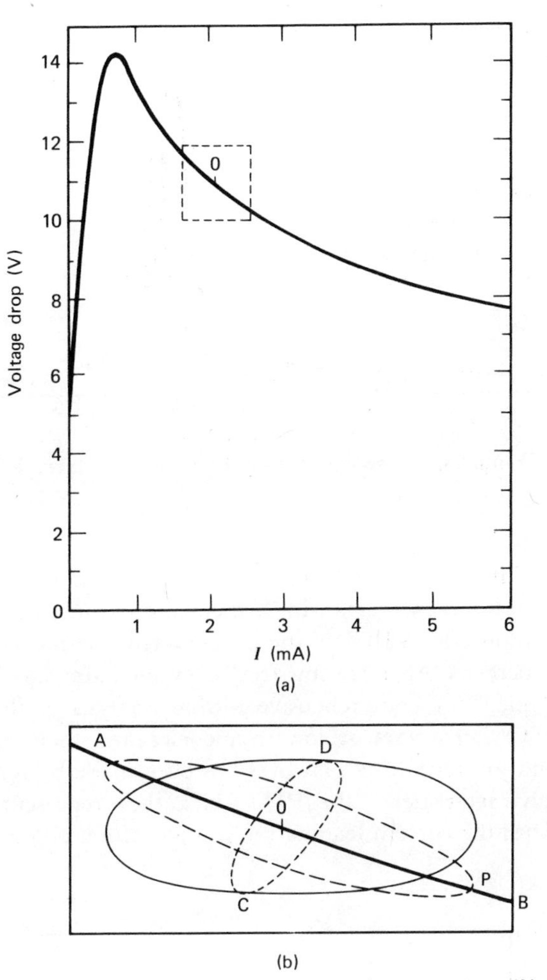

FIGURE 147. Thermistor characteristics (*a*) static characteristics, (*b*) dynamic characteristics. After Rasmussen.[682]

Rasmussen has used the apparent inductance of a Veco 51A4 thermistor (as shown in Figure 149) with a parallel capacitor and measured the transmission response of such a circuit, which acts as a band filter seen in Figure 148, for a range between 10^{-2} and 10^{-1} Hz with Q values of 14–24. An upper limit of 40 can be expected if the ambient temperature fluctuations cannot be completely eliminated. The apparent inductance of the thermistor is shown in Figure 149 as a function of frequency.

Figure 149 gives the entire impedance diagram of the thermistor starting at a negative resistance of 1600 Ω at 10^{-2} Hz, rising to a maximal reactance

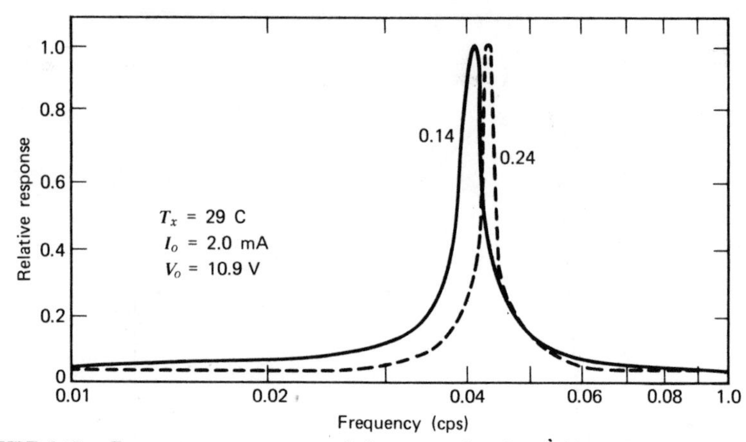

FIGURE 148. Transmission response of thermistor bandpass filter. Results of Rasmussen.[682]

of ~3000 Ω, which drops again to a much lower value at 0.7 Hz. The linearity of such filters increases with higher bias and smaller signal current. For a signal amplitude ≦10% of the bias current, the harmonic distortion is only a few percent. An ultra low-frequency oscillator has been designed on this principle.[652] A comprehensive formal analysis of the behavior of NTC and PTC thermistors at low frequencies and their application for integrating and phase-shifting networks has been made by Kraus.[653] Due to the inverse characteristic of the PTC units, they represent an apparent capacitance with the current leading the voltage with a phase angle depend-

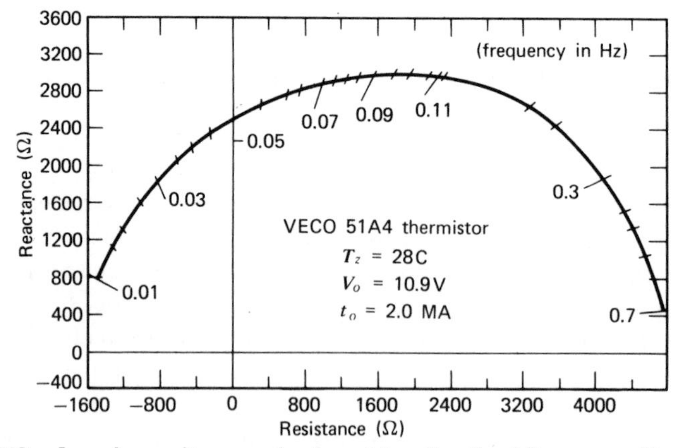

FIGURE 149. Impedance diagram of a thermistor. Results of Rasmussen.[682]

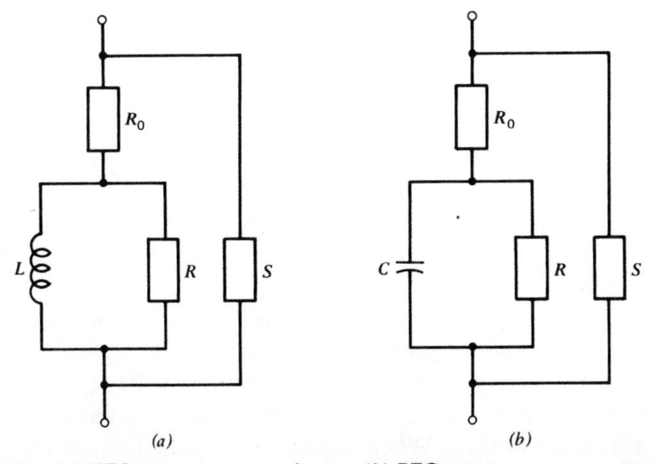

FIGURE 150. (*a*) NTC resistor as an inductor. (*b*) PTC resistor as a capacitor. (Equivalent circuits.) After Kraus.[653]

ing on frequency and their resistance. For comparison, the equivalent circuits of NTC and PTC thermistors are shown in Figures 150(*a*) and 150(*b*) in which R_∞ and R_0 are the resistance values at high and low frequency (dc). The corresponding impedance diagrams for Philips thermistors B 832003 P/1KS (NTC) and E 220ZZ/03 (PTC) are given in Figure 151 between dc and 8 Hz, the latter being nearly characteristic for R_∞.

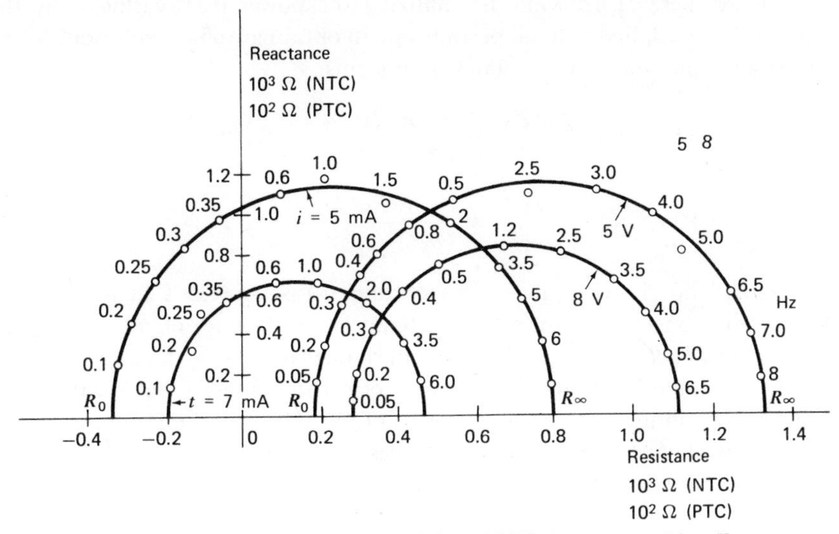

FIGURE 151. Impedance diagrams of NTC and PTC thermistors. After Kraus.[653]

PTC–NTC Networks as Low-Frequency Oscillators. Adopting the concept that a PTC and an NTC thermistor can be represented by the equivalent circuit of either a capacitance or an inductance [Figures 150(*a*) and 150(*b*)] for both in series with (Figure 152) a power source, "the well-known equation of the complete ohmic law for ac would apply." The peak-to-peak current value would be

$$I_{\text{peak-to-peak}} = \frac{2E}{[R^2 + (L\omega - 1/C\omega)^2]^{1/2}}$$

and for compensation of the opposing phase angles of inductance and capacitance, an oscillation, with the period $t = 2\pi(LC)^{1/2}$, would be possible.

A simple phenomenological explanation of the oscillation can be derived from Figure 153 when considering the temperature of the NTC and PTC units as a function of time. When voltage E is applied, the NTC has a high resistance and absorbs the major voltage drop and power input. This heats it up, increasing the total current in the circuit. At the same time shifting the major voltage drops gradually to the PTC unit, thus increasing its resistance, and decreasing the total current. This again raises the resistance of the NTC unit at the expense of the PTC unit, which has then a chance to cool and return nearly to its low initial resistance. Now the cycle can start again. The effect has already drawn attention from several investigators.[654,655] A rigorous mathematical analysis of this cycling taking in account the opposing terms of Joule heating and dissipation for variable temperatures of the NTC and PTC units has been made by Reenstra.[656]

Plotting these terms while normalizing the power by dividing it by the square of the applied voltage permits one to obtain graphical solutions to the algebraic equation that correlates both terms:

$$\text{NTC:} \qquad C_N \frac{dT_N}{dt} = \frac{E^2 R_N}{(R_p + R_N)^2} - D_N(T_N - T_0)$$

$$\text{PTC:} \qquad C_p \frac{dT_p}{dt} = \frac{E^2 R_p}{(R_p + R_N)^2} - D_p(T_p - T_0)$$

C, D, R, and T with the corresponding suffix represent heat capacity, dissipation constant, resistance, and temperature of both elements. For the static case $dT/dt = 0$. T_0 is the ambient temperature in both cases.

The dynamic behavior is, of course, of greater interest but creates difficulties in handling the nonlinear terms of the differential equations. Therefore a digital computer method was necessary to determine the currents and voltage as function of time. The solutions were verified by an experimental PTC–NTC network with known resistance-versus-temperature characteristics, dissipation constant, and heat capacity of both elements. The

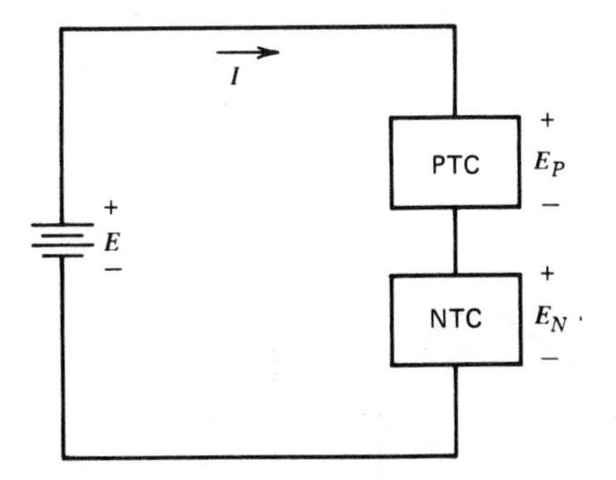

FIGURE 152. A two-element network that acts as a switch or oscillator depending on the thermal and electrical properties of the thermistors. After Reenstra.[656]

found boundary conditions for self-starting and stable oscillations agreed reasonably well with the computed data (Figures 154 and 155). The influence of the dissipation constant is asymmetrical. Its increase by 20% changes the period from 34 to 64 sec; its decrease has only a minor effect. Such an oscillator has been successfully used as proportional temperature controller.[708]

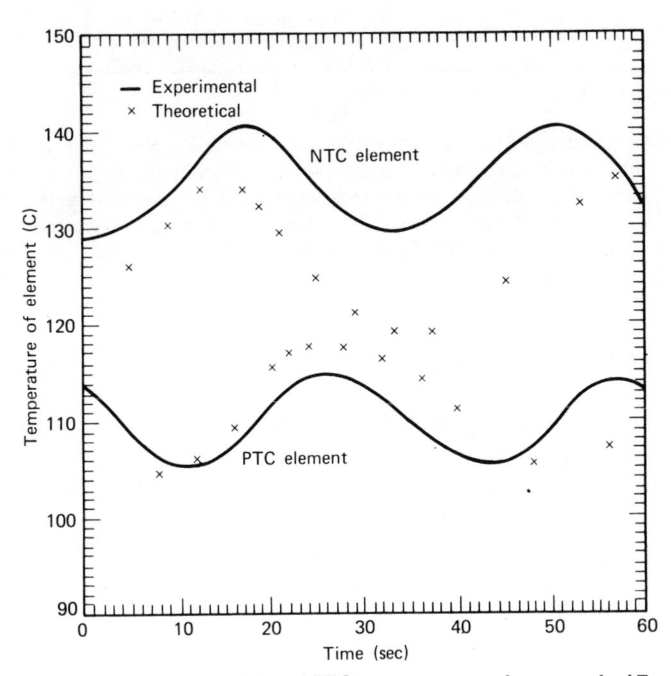

FIGURE 153. Temperature of NTC and PTC components in the network of Figure 152 as function of time. Results of Reenstra.[656]

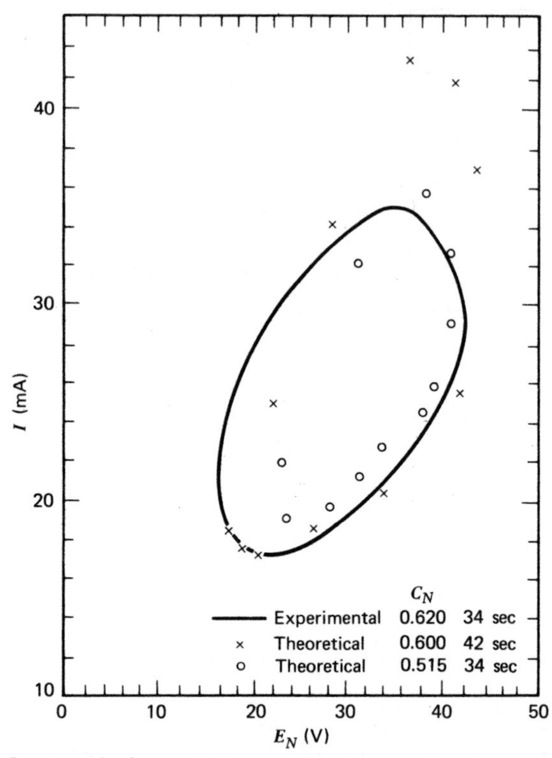

FIGURE 154. Limit cycle for oscillations in the I versus E_N plane. C_N is the thermal capacitance of the NTC thermistor; t is the period of oscillation. Improved agreement is obtained for an NTC thermal capacitance of 0.515 J K^{-1} or 86% of the measured 0.600 J K^{-1} value. The effective thermal capacitance appears to be reduced by current channeling in the NTC thermistor. Results of Reenstra.[656]

7.2.12. RF Microwave Power Meters

Thermistors have been found to be a useful substitute for the hot wire or thermocouple in power-measuring devices. They can be made very small, are less sensitive against short overloads, and their capacitance and inductance can be minimized by appropriate shape and design. The heating effect at low power input can be obscured by fluctuations of the ambient temperature. Assuming a dissipation constant of 1 mW K^{-1}, a minimum power input of 20 mW would be necessary to keep the signal-to-noise ratio above 10. A practical approach to increasing the signal-to-noise ratio at lower input is to operate the sensing thermistor in a dc or ac biased bridge at elevated temperature. This decreases its intrinsic sensitivity

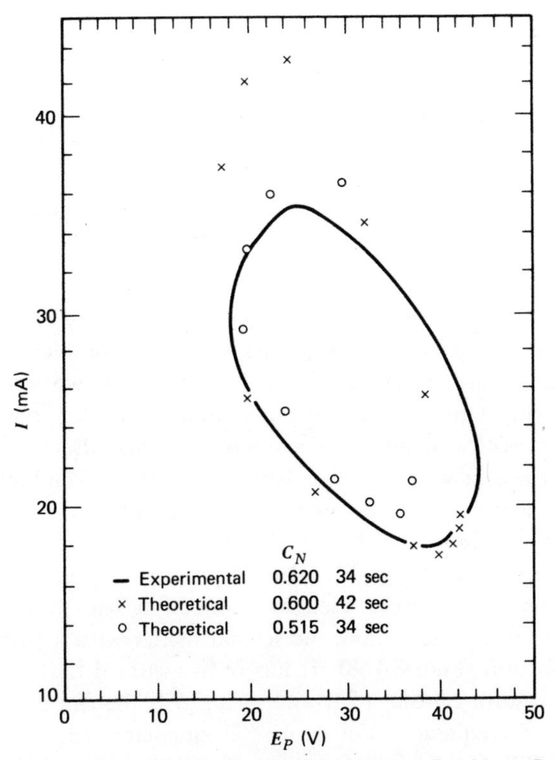

FIGURE 155. Limit cycle for oscillations in the I versus E_p plane. C_N is the thermal capacitance of the NTC thermistor, t is the period of oscillation. Improved agreement is obtained for an NTC thermal capacitance of 0.515 J K^{-1} value. The effective thermal capacitance appears to be reduced by current channeling in the NTC thermistor. Results of Reenstra.[656]

slightly ($\sim 1/T^2$), but reduces the influence of ambient fluctuation considerably. As an example, the intrinsic sensitivity decreases from 300 to 400 K by 45%; the influence of ambient fluctuations, however, is reduced to $\sim$1%, compounding the signal-to-noise ratio increase to a factor of 50.

Power measurements over a wide range of ambient temperatures require the addition of other thermistors as compensating circuits elements only responding to temperature changes. Indirectly and self-heated thermistors have been used in power meters. Thermistors power meters have found wide acceptance.[657–661]

A typical example is the Wood rms milliammeter with a frequency range from zero to 10 MHz, which has an accuracy of ±2% up to 7 MHz and an error of −4% at 10 MHz for full-scale deflection of 10 mA and a time

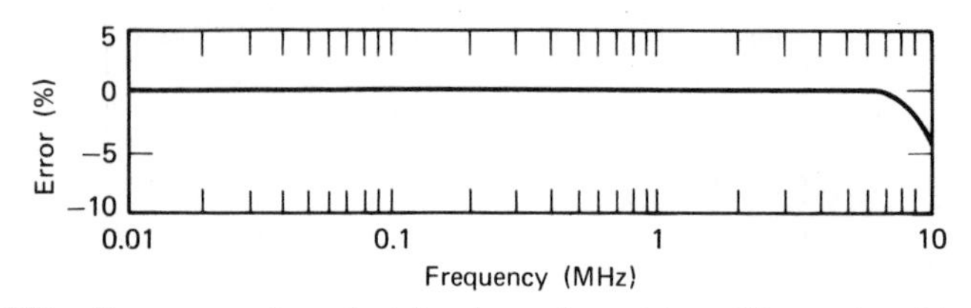

FIGURE 156. Frequency characteristic of rms thermistor milliammeter. After Wood.[659]

constant of ~2 sec (Figure 156). Its overload capacity is >400% and it lends itself to measure high frequencies with any complex wave form. In Figure 157 the basic circuit is shown. It consists of two indirectly heated thermistors with equal cold resistance and matching resistance-versus-power characteristics in adjacent arms of a bridge. Each of them is surrounded by a heater with 100 Ω resistance and 0.5 μH inductance. The ac current to be measured is fed into the heater of TH_1. The resulting unbalance of the bridge can either be measured directly or balanced by a known dc power fed to the heater of TH_2. In the practical design, thermistors with 120 mW power rating were used in connection with a high-gain direct coupled amplifier that is feeding the second heater with a current to restore bridge equilibrium (Figure 158). It has been assumed that the resistance of a thermistor resulting from a certain ac input is the same as for dc and is independent of frequency. For higher frequencies this condition is less likely to be fulfilled, especially for the oxide thermistors with polycrystalline structure where contact resistance cannot be ignored and grain boundary

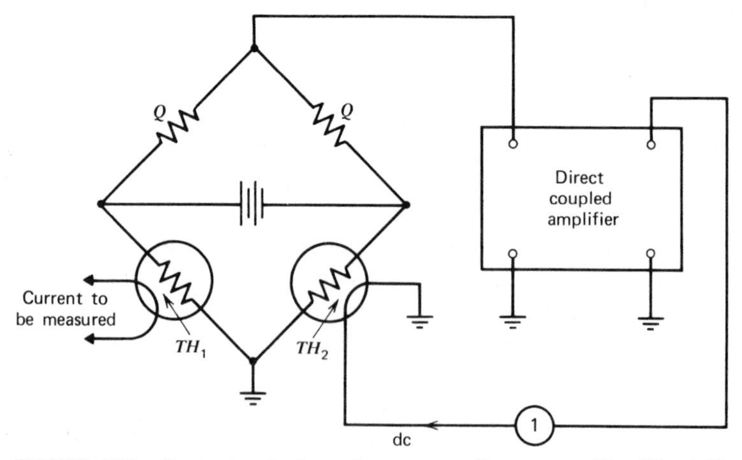

FIGURE 157. Basic circuit of rms thermistor milliammeter. After Wood.[659]

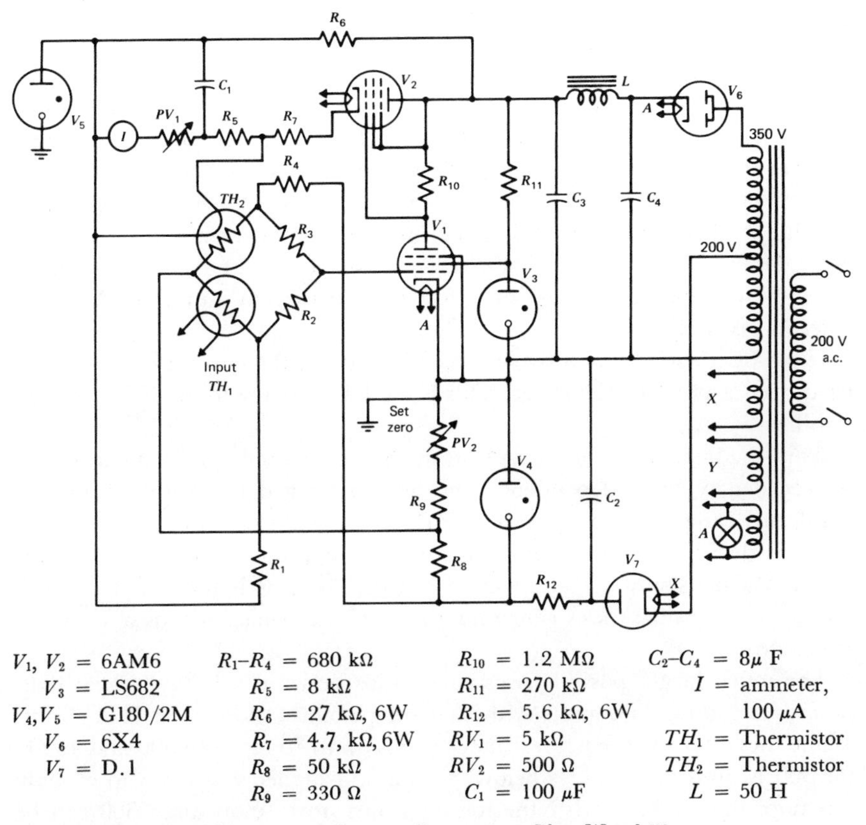

FIGURE 158. Complete circuit of rms milliammeter. After Wood.[659]

capacities can play an important role. The power fraction β dissipated in a bead thermistor is given by

$$\beta = \frac{1}{1 + \omega^2 C^2 R r}$$

where r is the lead resistance of the thermistor R using 25 μm 70 PT–30 Ir wire estimated as ~2.5 Ω. It enters into the consideration by its skin effect.[662] It was found that at frequencies above 10^4 MHz, β starts to drop. At 3×10^4 MHz only 50% of the total power is dissipated in the bead.

Thermal detection of EPR signals with bead thermistors of 800 Ω nominal resistance used in a differential system and introduced into the cylindrical microwave cavity at the site of maximal field was successful. The unbalance of the bridge operated at 100 KHz was used as input of a lock-in amplifier.[663]

7.2.13. Selected PTC Applications

Many PTC applications are analogs of those known for NTC units: temperature measurements and control, timing devices, level and flow control. Since the *V–I* characteristics of PTC and NTC are rotated 90° against each other (Figure 81), the voltage peak becomes a current peak and the negative branch of the voltage characteristic a slope of decreasing current. The conditions of nonuniform self-heating and large temperature gradients also exist. However, in this case the center heats faster and eliminates itself from the conduction process first, forcing the current into an surface skin (thermal skin effect).

If self-heating is limited by suitable dissipation, the current can be stabilized over a wide voltage range. This is the PTC analog to the NTC voltage stabilizer.

As in the latter case for current, in the first case a voltage limit exists. If it is exceeded, voltage dependence and even electrical breakdown limit the regulating range, especially since in PTC units the behavior of intergranular zones is decisive. Also in timing devices the opposite function of PTCs is apparent: in its simplest application as series resistor it turns a circuit off, not on. All considerations concerning time and dissipation constant remain the same.

The much larger possible resistance ratios between suitably chosen temperatures endow PTC units with switching ability, more than NTC units. While the latter may reach resistance ratios up to 15 between 300 and 400 K for 300 K resistance values below ~1 kΩ and higher values up to 60 only for more than 1 MΩ, with the former ratios up to more than 300 can be attained for the same temperature interval. This opens up possibilities far beyond those known for thermistors. Although they mainly fall under mode 2, they will be treated together with other PTC applications. To deviate in this case from the normal system appears to be excusable, considering the relative novelty of the PTC field.

Protective Devices

Overheating. If a rising ambient temperature would cause overheating of load resistance R under power (motor windings), the temperature influence on the PTC characteristics can be used to reduce the current through R. While the low temperature V characteristics have two stable intersections with the load line of R, at the high temperature only one stable point of intersection, P_2 exists, limiting the current to a permanent, rather a low value, which is retained even after the ambient temperature has dropped. To reactivate the circuit, it has to be switched off and on again. The PTC unit acts as a reusable fuse as in Figure 159(*b*). The technical details on this

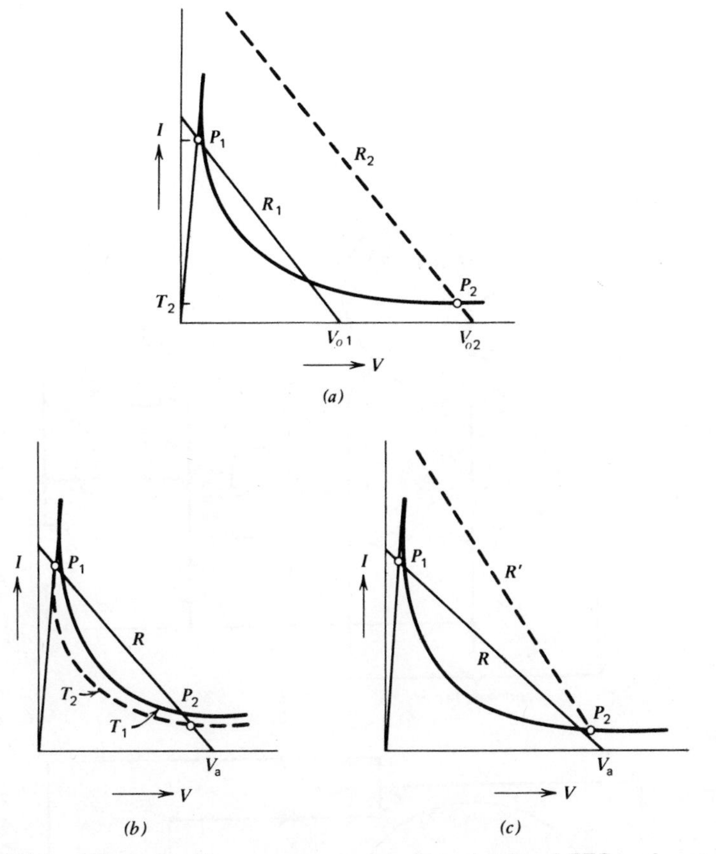

FIGURE 159. Effect of voltage versus current characteristic of PTC and series load resistance (*a*) by increasing voltage, (*b*) by raising the temperature ($T_2 > T_1$), (*c*) by reducing the load resistance. After Andrich and Haerdtl.[669]

application have been discussed for different levels of load, voltage (relative to rated value), and cooling, in running and locked rotor condition for 1- and 3-phase motors up to 10 hp.[664,665]

The obvious advantage of using these overheating sensors is that they directly monitor the temperature of the stator winding and not indirectly by changes of current as the thermal overload relays using the bimetal principle. Typical circuit applications are shown in the following examples (Figure 160).[666] Overheating raises the resistance of three PTC units in series with relay CR and an auxiliary dc source of 30 V supplied from a transformer and two silicon diodes. If the PTC resistance of $\approx 200\ \Omega$ at normal operating temperatures increases to $> 1500\ \Omega$, the series relays drops out and shuts

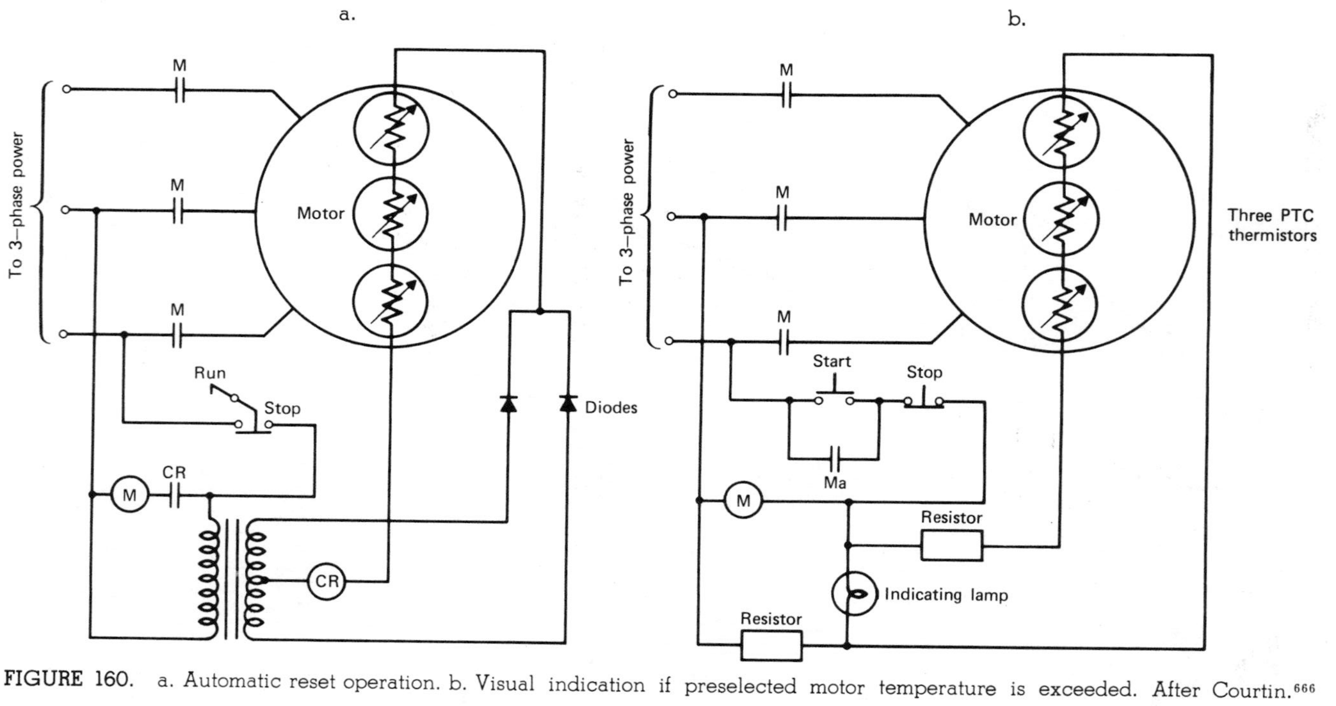

FIGURE 160. a. Automatic reset operation. b. Visual indication if preselected motor temperature is exceeded. After Courtin.[666]

down the motor at ~16 mA. The permanent load at the PTC units in this condition is ~0.4 W. In order to restart the motor by button, the windings must cool at least 15 K below the trip temperature. Early warnings of overheating can be accomplished simply by connecting a glow lamp across the PTC units and selecting the additional fixed resistors such that they fire at a desired overheating temperature safe enough to take prevention steps by reducing the load or by other means. The thermal inertia of the PTC units, although minimized by their small mass and good thermal contact with the winding, has still to be considered for momentary high overloads by stalling the motor. If this happens after a cold start of a 30-hp motor, 12 sec were necessary to trip the protective relay—compared to only 4 sec for a warm motor. The peak stator temperature principle using this protective (Guardistor) decreased from 450 to 410 K for hp ratings between 1 and 200.

The use of thermistors has been analyzed in comparison with other types of overload protectors with special attention to the inertia of thermistor-winding systems.[667]

Increased Line Voltage. A load resistance (motor, appliance) rated for 110 V requires a series resistance if connected to higher line voltages. Instead of manual connecting of this series resistance, it remains permanently in the circuit, but is shunted by a PTC unit. At the lower voltage it would permit a shunt current I_1 at the first intersection of the characteristic with the load line R_1. If the voltage is raised, the load line R_2 determines another intersection with the characteristic at a current I_2 ($I_2 \ll I_1$) and the operating current is mainly determined by the fixed series resistor that has been selected to absorb the excess line voltage.

Overload Resistance [*Figure* 159(*a*)]. If the load decreases by the operating conditions, for instance, mechanical braking of a motor, a PTC unit in series with the load normally operating with load resistance R_1, at a current I_1, would reduce its current to much smaller I_2, if the new load line R_2 is drawn [Figure 159(*c*)].

Other Self-Heating and Surge Applications. A PTC unit can be operated as a thermostat if it is self-heated by application of a suitable voltage to a temperature region where the resistance-versus-temperature characteristic is very steep. Fluctuations of the ambient temperature result in compensating variations of power input. Thermostats based on this principle have been developed with a stability of 0.5 K for a temperature change of 10 K. Better stability (±0.05 K between 240 and 330 K) has been accomplished by a thermostat using the semiconductor–metal transition of VO_2.[668]

Soldering irons have the unpleasant property of overheating when idling on line voltage, since the power input is designed to compensate for the heat dissipation during the soldering process. The power input can be automatically reduced and matched to the variable heat dissipation, if a PTC thermistor is used as heating element. Pb-substituted $BaTiO_3$ with higher Curie temperature ($Ba_{0.5}Pb_{0.5}TiO_3$) has been suggested for this purpose in the form of 0.1-cm-thick rectangular slabs.[669] For example, if the resistance of the PTC heater increases at a given voltage by a factor of 10, its power input drops by the same factor, thus minimizing the overheating risk. The initial power input is chosen in such a way that it heats the solder iron fast enough (in less than 20 sec) to soldering temperatures of 480–580 K. These temperatures will be maintained with a stationary input power of $\sim 50\%$ of the initial value, which of course has to be higher for a fast start. Soldering irons of this type have been developed for maximal heating powers of 30 W cm^{-2} with idling power of less than 30% and life times of >3000 hr. Similar applications have been considered for hot-plates operating continuously on line voltage. Although these and similar applications, where the PTC thermistor is used as solid-state switching element, look at first quite attractive, they also present some problems. Contrary to the NTC case, where the thermistor starts with low power, the PTC unit has to take in stride a power surge that not only imposes high internal heat stresses, but also results in heat expansion stresses against the contacts and their mechanical parts. To take a nonelectrical analog, it would be the same as quenching an ordinary ceramic with a linear thermal expansion of $\approx 10^{-5}$ from a temperature of several hundred degree into water.

Similar conditions exist in another switching operation: automatic degaussing of color television sets, which has become a favorite application for PTC thermistors. In this case conditions are still more severe than for the self-regulated soldering irons. While for the latter a residual load of ≈ 15–30% was acceptable to serve this purpose, in the degaussing case the residual input power of the thermistor should not exceed 3–4 W, while the input surge is of the order of 600 W. This represents a power ratio of ≈ 200 between peak and idling compared to 3–6 for thermal appliances. This ratio could be reduced by the selection of a degaussing coil with higher impedence and a large number of turns, thus reducing the required demagnetizing current peak from 6 to 8 and to lower values. However, economical considerations shifted the burden of the degaussing on the PTC thermistor.

For low-frequency oscillations generated in PTC- NTC-networks, see Section 7.2.11.

Among the many other possible PTC applications in mode 1 the long-time stabilization crystal-type generators used in carrier-frequency systems deserve special attention. The annual crystal-thermostat drift could be reduced to ≤ 50 mK.[670]

7.3. APPLICATIONS WITH INDIRECTLY HEATED THERMISTORS

7.3.1. Automatic Gain Control in Lamp-Photocell Transducers Systems

It is important to compensate the inevitable downdrift of lamp intensity, especially if mechanical vibrations are measured with a modulated light beam. Using up to six series connected indirectly heated thermistors as the load resistor of a photocell KMV6 stabilizes the transducer system within 0.03% for symmetrically modulated light. This is accomplished by comparing the mean component of voltage with a reference voltage and using the difference, after amplification, to control the heater current in the indirectly heated bead thermistors in the evacuated glass envelopes (Types B 5513/60 and B 2552/60 Standard Tel. Cables, England). Compensation takes place simultaneously for lamp drift and changes of photocell characteristics.[671]

7.3.2. DC–AC Converter

Externally heated thermistors can be used in circuits, where conversion of dc into ac voltage or stepping up a dc voltage is desirable

Figure 161 shows how this purpose can be accomplished. The ac power is feeding a bridge circuit R_1–R_4. R_2 is variable and R_3 is a thermistor externally heated by a heating element controlled with a dc input. The bridge, initially balanced with dc input zero, produces an ac signal as a function of the dc input at the heater of R_3. The lower amplitude of this ac signal compared to that of the ac power supply can be compensated by a step-up transformer. Introducing a rectifier with filter into the converter output makes this circuit a dc–dc converter. Even thermistors without external heater can

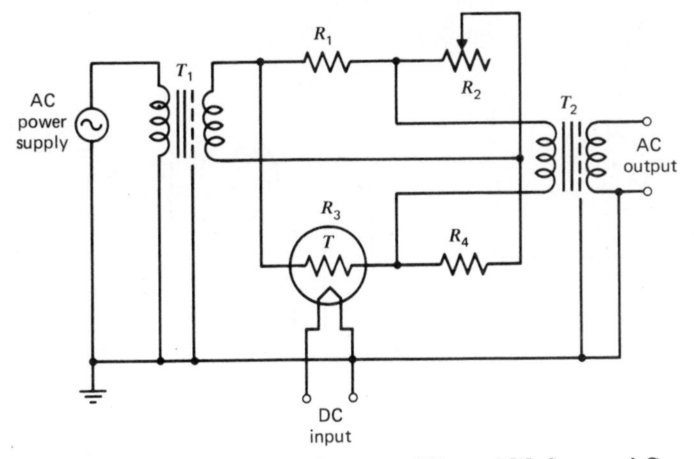

FIGURE 161. Dc versus ac converter. Courtesy of Howard W. Sams and Co.

be used as converter, for instance by capacitive coupling to an rf supply. Bell[672] used a VE 43 Al thermistor immersed into mercury to obtain a time constant of the order of 10 msec and a total resistance decrease by a factor of 8 during rf application and suggested how to apply this principle to convert small dc signals to ac in remote locations.

7.3.3. Phase Shifting

The phase angle θ of an ac input can be shifted by a single or cascaded RC network as shown in Figure 162 according to

$$\tan \theta = \frac{1}{2\pi f RC}$$

Using NTC or PTC thermistors as the resistive component, the phase angle θ changes with their temperature-dependent resistance and can therefore be controlled either by ambient, self-, or indirect heating. Whenever phase conditions in an ac circuit are important, for instance, in an ignitron, thyratron, or SCR, this method of phase shifting has merits. For rf circuits it could have adverse effects that are undesirable because of their relation to random temperature fluctuations and their effect on the relative gain in amplifiers.

For all practical purposes, temperature-controlled phase shifting would be accomplished with indirectly heated thermistors—except for cases where the phase shift is used as the operating principle of a temperature transducer. The quantitative relationship between phase shifting angle and temperature is easily derived by introducing the basic equation correlating resistance and temperature into the definition of the equation for tan θ.

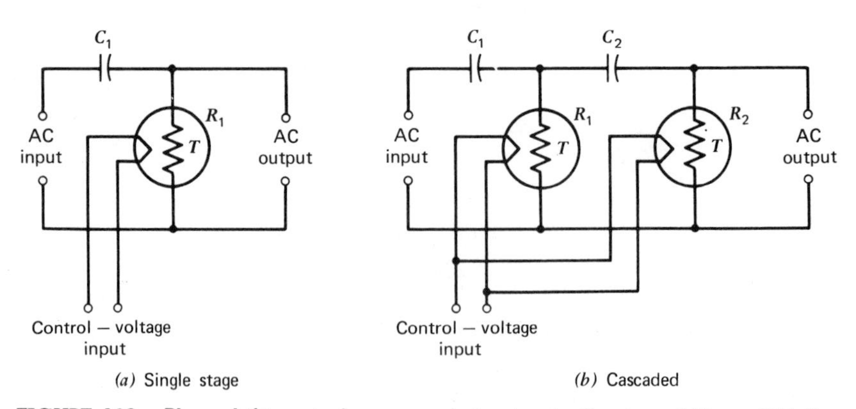

FIGURE 162. Phase shifting single or cascaded network. Courtesy of Howard W. Sams and Co.

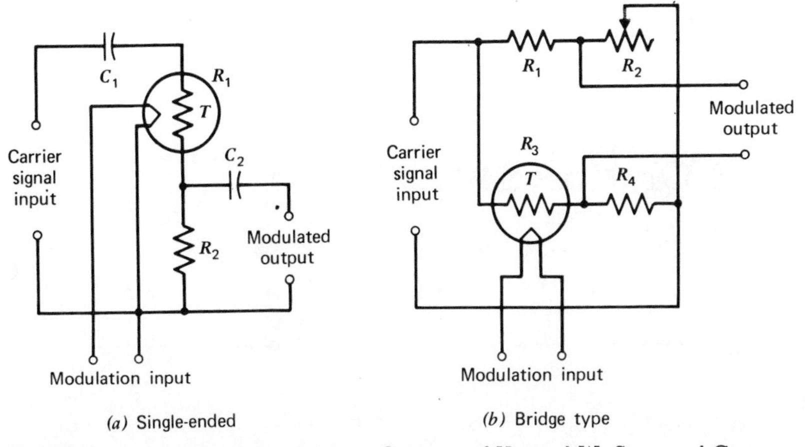

(*a*) Single-ended (*b*) Bridge type

FIGURE 163. Amplitude modulators. Courtesy of Howard W. Sams and Co.

7.3.4. Amplitude Modulation

It is possible to modulate a carrier frequency with a modulating frequency low enough so that it can produce temperature (i.e., resistance) fluctuations of a thermistor in the modulator. The modulation can be accomplished by applying the modulating input to the heater of an indirectly heated thermistor (Figure 163). In this case not only the inertia of the heater but also of the heat transfer to the thermistor limits the modulating frequency. Somewhat better would be a direct (self-heating) of the thermistor by the modulating input. Since the carrier current also flows through the thermistor, the carrier and modulating frequencies must be different enough that their circuits can be separated by suitable blocking capacitors. It is always desirable to minimize the carrier amplitude if no modulation occurs. This can easily be accomplished in a modulating bridge that is balanced for modulation input at zero and therefore results in an output of zero. With increasing modulating input, the bridge becomes unbalanced thus permitting increased carrier output with increasing modulation amplitude. The options for indirect or direct thermal coupling of both circuits remain the same.

List of Patents

No.	Dates		Topics
	Priority	Issue	
I. U.S.A.			
2,082,102		June 1, 1937	Sintered silver sulfide (Ag_2S)
2,081,894		May 25, 1937	Sintered Co-Ni oxide
2,091,259		Aug. 31, 1937	Ag_2S produced by reaction of S with Ag
2,079,690		May 11, 1937	Ag_2 thermistor film on substrate
2,178,548		Nov. 7, 1940	Indirectly heated Ag_2S rod
2,184,847		Dec. 26, 1940	Boron with 2 Pt wire leads
2,219,365		Oct. 29, 1941	Sintered Ni-Mn oxide
2,258,646		Oct. 14, 1942	Sintered Mn-Fe-Ni oxide
2,264,834		Dec. 2, 1942	Tellurium
2,264,073		Nov. 25, 1942	Ag-Te alloy
2,278,072		Mar. 31, 1942	Sintered Cr oxide
2,282,944		May 12, 1942	Sintered Mn-Ni-Cu oxide
2,294,756		Sept. 1, 1943	CuO and Cr_2O_3 sintered in air or O_2
2,298,679		Oct. 13, 1943	Sintered Mn-Ni-Co-Cu oxide
2,276,864		Mar. 17, 1942	Uranium oxide film thermistor
2,329,511		Sept. 14, 1944	Sintered Mn-Cu oxide
2,414,793		Jan. 28, 1947	Sintered oxide flake

No.	Dates		Topics
	Priority	Issue	
2,462,162		Feb. 22, 1949	Metallic oxides sintered in controlled O_2 atmosphere
2,475,864		July 12, 1949	Sintered MgO-AC_2O_3 spinel
2,487,279	Dec. 24, 1946	Nov. 8, 1949	Bead thermistor with large heat dissipation to generate alternating currents between 1 and 100 Hz
2,492,543	July 24, 1946	Dec. 27, 1949	Spinel phase consisting of MgO + FeO: Al_2O_3 + Fe_2O_3 with 5 and 10% MgO and FeO, respectively
2,524,611		Oct. 3, 1950	FeO, Fe_2O_3, MgO, and Al_2O_3 in which the total amount of Fe oxides is 50 mol. % or less
2,511,216		June 13, 1950	O-deficient TiO_2
2,616,859		Nov. 4, 1952	Fe-Ti oxide sintered in N_2
2,633,521		Mar. 31, 1953	Films of Mn-Ni-Co, Cu, Fe and Zn oxide
2,636,012		Apr. 21, 1953	Flake thermistors by oxidation of Ni-alloys
2,645,700		July 14, 1953	Ni-Mn-Fe oxide sintered at 1473 K
2,649,424		Aug. 18,1953	Organic thermistor
2,694,050	Sept. 1, 1949	Nov. 9, 1954	56 Mn, 24 Fe, 20 NiC at. %, sintered in air 1573 K
2,697,028		Nov. 11, 1954	Sintered steatite impregnated with Cr, Mn, Fe, Ni salts
2,700,720	Dec. 15, 1948	Jan. 25, 1955	Cr oxide with specified upper limits of other oxides Ba 8, Fe 32, Ni 7, Th 50, Ti 8, W 17, V 8, Zr 16 (%)
2,703,354	Feb. 23, 1950	Mar. 1, 1955	Alkaline earth manganate $x\,MnO_3$ (x = Mg, Ca, Sr, Ba)

No.	Dates		Topics
	Priority	Issue	
2,714,054		July 26, 1955	Cr_2O_3 sintered in N_2 and O_2
2,717,299		Sept. 6, 1955	Temperature-dependent resistor
2,720,573	June 27, 1951	Oct. 11, 1955	Cu-Mn, Ni-Fe oxide sintered between alloyed Fe electrodes
2,720,471		Oct. 11, 1955	Mn, Ni, Co, Fe, Cu oxide with Ti, V_1-Cr, Al and Si oxide
2,728,836	June 7, 1951	Dec. 27, 1955	Conductive glass
2,735,824	Dec. 19, 1947	Feb. 21, 1956	Fe_2O_3 + 1 mole % TiO sintered in air
2,736,784		Feb. 21, 1956	Thermistor probe
2,736,858		Feb. 28, 1956	Intermetallic crystals as temperature sensor
2,740,031		Mar. 17, 1956	Bead thermistor
2,761,849	Dec. 27, 1950	Sept. 4, 1956	Warm pressed plastic composite resistor containing semiconducting metal oxides
2,806,200	Dec. 17, 1952	Sept. 10, 1957	Temperature compensation with thermistors
2,816,023	Nov. 2, 1955	Dec. 10, 1957	Al-Sb doped with Ta
2,837,618		June 3, 1958	Sb_2Se_3 + As_2Se_3 from melt
2,841,508		July 1, 1958	Ba, Sr, Ca-titanate Ba, Mg-cerconate doped with rare earth
2,940,941	May 20, 1953	June 14, 1960	Semiconducting, O-deficient TiO_2
2,970,411	July 1, 1958	Feb. 7, 1961	Thermistor standardization
2,976,505		Mar. 21, 1961	Ba(SrPb) TiO_3 Ce- or Y-doped
2,977,558	June 19, 1958	Mar. 28, 1961	Indirectly heated thermistor
2,981,699	Dec. 28, 1959	Apr. 25, 1961	Y- or Ce-doped $Ba(Ti_1ZrO_3)$
2,989,482		June 20, 1961	Ba(Pb)-titanate with Li additions

No.	Dates		Topics
	Priority	Issue	
2,989,483		June 20, 1961	Titanate ceramic
3,016,506	Feb. 1, 1960	Jan. 9, 1962	Mn-Co oxide bead with Ni Cr-Fe alloy leads
3,065,532	Apr. 22, 1958	Nov. 27, 1962	Contacting for high temperatures
3,068,438	Feb. 17, 1960	Dec. 11, 1962	Multiple-resistance thermistor consisting of Mn-Cu oxide compositions
3,071,522	Oct. 30, 1958	Jan. 1, 1963	Contacting by electroless plating
3,109,097	Sept. 28, 1961	Oct. 29, 1963	Thermistor, bolometer
3,140,531	Feb. 20, 1964	July 14, 1964	Au-doped Si
3,198,012	Mar. 29, 1961	Aug. 3, 1965	Ga As with 0.75 eV band gap
3,209,435	Feb. 23, 1962	Oct. 5, 1965	PTC thermistor
3,219,480	June 29, 1961	Nov. 23, 1965	Thick-film thermistor of Mn-Co-Ni-Cu oxide
3,221,393	Sept. 5, 1961	Dec. 7, 1965	Manufacture of bead thermistor
3,231,522	Sept. 26, 1963	Jan. 25, 1966	(BaPb)-titanate doped with rare earth
3,239,785		Mar. 3, 1966	Organic semiconductor
3,267,403	Sept. 4, 1963	Aug. 16, 1966	Infrared Ge sensor
3,284,418	Feb. 1, 1963	Nov. 8, 1966	Organic thermistor
3,292,129	Oct. 7, 1963	Dec. 12, 1966	*N*-type Au-doped Si
3,312,572	June 7, 1963	Apr. 4, 1967	Thin-film Ge or Si bolometer
3,313,004	June 14, 1965	Apr. 11, 1967	Device to measure resistance of metal oxides at high temperature and pressures
3,316,765		May 2, 1967	Temperature sensing with linear ranges
3,317,354	May 28, 1964	May 2, 1967	Doping of natural diamonds in gaseous electrical discharge
3,329,917	July 26, 1965	July 4, 1967	Boron chip with B-alloyed Pt wire

No.	Dates		Topics
	Priority	Issue	
3,341,473	Feb. 16, 1966	Sept. 12, 1967	SnO_2-TiO_2 doped with Sb_2O_3 or Ta_2O_5
3,351,568	Apr. 13, 1964	Nov. 7, 1967	PTC material La-doped $BaTiO_3$
3,359,632	Feb. 10, 1965	Dec. 26, 1967	Thick-film flake thermistor
3,364,565	Feb. 10, 1965	Jan. 23, 1968	Film thermistor on Ni strip
3,372,469	Oct. 28, 1963	Mar. 12, 1968	Low-resistance bonds
3,377,561	Oct. 15, 1964	Apr. 9, 1968	PTC $BaTiO_3$ thermistor
3,393,448	Dec. 22, 1965	July 23, 1968	Molten bead of $FeTiO_3$ + TiO_2
3,402,131	July 27, 1965	Sept. 9, 1968	Doped VO_2 bead with resistivity jump
3,408,311	Sept. 29, 1966	Oct. 29, 1968	Low temperature coefficient material (Co-Mn oxide containing Pd and Ag)
3,414,861	Mar. 30, 1966	Dec. 3, 1968	Bead thermistor fast response
3,417,032	Dec. 10, 1965	Dec. 12, 1968	Alkali niobate or tantalate doped with Ba, Sr, Ca, Pb
3,430,336	Sept. 15, 1965	Mar. 4, 1969	Attaching leads using enamel
3,435,398	Apr. 19, 1966	Mar. 25, 1969	Semiconducting boron nitride
3,435,399	Apr. 19, 1966	Mar. 25, 1969	Semiconducting synthetic diamond
3,455,724	Apr. 7, 1964	July 15, 1969	Vanadium suboxide film on substrate
3,469,224	Dec. 8, 1966	Sept. 23, 1969	Film thermistor with fast response
3,479,631	Dec. 22, 1965	Nov. 18, 1969	Fused bead of Fe_2O_3-TiO_2
3,483,140	Mar. 18, 1966	Dec. 9, 1969	Solid solution of AgI + CuI in Ag_2S + Au_2S
3,503,030	Nov. 11, 1966	Mar. 24, 1970	Indirectly heated thermistors
3,511,786	Aug. 22, 1967	May 12, 1970	Sr-titanate ferrates
3,526,541	Dec. 23, 1966	Sept. 1, 1970	Thin-film contacts
3,547,835	June 1, 1965	Dec. 15, 1970	Conductive silver coating

No.	Dates		Topics
	Priority	Issue	
3,568,125	Oct. 20, 1967	Mar. 2, 1971	Indirectly heated Ge thermistor
3,586,534	Dec. 5, 1966	June 22, 1971	Electroless Ni contact
3,598,764	Dec. 9, 1968	Aug. 10, 1971	CeO_2-ZrO_2 high-temperature thermistors
3,610,023	Nov. 14, 1969	Oct. 5, 1971	Gas analysis with thermistor sensor
3,642,527	Dec. 30, 1968	Feb. 15, 1972	Surface doping of $BaTiO_3$
3,645,785		Feb. 29, 1972	Laminated contact
3,652,463	Jan. 26, 1970	Mar. 28, 1972	Co-Zn-Mn spinel of improved stability
3,654,785	Nov. 12, 1969	Feb. 29, 1972	Multilayer ohmic contact on semiconducting on $BaTiO_3$ using Ge, Au, Ni, and Pd
3,656,029	Dec. 31, 1970	Apr. 11, 1972	Bistable resistor made of europium oxide, europium sulfide, or europium selenide doped with a 3d transition or Group VA element
3,660,155	Apr. 15, 1970	May 2, 1972	Controlling the transition temperature of V oxides
3,679,606	Oct. 3, 1966	July 25, 1972	Printable glaze thermistors with low temperature coefficient
3,684,930	Dec. 28, 1970	Aug. 15, 1972	Ohmic contacts on Ge or Si
3,704,266	Nov. 18, 1968 Japan appl.	Nov. 28, 1972	Barium titanate zirconate semiconducting ceramic compositions
3,716,407	Sept. 23, 1969	Feb. 13, 1973	Solderable ohmic contact to semicontacting
3,735,321	June 18, 1971	May 22, 1973	Diamond, cubic boron nitride, silicon carbide particles sintered above 1570 K under pressure, doping with B or Bi.

No.	Dates: Priority	Dates: Issue	Topics
II. U.K.			
945,566		Jan. 2, 1964	Thermistor bolometer of Ge or Si films
922,491	Apr. 3, 1959	Apr. 3, 1963	Mn-Ni-Cu oxide with PbO addition
1,040,072	Jap. appl. Mar. 22, 1962	Aug. 24, 1966	V_2O_4 thermistor
1,054,525		Jan. 11, 1967	Thick-film thermistor Ni-Mn-Cu oxide
1,083,314	Aug. 22, 1963 Jap. appl.	Sept. 13, 1967 Issue	V-oxide thermistors for hf application
1,084,879	Nov. 7 ,1964 Jap. appl.	Sept. 27, 1967	Glass thermistor
1,115,937	Feb. 25, 1965	June 6, 1968	Sputtering film resistors
1,133,807	Mar. 4, 1966 U.S. appl.	Nov. 20, 1968	SiC sensor
1,136,198	May 16, 1966 U.S. appl.	Dec. 11, 1968	Co-oxide paste fired on ceramic substrate
1,214,282	Jan. 6, 1969	Dec. 2, 1970	CeO_2-ZrO_2 for high temperatures
1,244,422	Jan. 22, 1969	Mar. 10, 1971	V-oxides containing Co, Fe, Ni, Cu, Ti, Mo, W
1,226,789	Dec. 22, 1967 U.S. appl.	Mar. 31, 1971	Printed thermistor of Co-Mn oxide on Al_2O_3 substrate
1,267,107	Nov. 25, 1969	Mar. 15, 1972	Thin-film thermistor
1,093,073	Nov. 20, 1963	Nov. 29, 1967	High-temperature thermistor SnO_2, TiO_2 doped with Sb and Zn
1,287,930	Dec. 31, 1968	Sept. 6, 1972	Laminated thermistor by cathode sputtering of oxide
1,238,261	June 21, 1968 U.S. appl.	July 7, 1971	La-doped (BaSr) or (BaPb)-titanate motor protector
1,320,111	Jan. 8, 1971	June 13, 1973	Multielectrode body of intrinsic Ge

No.	Dates: Priority	Dates: Issue	Topics
III. Germany			
Ger. Offen.			
815,062		Sept. 21, 1951	Electrical resistors
972,851		Oct. 20, 1948	Contact metals with higher ionization potential
1,056,243		Apri 30, 1959	Surface treatment to produce different surface resistivity
1,067,918		Oct. 29, 1959	SiC with PbO, BaO, TiO_2
1,082,333		May 25, 1960	Si-metal oxides (Pb, Ba, Sn) to produce *n-p* junction
1,954,225	Russ. appl. Oct. 31, 1968		
	Oct. 14, 1969	Jan. 14, 1971	Organic thermistor
2,012,031	Mar. 13, 1970	Sept. 23, 1971	Cr or Mo contact on semiconductors.
	This is an addition to 1,963,514		
Ger. Offen.			
2,135,916	Jap. appl. July 15, 1970	Mar. 16, 1972	High-temperature thermistor of ZnO, PbO-Sb_2O_3, or BaO oxide
Ger. Offen.			
2,161,517	U.S. appl. Dec. 10, 1970	June 29, 1972	Silver powder of small particle size
Ger. Offen.			
2,300,199	U.S. appl. Jan. 3, 1972	July 19, 1973	Film thermistor consisting of 5–55% V 60, Al, Ge, Ti, W, V, Zr, Te, or Se 5 inorg. binder 15, and inert. liq. 20%
1,490,498	Dec. 14, 1963	July 5, 1973	Magnetically controlled thermistor made from Group III-V semiconducting compounds Zn-doped InSb

No.	Dates: Priority	Dates: Issue	Topics
2,117,446		Apr. 8, 1971	PTC units sintered in N_2 with low O_2 contents
2,055,657		May 18, 1972	Activated solder alloys containing small concentrations of Li or of strongly reducing elements

IV. France

No.	Dates: Priority	Dates: Issue	Topics
991,891		Oct. 11, 1951	Al_2O_3-Cr_2O_3-ThO_2 TiO_2, V_2O_5
1,052,015		Jan. 20, 1954	Mn-Ni-Fe oxide with 2% Al_2O_3 or SiO_2
Addition 65,137		Jan. 26, 1956	Mn-Co-Ni-Cu oxide 3.5–5% Al_2O_3
1,112,965		Mar. 21, 1956	Addition of V_2O_5 to preceding composition
1,165,582		Oct. 27, 1958	CoO-CuO-Li_2O sintered at 1350 K
Addition 75,622		Nov. 3, 1961	CuO replaced by PbO
1,370,979	Jap. appl. Apr. 6, 1963	Aug. 28, 1966	Ferrite Mn-Ni-Zn ferrite
1,367,253	U.S. appl. Aug. 23, 1962	July 17, 1967	Semiconducting glass containing Mn, Co, Mo oxide
1,403,256	U.S. appl. July 10, 1963	June 18, 1965	Coating of ir-detecting thermistors
1,443,121	June 24, 1966	Jan. 20, 1967	Organic thermistor complexes with metal salts
Addition 88,337			
Addition 89,238		Aug. 30, 1967	

No.	Dates: Priority	Dates: Issue	Topics
Addition 91,925		May 26, 1967	Organic thermistor
1,491,529	July 1, 1966	Aug. 11, 1967	Ti-Cr oxide rods
1,520,016		Apr. 5, 1968	BN crystal sealed in glass
1,520,015			Synthetic diamond thermistor sealed in glass
1,510,776	U.S. appl. Feb. 11, 1966	Jan. 19, 1968	Thermistor doped (Ba Sr) TiO_3
2,009,536	Jap. appl. May 29, 1968	Feb. 6, 1970	Stabilized PTC thermistor for use under elastic discharge
1,599,192	Dec. 10, 1968	Aug. 21, 1970	CeO_2-ZrO_2 for high temperature
1,601,788	Ger. appl. May 27, 1968	Oct. 23, 1970	CeO_2 with 2–15% ZrO_2
Fr. Demande 2,123,402	Jan. 25, 1971	Oct. 13, 1972	Boron Ni high-temperature thermistor with B = 9000 K between 470 and 1225 K

V. Japan

No.	Priority	Issue	Topics
69 17,023	Nov. 10, 1964	July 28, 1969	Intermetallic thermistor
71 26,228	Oct. 26, 1966	July 29, 1971	Containing material for semiconducting $BaTiO_3$
70 29,576	Apr. 30, 1965	Sept. 25, 1970	VO_2 film thermistor
73 53,291	Nov. 5, 1971	July 26, 1973	MnO_2 75–85 wt % and NiO 15–25 wt % mixed with 1–5 wt. % Cu oxide, B = 3000 K

VI. Netherlands

No.	Priority	Issue	Topics
83,997		Jan. 15, 1957	Contacting with TiO_2 or ZrO_2 with Ag_2O-Cu_2O in reducing atmosphere

No.	Dates: Priority	Dates: Issue	Topics
292,542		May 11, 1964	Ceramic thermistor
6,516,534	Dec. 18, 1965	June 19, 1967	Reaction of $BaTiO_3$ with earth alkali fluorides
6,614,015	Oct. 5, 1966	Apr. 8, 1968	Spinel structure of Mn and Fe, together with one or more of the group Mg, Co, Ni, the MgO eventually replaced to 50% by ZnO
6,710,691	Aug. 2, 1967	Feb. 4, 1969	Mn-, Fe-, Mg-Co, Zn and N oxide
6,412,123	Oct. 17, 1964	Apr. 18, 1966	Improving PTC characteristics by haloid treatment

VII. U.S.S.R.

No.	Priority	Issue	Topics
134,307		Dec. 25, 1960	TiO_2-CoO sintered between 1725 and 1825 K suitable for application 870 K
235,146	Oct. 16, 1967	Jan. 16, 1969	Thermistor manufacture for high temperature
288,093	June 15, 1967	Dec. 3, 1970	ZrO_2 composition
338,924	May 11, 1970	May 15, 1972	Thick-film thermistor of Mn, Co, Cu oxides
337,966	July 23, 1969	May 5, 1972	MgO-NiO doped with Li_2O

VIII. Czechoslovakia

No.	Priority	Issue	Topics
141,639	June 9, 1969	June 15, 1971	Mn-oxides doped with Sc_2O_3

IX. India

No.	Priority	Issue	Topics
53,608		Jan. 18, 1956	TiO_2 wetted with H_2O, granulated to 20 mesh dried to 10% moisture
113,404	Dec. 4, 1967	Feb. 27, 1971	$BaTiO_3$ doped with rare earth

No.	Dates: Priority	Issue	Topics
X. Poland			
48,183	Mar. 18, 1963	Aug. 27, 1964	Co-Mn oxide, cooled 600°C show good stability
60,173	Jan. 30, 1969	June 30, 1970	Mn-Ni-Co oxide thermistors quenched
XI. Spain			
363,566	Feb. 12, 1969	Jan. 1, 1971	Printed thermistor on Al_2O_3 substrate
XII. Switzerland			
Nr.280,644	Nov. 28, 1949	Jan. 31, 1952	Reactive contact mixture for Fe_3O_4 and Zn_2TiO_4

Useful Charts and Diagrams to Select Temperature Sensors

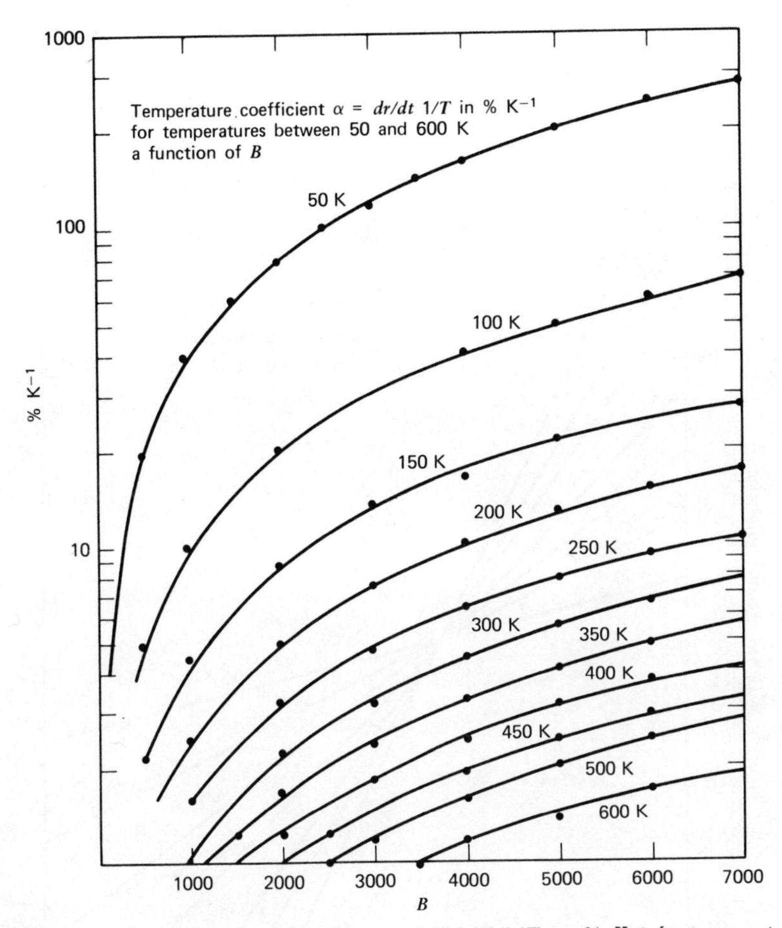

FIGURE 164. Temperature coefficient $\alpha = (dR/dT)(1/T)$ in % K^{-1} for temperatures between 50 and 600 K as function of B.

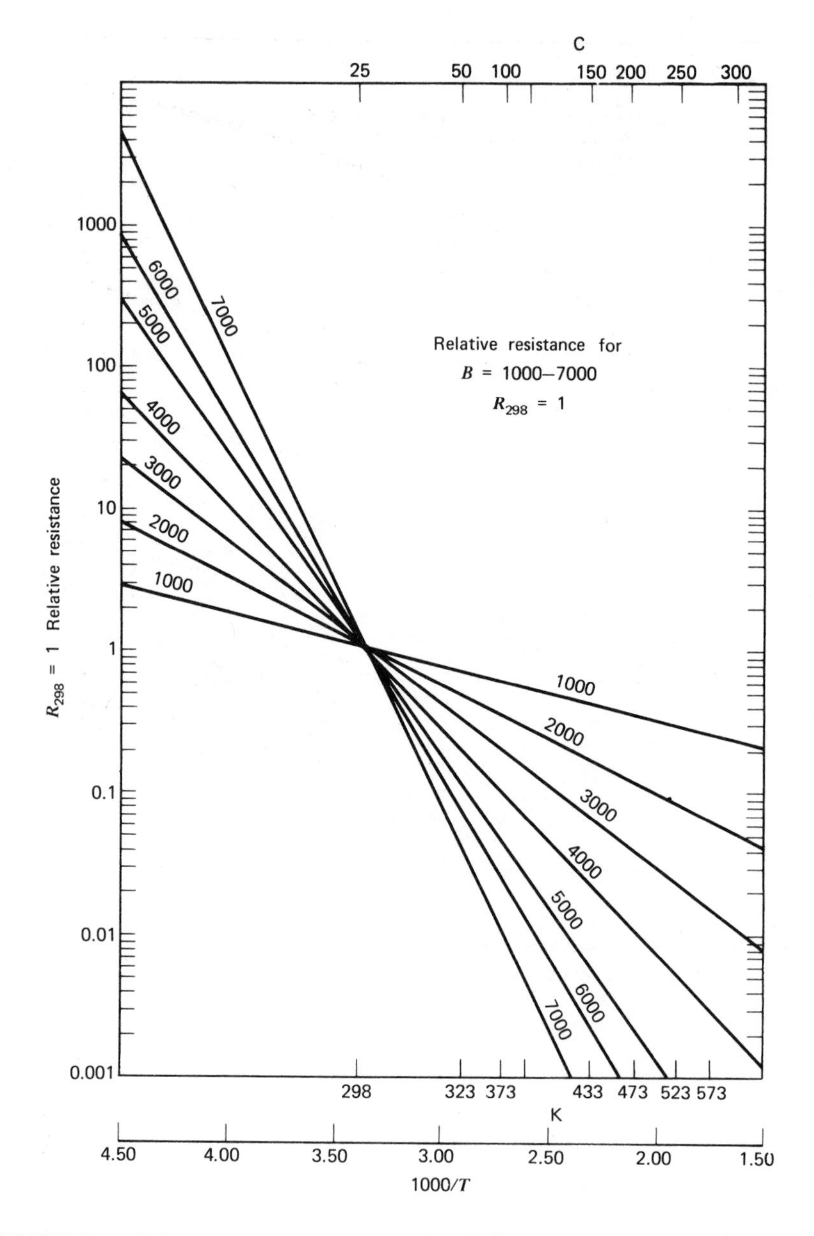

FIGURE 165. Relative resistance value RT/R 298 for material constants from 1000 to 7000 K.

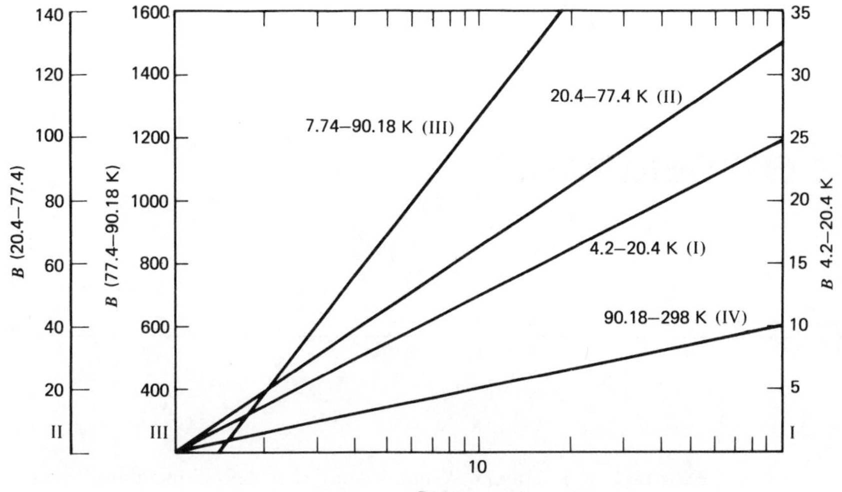

FIGURE 166. Material constant B as function of the resistance ratio for temperature intervals. I. 4.2–20.4 K, II. 20.4–77.4 K, III. 77.4–90.18 K, IV. 90.18–273.15 K.

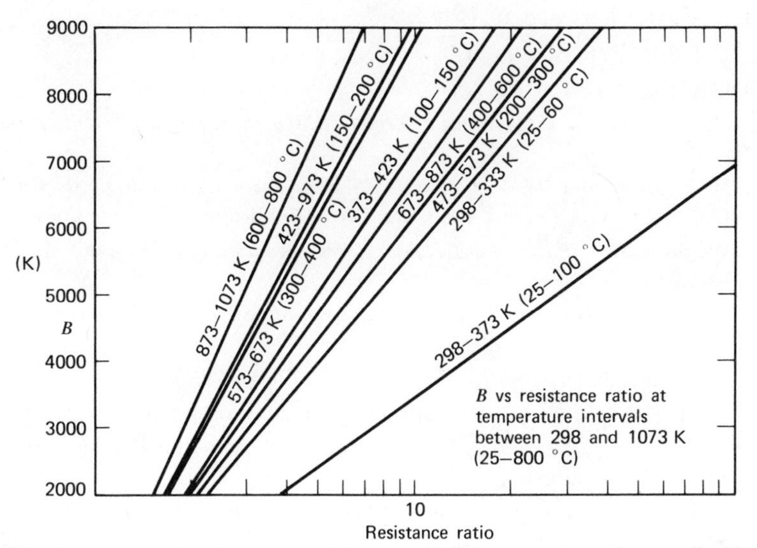

FIGURE 167. Material constant B as function of the resistance ratio for the temperature intervals: 298–333 K, 373–423 K, 423–473 K, 473–573 K, 573–673 K, 673–873 K, 873–1073 K. (temperatures in parenthesis in °C.)

References

1. D. A. Grant and W. F. Hickes, "Temperature: Its Measurement and Control in Science and Industry," Vol. 3, Part 2, Reinhold Publishers, Inc., New York, 1962, pp. 305–315.
2. H. B. Sachse, *Siemens Zeitschrift*, **19,** (5), 214–218 (1939).
3. G. L. Pearson, *Bell Lab. Record*, **19,** 106–111 (December 1940).
4. U.S. 2,091,259, Aug. 31, 1937.
5. U.S. 2,178,548, Nov. 7, 1940.
6. U.S. 2,184,847, Dec. 26, 1940.
7. C. Tubandt and H. Reinhold, *Zeitschr. Elektrochem. Bd.*, **37,** No. 8/9, 589–593 (1931).
8. N. F. Mott and R. W. Gurney, *Electronic Processes in Ionic Crystals*, Oxford University Press, Oxford, 1940, and Dover Publications, Inc., New York, 1964.
9. W. Shockley, *Electrons and Holes in Semiconductors*, Van Nostrand, Inc., New York, 1950.
10. A. F. Ioffe *Semiconductors in Contemporary Physics*, Izdatelstuo Akad. Nauk SSSR, Moscow, 1954.
11. E. Spenke, *Elektronische Halbleiter*, Springer Verlag, Berlin, 1955.
12. N. B. Hannay, *Semiconductors*, Reinhold Publishing Corp., New York, 1959.
13. A. F. Ioffe, *Physik der Halbleiter*, translated from the Russian, Akademie Verlag, Berlin, 1960.
14. A. F. Ioffe, *Physics of Semiconductors*, Academic Press, New York, 1961.
15. J. P. Suchet, *Chemie physique des semiconducteurs*, H. Dunod, Paris, 1962.
16. Charles Kittel, *Introduction to Solid State Physics*, 2nd edition, John Wiley and Sons, Inc., New York, 1963.
17. F. Seitz, *Modern Theory of Solids*, McGraw-Hill Book Co., Inc., New York, 1940.
18. R. Peierls, *Quantum Theory of Solids*, Clarendon Press, Oxford, England, 1955.
19. A. H. Wilson, *Theory of Metals*, Cambridge University Press, Cambridge, 1953.

20. P. P. Debye and E. M. Conwell, *Phys. Rev.*, **93,** 693 (1954).
21. K. Baedeker, *Ann. Phys.*, **22,** 749 (1907); **29,** 566 (1909).
22. M. LeBlanc and H. B. Sachse, *Berichte der Math. Phys. Klasse Sachs. Akad. Wiss.*, **82,** 153–158 (1930).
23. M. LeBlanc and H. B. Sachse, *Ann. Phys.*, **5,** F.11 (6) 727–735 (1931).
24. C. Wagner and W. Schottky, *Z. phys. chem.*, **B.11,** 163–210 (1931).
25. A. C. Switendick, *Quart. Progr. Rept.*, **49,** 41 (1963).
26. J. C. Slater, *Phys. Rev.*, **82,** 538 (1951).
27. D. Adler and H. Brooks, *Phys. Rev.*, **155,** 826–840 (1967).
28. W. Marshall, *Informal Proc. Buhl Conf. on Materials in Pittsburgh*, Gordon and Breach, Science Publishers, Inc., New York, 1964.
29. D. Adler, *Phys. Rev. Letters*, **17,** 139 (1966).
30. E. Burstein, E. E. Bell, J. W. Davisson, and M. Lax, *J. Phys. Chem.*, **57,** 849–852 (1953).
31. W. Heywang, *Proc. Solid State Phys. in Electronics Telecommun., Brussells*, 1958, **4,** 877–882 (1960).
32. N. F. Mott and R. W. Gurney, *Electronic Processes in Ionic Crystals*, 2nd Edition, Oxford University Press, pp. 80–83 (1950).
33. (a) W. Heywang, *Angew. Phys.*, **16,** 1–5 (1963); (b) *J. Am. Ceram. Soc.*, **47,** (10), 484–490 (1964).
34. W. Heywang and E. Fenner, *Zeitschr. f. Angew. Phys.*, **18,** 316–318 (1965).
35. G. H. Jonker, *Solid State Electronics*, **7,** 895–903 (1964).
36. F. A. Kröger, *The Chemistry of Imperfect Crystals*, North Holland Publishing Co., Amsterdam, 1964.
37. T. Y. Tien and W. G. Carlson, *J. Am. Chem. Soc.*, **46,** 297–298 (1963).
38. I. Ueda and S. Ikegami, *J. Phys. Soc. (Japan)*, **20,** (4), 546–552 (1965).
39. Netherland Patent: 64.12123, N. V. Philips Gloeilampen fabrieken, Eindhoven, Netherlands.
40. G. H. Jonker, *Materials Res. Bull.*, **2,** (4), 401–407 (1967).
41. O. Saburi, *J. Phys. Soc. (Japan)*, **14,** 1159–1174 (1959).
42. G. Goodman, *J. Am. Ceram. Soc.*, **46,** 48–54 (1963).
43. C. Fielding Brown and C. E. Taylor, *J. Appl. Phys.*, **35,** 2554–2556 (1964).
44. R. J. Mytton and R. K. Benton, *Phys. Letters A*, **39,** (4), 329–330 (1972).
45. J. Holt, *Solid State Electronics*, **9,** 813–818 (1966).
46. B. M. Kulwicki and A. J. Purdes, *Ferroelectrics*, **I**, 253–263 (1970).
47. W. T. Peria, W. R. Bratschun, and R. D. Fenitey, *J. Am. Ceram. Soc.*, **44,** 249–250 (1961).
48. A. F. Devonshire, *Phil. Mag.*, **4,** 1040–1063 (1949); Suppl. **3,** 86–130 (1954).
49. P. Gerthsen and K. H. Härdtl, *Z. Naturforsc.* **18a,** 484 (1963).
50. H. Rehme, *Phys. State Solid*, **18,** K101–102 (1966).

51. W. Heywang and H. Brauer, *Solid State Electronics*, **8,** 129–135 (1965).
52. H. Brauer, *Z. Angew. Phys.*, **23,** 373–376 (1967).
53. W. Heywang and E. Fenner, *Siemens Zeitschrift*, **41,** 878–886 (1967).
54. O. Eberspächer, *Naturwissenschaften*, **49,** 155–156 (1962).
55. H. Schmelz, *Phys. State Solid*, **35,** 219–225 (1969).
56. H. Schmelz, *Physica Status Solidi*, **31,** 121–128 (1969).
57. W. Heywang, *Ferroelectrics*, **I**, 177–178 (1970).
58. N. G. Eror and D. M. Smyth, *The Chemistry of Extended Defects in Nonmetallic Solids*, North Holland Publishing Co., Amsterdam, 1970, pp. 62–74 (conference proceedings).
59. J. Novak and H. Arend, *J. Am. Ceram. Soc.*, **47,** (10), 530 (1964).
60. W. D. Johnston and D. Sestrich, *J. Inorg. and Nucl. Chem.*, **20,** 32–38 (1961)
61. Yu. I. Gol'tsov, A. S. Bogatin, and O. I. Prokopalo, *Izv. Akad. Nauk. SSSR, Ser. Fiz.*, **31,** (11), 1821–1823 (1967).
62. R. C. Kell and N. J. Hellicar, *Acustica*, **6,** 235 (1956).
63. C. I. Moratis and R. I. Bratton, *Japan J. Appl. Phys.*, **10,** 421–426 (1971).
64. H. A. Sauer and S. S. Flaschen, *Proc. Electronics Comp. Symp. Washington, D.C.*, Engineering AIEE, 41–46, 1956, Publishers, New York, 1956.
65. Y. Ichikawa and W. G. Carlson, *Ceram. Bull.*, **42,** C 5, 312–316 (1963) (see also U.S. Patent 2,976,505).
66. G. S. Gruintjes, G. J. Oudemous, J. B. Huffadine, A. J. Whitehead, M. J. Latimer, R. M. Spriggs, L. Atheraas, and R. B. Runk, *Proc. Br. Ceram. Soc.*, No. 12, Fabrication Science, pp. 83–98, 165–182, 201–209 (1970).
67. G. Haertling, *Bull. Am. Ceram. Soc.*, **43,** 875–879 (1964).
68. A. J. Mountvala, *Bull. Am. Ceram. Soc.*, **42,** 120–121 (1963).
69. G. N. Tekster-Proskuryakova and I. T. Sheftel, *Soviet Phys. Solid State*, **5,** (12), 2542–2548 (1964).
70. Yu. I. Gol'tsov, O. I. Prokopalo, and L. A. Belova, *Soviet Phys. Solid State*, **14** (3), 805–806 (1972).
71. Makoto Kuwabara and Hiroaki Yanagida, *Japan J. Phys.*, **10,** 805 (1971).
72. Kazuyuki Ohe and Yoshihide Naito, *Japan J. Appl. Phys.*, **10,** (1), 99–108 (1971).
73. P. K. Gallagher, Frank Schrey, and F. V. DiMarcello, "Preparation of Semiconducting Titanates by Chemical Methods," *J. Am. Ceram. Soc.*, **46,** (8), 359–365 (1963).
74. E. M. Swiggard and W. St. Clabaugh, *Am. Ceram. Bull.*, **45,** (9), 777–781 (1966).
75. M. Foex, *J. Rech. CNRS*, 238–246 (1952).
76. M. Foex, *Compt. Rend.*, **223,** 1126–1128 (1946).
77. M. Foex and J. Wucher, *Compt. Rend.*, **241,** 184–186 (1955).
78. F. J. Morin, *Phys. Rev. Letters*, **3,** 34–36 (1959).

79. R. F. Janninck and D. H. Whitmore, *J. Phys. Chem. Solids*, **27,** 1183–1187 (1966).
80. J. B. Goodenough, *Phys. Rev.*, **117,** 1442–1451 (1960); *J. Appl. Phys. Suppl.*, **31,** 359 S (1960).
81. D. Adler, J. Feinleib, H. Brooks, and W. Paul, *Phys. Rev.*, **155,** 851–860 (1967).
82. S. Kachi, T. Takada, and K. Kosuge, *J. Phys. Soc. (Japan)*, **18,** 1839–1840 (1963).
83. T. Mitsuishi, *J. Appl. Phys. (Japan)*, **6,** 1060–1071 (1967).
84. S. Koide and H. Takei, *J. Phys. Soc. (Japan)*, **22,** 956 (1965).
85. Sh. Minomura and H. Nagasaki, *J. Phys. Soc. (Japan)*, **19,** 131–132 (1964).
86. J. B. MacChesney and H. J. Guggenheim, *Phys. Chem. Solid*, **30,** 225–234 (1969).
87. M. Rubinstein, *Solid State Comm.*, **8,** 1469–1472 (1970).
88. V. N. Novikov, *Izv. Akad. Nauk. SSSR. Neorganicheskie Mat.*, **6,** (1), 96–99 (1970).
89. Japan Patent: Cl. 62 A, No. 39-8583 (1964).
90. T. Okashi and A. Watanabe, *J. Am. Ceram. Soc.*, **49,** 519 (1966).
91. A. Watanabe, I. Kitahiro, and H. Sasaki, *J. Phys. Soc. (Japan)*, **21,** 196 (1966).
92. I. Kitahiro and A. Watanabe, *J. Appl. Phys.*, **6,** 1023–1024 (1967).
93. H. Sasaki and A. Watanabe, *J. Phys. Soc. (Japan)*, **19,** 1748 (1964).
94. M. Guntersdorfer, *Solid State Electronics*, **13,** 369–379 (1970).
95. I. Kitahiro, A. Watanabe, and H. Sasaki, *J. Phys. Soc. (Japan)*, **21,** 196 (1966).
96. K. van Steensel, F. van der Burg, and C. Koog, *Philips Res. Rep.*, **22,** 170–177 (1967).
97. P. F. Bongers and U. Enz, *Philips Res. Rep.*, **21,** 387–389 (1966).
98. C. E. N. Fuls, D. H. Hensler, and A. R. Ross, *Appl. Phys. Letters*, **10,** (7), 199–201 (1967).
99. G. A. Rozgony and D. H. Hensler, *J. Vac. Sci. Tech.*, **5,** (6), 194–199 (1968).
100. A. G. Zhden, A. S. Darevski, M. E. Chuganova, and I. A. Serbinov, *Soviet Phys. Solid State*, **14,** (7), 1879–1880 (1973).
101. I. G. Austin, *Phil. Mag.*, **7,** 961–967 (1962).
102. Y. Shapira, S. Fosser, and T. B. Reed, *Phys. Rev. B*, **8,** 2299–2315 (1973).
103. C. Wagner, *Diffusion and High Temperature Oxidation*, American Society of Metals, Cleveland, Ohio, 1951, pp. 153–173.
104. W. A. Weyl, *Ceramic Age*, **60,** (5), 28–38 (1952).
105. K. Hauffe, *Reaktionen in und an festen Stoffen*, Springer-Verlag, Berlin, 1966.
106. C. Wagner, *Z. phys. Chem. B*, **22,** 181–194 (1933).
107. K. Hauffe, *Reaktionen in und an festen Stoffen*, Springer-Verlag, Berlin, 1955, pp. 180–242.
108. R. Fricke and G. Weitbrecht, *Z. Elektrochem.*, **48,** 87–110 (1942).
109. H. Schmalzried, *Z. Physik. Chem. (Frankfurt)*, **25,** 178–192 (1960).

110. B. R. Puri and V. Bhushan, *J. Sci. Ind. Res.* (*India*), **11B,** 504–505 (1952).

111. F. Bertaut and C. Delorme, *Compt. Rend.*, **238,** 1829–1830 (1954).

112. W. D. Johnston, R. C. Miller, and R. Mazelsky, *J. Phys. Chem.*, **63,** 198–202 (1959).

113. E. J. W. Verwey, *Z. Kristallog.*, **91,** 65–69 (1935).

114. E. J. W. Verwey and M. G. van Bruggen, *Z. Kristallog.*, **92,** 136–138 (1935).

115. A. E. van Arkel, E. J. W. Verwey, and M. G. van Bruggen, *Rec. Trav. Chim.*, **55,** 331–339 (1936).

116. E. E. Vainshtein, R. M. Ovrutskava, B. I. Kotlyar, and V. R. Linde, *Soviet Phys. Solid State*, **5,** (10), 2150–2153 (1964).

117. T. Bedynska, *Acta Phys. Polon.*, **18,** 199–203 (1959).

118. L. Azaroff, *Z. Kristallog.*, **112,** 33–43 (1959).

119. E. G. Larson and R. J. Arnott, and D. G. Wickham, *J. Phys. Chem. Solids*, **23,** 1771–1781 (1962).

120. Shigeharu Naka, Michio Inagaki, and Tatsuo Tanaka, *J. Mater. Sci.*, **7,** 441–444 (1972).

121. E. V. Kavalenko, V. A. Lyntsedarskii, and T. A. Ryshkina, *Zh. Prikl. Khim.*, **34,** 1880–1883 (1961).

122. B. T. Kolomiets, I. T. Sheftel, and E. V. Kurlina, *Zh. Thekh. Fiz.*, **27,** (1), 51–72 (1957).

123. B. N. Naik and A. P. B. Sinha, *Ind. J. Pure Appl. Phys.*, **7,** 170–174 (1969).

124. G. Blasse, *J. Phys. Chem. Solids*, **27,** (2), 383–389 (1966).

125. C. D. Sabane, A. P. B. Sinha, and A. B. Biswas, *Ind. J. Pure Appl. Phys.*, **4,** (5), 187–190 (1966).

126. M. Rosenberg, P. Nicolau, R. Manaila, and P. Pausescu, *J. Phys. Chem. Solids.*, **24,** 1419–1434 (1963).

127. J. S. Smart and S. Greenwald, *Phys. Rev.*, **82,** 113–114 (1951).

128. W. L. Roth, *Acta Cryst.*, **13,** 140–149 (1960).

129. G. H. Jonker, *J. Phys. Chem. Solids*, **9,** 165–175 (1959).

130. F. K. Lotgering, *J. Phys. Chem. Solids*, **25,** (1), 95–103 (1964).

131. N. Reslescu, *Compt. Rend.*, **268B,** 136 (1969).

132. H. F. McMurdic, B. M. Lullivan, and F. A. Mauer, *J. Res. Natl. Bur. Std.*, **45,** 35–41 (1960); Res. Paper No. 2111.

133. A. Eucken, *Energie and Wärmeinhalt*, Bd. 8, Teil 1, *Handbuch der Experimentalphysik*, V. Wiens und F. Harms, editors, Akademie Verlag, Leipzig, 1929, p. 230.

134. Landolt Börnstein, Technik Bd 1, 702–707 (1927).

135. I. Licea, *Phys. Status Solidi*, **10,** (2), K115–K118 (1965).

136. H. van Dijk and W. H. Keesom, *Physica*, **7,** 970–984 (1940).

137. R. P. Hudson, Thesis: "Thermal conductivity of iron ammonium alum compressed below 0.2°K." Oxford, 1949 (unpublished).

138. H. W. Godbee and W. T. Ziegler, *J. Appl. Phys.*, **37,** (1), 40–55 (1966).

139. A. Eucken and G. Kuhn, *Z. Phys. Chem.*, **134,** 193–219 (1928).

140. R. Berman, F. E. Simon, and J. M. Ziman, *Proc. Roy. Soc. London, Ser. A*, **220,** 171–183 (1953).

141. M. Remoissenet and L. Godefrey, *Compt. Rend.*, **262,** (II), 56–59 (1966).

142. A. J. H. Mante and J. Volger, *Phys. Letters A*, **24,** (3), 139–140 (1967).

143. P. V. Tamarin and S. S. Shalyt, *Soviet Phys., Semiconductors*, **5,** (6), 1097–1098 (1971).

144. E. Grüneisen, *Handbuch der Physik*, **10,** 1–59 (1926).

145. T. H. K. Barron, *Phil. Mag.*, **46,** 720–734 (1955).

146. Alexander Goldsmith, Thomas E. Waterman, and Harry J. Hirschhorn, editors, *Handbook of Thermophysical Properties of Solid Materials* Macmillan (vol. 5), New York, 1961.

147. H. B. Sachse, *Anal. Chem.*, **32,** 529–530 (1960).

148. A. Smakula, "A study of the physical properties of high temperature single crystals," 1967. Technical Report AD 663 734, Clearinghouse for Federal Scientific and Technical Information.

149. H. R. Thornton, Thesis, Univ. of Illinois, 1963 (unpublished).

150. C. M. Osburn and R. W. Vest, *J. Phys. Chem. Solids*, **32,** 1355–1363 (1971).

151. B. T. Kolomiets, I. T. Sheftel, and E. V. Kurlina, *Soviet Phys.-Tech. Phys.*, 40–58 (1957) (English translation).

152. H. Walch, *Siemens Zeitschrift*, **47,** 65–67 (1973).

153. E. J. W. Verwey, P. W. Haaijman, F. C. Romeijn, and G. W. van Oosterhout, *Philips Res. Repts.*, **5,** 173–187 (1950).

154. E. J. W. Verwey, *Bull. Soc. Chim. France 1949*, Mises au point D 122.

155. W. D. Johnston and R. R. Heikes, *J. Am. Chem. Soc.*, **78,** 3255–3260 (1956).

156. R. R. Heikes and W. D. Johnston, *J. Chem. Phys.*, **26,** 582–587 (1957).

157. W. D. Johnston, R. R. Heikes, and D. Sestrich, *J. Phys. Chem. Solids*, **7,** 1–13 (1958).

158. R. C. Miller, R. R. Heikes, and W. D. Johnston, *J. Chem. Phys.*, **31,** 116–118 (1959).

159. R. R. Heikes, A. A. Maradudin, and R. C. Miller, *Ann. Phys. (France)*, **8,** (11–12), 733–746 (1963).

160. K. Hauffe and J. Block, *Z. phys. chem.*, **196,** 43–46 (1951).

161. V. P. Zhuze and A. I. Shelykh, translated from *Fiz. Tverd. Sov. Phys. Solid State*, **5,** 1278–1280 (1963).

162. Ya. M. Ksendzov, L. N. Ansel'm, L. L. Vasil'eva, and V. M. Latysheva, *Sov. Phys. Solid State*, **5,** (6), 1116–1123 (1963).

163. I. G. Austin, A. J. Springthorpe, B. A. Smith, and C. E. Turner, *Proc. Phys. Soc.*, **90,** 157–174 (1967).

164. I. G. Lang and Yu. A. Firsov, *Sov. Phys. Solid State, Eng. Transl.*, **5,** (10), 2029 (1964).
165. H. J. Van Daal and A. J. Bosman, *Phys. Rev.*, **158,** (3), 736–747 (1967).
166. F. E. Maranzana, *Phys. Rev.*, **160,** (2), 421–430 (1967).
167. A. J. Bosman and C. Crevecoeur, *Phys. Rev.*, **144,** (2), 763–770 (1966).
168. B. Fisher and D. S. Tannhauser, *J. Chem. Phys.*, **44,** (4), 1663–1672 (1966).
169. F. J. Morin, *Phys. Rev.*, **93,** 1195–1204 (1954).
170. A. J. Springthorpe, I. G. Austin, and B. A. Austin, *Solid State Comm.*, **3,** 143–146 (1965).
171. P. Nagels and M. Denayer, *Solid State Comm.*, **5,** 193–197 (1967).
172. B. Fisher and J. B. Wagner, Jr., *J. Appl. Phys.*, **38,** 3838–3842 (1967).
173. B. Fisher and J. B. Wagner, Jr., *Phys. Letters*, **21,** 606–607 (1966).
174. A. J. Bosman, H. J. Van Daal, and G. F. Knuvers, *Phys. Letters*, **19,** 372–373 (1965).
175. C. Greskovich and H. Schmalzried, *J. Phys. Chem. Solids*, **31,** 639–646 (1970).
176. T. F. W. Barth and E. Posnjak, *Z. Kristallog*, **88,** 271–280 (1934).
177. F. J. Morin, *Phys. Rev.*, **83,** 1005–1010 (1951).
178. C. Wagner and E. Koch, *Z. Phys. Chem.*, **B32,** 439–446 (1936).
179. K. Hauffe, H. Grunewald, and R. Tränckler-Greese, *Z. Elektrochem.*, **56,** 937–944 (1952).
180. H. Grunewald, *Ann. Phys. 6. Folge*, **14,** 121–124 (1954).
181. C. G. Koops, *Phys. Rev.*, **83,** 121–124 (1951).
182. K. W. Wagner, *Ann. Phys.*, **40,** 817 (1913).
183. Yootarou Yamazaki and Minoru Satou, *Japan J. Appl. Phys.*, **12,** (7), 998–1000 (1973).
184. K. V. Rao and A. Smakula, *J. Appl. Phys.*, **36,** (6), 2031–2038 (1965).
185. A. K. Chaudhury and K. V. Rao, *Phys. Status Solidi*, **32,** (2), 731–739 (1969).
186. Shigeharu Kabashima, *J. Phys. Soc. (Japan)*, **26,** (4), 975–978 (1969).
187. E. Schleicher, *Hermsdorfer Techn. Mitteilungen*, Heft 22, 681–690 (1968).
188. R. W. Vest, N. M. Tallan, and W. C. Tripp, *J. Am. Ceram. Soc.*, **47,** 635–640 (1964).
189. R. W. Vest and N. M. Tallan, *J. Am. Ceram. Soc.*, **48,** 472–475 (1965).
190. N. M. Tallan, R. W. Vest, and H. C. Graham, *Mater. Sci. Res.*, **2,** 33–37 (1965).
191. P. J. Freud, *Phys. Rev. Letters*, **29,** (17), 1156–1159 (1972).
192. M. R. Notis, R. M. Spriggs, and W. C. Hahn, Jr., *J. Appl. Phys.*, **44,** (9), 4165–4171 (1973).
193. St. J. Fonash, *J. Appl. Phys.*, **44,** (10), 4607–4613 (1973).
194. H. B. Sachse, *Electronic Ind.*, **16,** 55–56 (January 1957).
195. H. B. Sachse, *Electronic Design*, **31,** 30–33 (April 1958).

196. H. B. SACHSE, *Bulletin de l'Institut International du Froid Delft*, **1958-1,** 145–154 (1958).

197. H. B. SACHSE and G. W. VOLLMER, *Electronic Ind.*, **18,** 67–68 (Feb. 1959).

198. H. B. SACHSE, in *Temperature—Its Measurement and Control in Science and Industry*, Instrument Society of America, Pittsburgh, 1962, Vol. 3, Part 2, Reinhold, New York, pp. 347–353.

199. H. B. SACHSE, *Z. Angew. Phys.*, **15,** Heft 1, 4–7 (1963).

200. C. T. N. PALUDAN, *Aerospace Eng.*, 107, 108, 111, 114, 118 (Sept. 1959).

201. W. F. SCHLOSSER and R. H. MUNNINGS, *Rev. Sci. Instr.*, **40**, 1356–1360 (1969).

202. W. F. SCHLOSSER and R. H. MUNNINGS, in *Temperature, Its Measurement and Control in Science and Industry*, Vol. 4, Part 2, Instrument Society of America, Pittsburgh, 1973, pp. 795–801.

203. L. J. NEURINGER and Y. SHAPIRA, *Rev. Sci. Instr.*, **40,** 1314–1321 (1969).

204. L. J. NEURINGER, A. J. PERLMAN, L. G. RUBIN, and Y. SHAPIRA, *Rev. Sci. Instr.*, **42,** 9–14 (1971).

205. L. J. NEURINGER and L. G. RUBIN, in *Temperature, Its Measurement and Control in Science and Industry*, Vol. 4, Part 2, Instrument Society of America, Pittsburgh, 1973, pp. 1085–1094.

206. H. B. SACHSE, *Electronic Industries*, **20,** (1), 96–99 (Jan. 1961).

207. I. ESTERMANN, *Phys. Rev.*, **78,** 83–84 (1950).

208. J. E. KUNZLER, T. H. GEBALLE, and G. W. HULL, *Rev. Sci. Instr.*, **28,** 96–98 (1957).

209. H. FRITZSCHE, *Phys. Rev.*, **119,** 1899–1900 (1960).

210. J. S. BLAKEMORE, J. M. SCHULTZ, and J. W. MYERS, *Rev. Sci. Instr.*, **33,** 545–551 (1962).

211. J. S. BLAKEMORE, *Rev. Sci. Instr.*, **33,** 106–112 (1962).

212. G. CATALAND and H. H. PLUMB, *J. Res. Natl. Bur. Std.*, **70A,** 243–252 (1966).

213. J. BLAKEMORE, J. WINSTEL, and R. V. EDWARDS, *Rev. Sci. Instr.*, **41,** 835–842 (1970).

214. J. G. COLLINS and W. R. G. KEMP, in *Temperature, Its Measurement and Control in Science and Industry*, Vol. 4, Part 2, Instrument Society of America, Pittsburgh, 1973, pp. 835–842.

215. M. P. ORLOVA, D. N. ASTROV, and L. A. MEDVEDEVA, *Prib. Tekh. Eksp.*, **5,** 231 (1964) [translation, *Cryogenics*, 165–66 (June 1965)].

216. C. A. SVENSON and P. C. F. WOLFENDALE, *Rev. Sci. Instr.*, **44,** 339–341 (1973).

217. W. V. JOHNSTON and G. W. LINDBERG, *Bull. Am. Phys. Soc.*, **10,** 719 (1965).

218. T. H. HERDER, R. O. OLSON, and J. S. BLAKEMORE, *Rev. Sci. Instr.*, **37,** 1301–1305 (1966).

219. R. BACHMANN, H. C. KIRSCH, and T. H. GEBALLE, *Rev. Sci. Instr.*, **41,** 547–549 (1970).

220. H. C. Kirsch, R. Bachmann, and T. H. Geballe, in *Temperature, Its Measurement and Control in Science and Industry*, Vol. 4, Part 2, Instrument Society of America, Pittsburgh, 1973, pp. 843–846.

221. B. G. Cohen, W. B. Snow, and A. R. Tretola, *Rev. Sci. Instr.*, **34,** (10), 1091–1093 (1963).

222. H. Harris, *Sci. Am.*, **240,** (6), 192 (1961).

223. A. G. McNamara, *Rev. Sci. Instr.*, **33,** 330–333 (1962).

224. NASA Report CR 54962 Contract NAS 36 219.

225. J. M. Swartz and J. R. Gaines, in *Temperature, Its Measurement and Control in Science and Industry*, Vol. 4, Part 2, Instrument Society of America, Pittsburgh, 1973, pp. 1117–1124.

226. K. F. Pavese and S. Limbarinu, in *Temperature, Its Measurement and Control in Science and Industry*, Vol. 4, Part 2, Instrument Society of America, Pittsburgh, 1973, pp. 1103–1116.

227. O. V. Emel'yanenko, D. N. Nasledov, E. I. Nikulin, and I. N. Timchenko, *Soviet Phys. Semiconductors*, **6,** (11), 1926–1927 (1973).

228. T. C. Verster, in *Temperature, Its Measurement and Control in Science and Industry*, Vol. 4, Part 2, Instrument Society of America, Pittsburgh, 1973, pp. 1125–1134.

229. W. Wlodarski, *Rev. Sci. Instr.*, **42,** 260–261 (1971); *Elektronika*, **11**, 10 (1970) (Polish).

230. J. R. Clement and E. H. Quinnell, *Rev. Sci. Instr.*, **23,** 213–216 (1952).

231. H. H. Plumb and M. H. Edlow, *Rev. Sci. Instr.*, **30,** 376–377 (1959).

232. R. Guenther, H. Weinstock, and R. W. Schleicher, *Cryogenic Tech.*, **6,** 13 (1970).

233. H. Weinstock and J. Parpia, in *Temperature, Its Measurement and Control in Science and Industry*, Vol. 4, Part 2, Instrument Society of America, Pittsburgh, 1973, pp. 785–790.

234. W. C. Black, Jr., W. R. Roach, and J. C. Wheatley, *Rev. Sci. Instr.*, **35,** 887–891 (1964).

235. A. S. Edelstein and K. W. Mess, *Physica*, **31,** 1707–1712 (1965).

236. L. Gordy and H. Fritzsche, *J. Appl. Phys.*, **41,** 3546–3547 (1970).

237. W. F. Schlosser and R. H. Munnings, *Cryogenics*, **12,** 225–226 (1972).

238. P. Lindenfeld, *Rev. Sci. Instr.*, **32,** 9–11 (1961).

239. J. R. Clement, E. H. Quinnell, M. C. Steele, R. A. Hein, and R. L. Dolecek *Rev. Sci. Instr.*, **24,** 245–246 (1953).

240. N. H. Edlow and H. H. Plumb, *Adv. Cryogenic Eng.*, **6,** 542–547 (1961).

241. E. H. Schulte, *Cryogenics*, **6,** 321–323 (1966).

242. D. C. Pearce, A. H. Markham, and J. R. Dillinger, *Rev. Sci. Instr.*, **27,** 240 (1956); **28,** 382 (1957).

243. J. E. Robichaux and A. C. Anderson, *Rev. Sci. Instr.*, **40,** 1512–1513 (1969).

244. B. L. Booth and A. W. Ewald, *Rev. Sci. Instr.*, **40,** 1354–1355 (1969).

245. L. P. Mezhov-Deglin and A. I. Shalnikov, *Prib. Tekh. Exp.*, **1,** 209 (1968).
246. R. Miller and C. W. Ulbrich, *Cryogenics*, **12,** 173–175 (June 1972).
247. K. J. Komatsu, *Phys. Chem. Solids*, **6,** 380 (1958); **25,** 707 (1963).
248. L. P. Mezhov-Deglin and A. I. Shalnikov, *Cryogenics*, **9,** 60 (1969).
249. W. F. Giauque, *Ind. Eng. Chem.* (*Intern. Ed.*), **28,** 743 (1936).
250. T. H. Geballe, D. N. Lyon, J. M. Whelan, and W. F. Giauque, *Rev. Sci. Instr.*, **23,** 489–492 (1952).
251. L. Shen and D. C. Heberlein, in *Temperature, Its Measurement and Control in Science and Industry*, Vol. 4, Part 2, Instrument Society of America, Pittsburgh, 1973, pp. 791–794.
252. G. A. Saunders, *Appl. Phys. Letters*, **4,** (8), 138–140 (1964).
253. N. N. Mikhailov and A. Ya. Kaganovski, *Cryogenics*, 98–100 (December 1961).
254. B. Lalevic, *Rev. Sci. Instr.*, **33,** 103–105 (1962).
255. G. A. Zaytsev and I. A. Khrebtov, *J. Opt. Techn.*, **33,** (1), 35–37 (1966).
256. I. Giaever and K. Megerle, *Phys. Rev.*, **122,** (4), 1101–1111 (1961).
257. J. Bardeen, L. N. Cooper, and J. R. Schrieffer, *Phys. Rev.*, **108,** (5), 1175–1204 (1957).
258. J. W. Bakker, H. van Kempen, and P. Wyder, in *Temperature, Its Measurement and Control in Science and Industry*, Vol. 4, Part 2, Instrument Society of America, 1973, pp. 1097–1102.
259. C. M. Ignat'ev and V. Yu. Tarenkov, *Societ Phys. Semiconductors*, **4,** (11), 1895–1896 (May 1971).
260. A. F. G. Wyath, *Phys. Rev. Letters*, **13,** 401 (1964).
261. C. R. Tallman, in *Temperature, Its Management and Control in Science and Industry*, Vol. 4, Part 2, Instrument Society of America, Pittsburgh, 1973, pp. 1071–1076.
262. S. Okude, S. Takamura, and H. Maeta, *Intern. Conference on Vacancies and Interstitials in Metals*, Aachen, Germany, 1968, pp. 317–326.
263. J. E. Zimmerman and A. H. Silver, *J. Appl. Phys.*, **39,** (6), 2679–2682 (1968).
264. J. T. Harding and J. E. Zimmerman, *J. Appl. Phys.*, **41,** (4), 1581–1588 (1970).
265. M. O'Keeffe and M. Valigi, *J. Phys. Chem. Solids*, **31,** 947–962 (1970).
266. V. Croatto and A. Mayer, *Gazz. chim. ital.*, **73,** 199–210 (1943).
267. W. Noddack, H. Walch, and W. Dobner, *Z. Phys. Chem. Leipzig*, **211,** 181–193 (1959).
268. W. Noddack and H. Walch, *Z. Phys. Chem. Leipzig*, **211,** 194–207 (1959).
269. W. Noddack and H. Walch, *Z. Elektr.*, **63,** (2), 269–274 (1959).
270. V. B. Tare and H. Schmalzried, *Z. Phys. Chem. NF*, **43,** 30–32 (1964).
271. G. V. Subba Rao, B. Ramdas, P. N. Mehrotra, and C. N. R. Rao, *J. Solid State Chem.*, **22,** (2), 377–384 (1970).
272. G. R. Hyde, E. E. Moust, Jr., and LeRoy Furlong, *Bur. Mines Rep Invest.*, No. 7458 (1970).
273. N. M. Tallan and R. W. Vest, *J. Am. Ceram. Soc.*, **49,** (8), 401–404 (1966).

274. E. C. SUBBARO, P. H. SUTTER, and J. HRIZO, *J. Am. Ceram. Soc.*, **48,** (9), 443 (1965).

275. C. WAGNER, *Naturwissenschaften*, **31,** 265–268 (1943).

276. D. W. STRICKLER and W. G. CARLSON, *J. Am. Ceram. Soc.*, **48,** 286–289 (1965).

277. F. HUND, *Z. Phys. Chem.*, **199,** (1–3), 142–151 (1952).

278. W. D. KINGERY, J. PAPPIS, M. E. DOTY, and D. C. HILL, *J. Am. Ceram. Soc.*, **42,** (8), 393–398 (1959).

279. J. WEISSBART and R. RUKA, *J. Electrochem. Soc.*, **109,** (8), 723–726 (1962).

280. D. T. BRAY and U. MERTEN, *J. Electrochem. Soc.*, **109,** 447–452 (1964).

281. F. HUND, *Z. anorg. allg. Chem.*, **274,** 105–113 (1953).

282. F. HUND and R. MEZGER, *Z. phys. Chem.*, **201,** 268–277 (1952).

283. N. T. PLASHINSKY and I. T. SHEFTEL, U.S. 3,598,764, August 10, 1971; appl. Dec. 9, 1968.

284. J. L. LUMLEY, *Symp. on Measurement in Unsteady Flow*. Worcester, Mass., 1962, p. 75.

285. P. R. MALMBERG and C. G. MATTLAND, *Rev. Sci. Instr.*, **27,** 136–139 (1956).

286. H. R. WISELY, P. D. FREEZE, and E. F. FIOCH, WADC Tech. Report 54-388, 1954.

287. H. R. WISELY, *Am. Ceram. Soc. Bull.*, **36,** 133–136 (1957); H. R. WISELY, P. FREEZE, and E. FIOCH, *A Study of Thermistor Materials for Use as Temperature Sensing Elements in the High-Velocity Exhaust Gases of Jet-Type Engines*. WADC Tech. Rept. 54-388 (1954).

288. G. BLACKBURN, D. FREEZE, and F. R. CALDWELL, WADC-TR. 54-388, Suppl. 1, AD 97636 (April 1965).

289. A. R. ANDERSON and T. M. STICKNEY, in *Temperature, Its Measurement and Control in Science and Industry*, Vol. 3, Part 2, Reinhold Publishers, Inc., New York, 1962, pp. 361–367.

290. E. G. WOLFF, *Rev. Sci. Instr.*, **40,** 544–549 (1969).

291. A. M. ANTHONY, *Compt. Rend.*, **260,** No. 7, 1936–1939 (1965).

292. J. ELSTON, Z. MIHAILOVIC, and MADELEINE ROUX, *Proc. Symp. MHD Salzburg*, **3,** 389 (1966).

293. M. ROUX, Rapport CEA-4353 (1972),C.E.N. Saclay B.P. N2, 91-GIF-sur Yvette, France.

294. C. H. MCMURTRY, W. T. TERRELL, and W. T. BENECKI, *IEEE Trans. Industry and Gen. Appl.*, **IGA-2,** (6), 461–464 (1966).

295. U.S. 3,341,473, Sept. 12, 1967.

296. E. G. WOLFF and F. M. VILES, AFFDC-TL, 65-134 (July 1965).

297. CH. E. RYAN, *Mater. Res. Bull.*, **4,** 1–12 (1969).

298. G. A. SLACK, *J. Chem. Phys.*, **42,** 805 (1965).

299. C. J. KROBO and A. G. MILNES, *Sol. State Electronics*, **9,** 1125 (1965); **8,** 829–830 (1965).

300. E. L. Kern, D. W. Hamill, H. W. Deem, and H. D. Sheets, *Mater. Res. Bull.*, **4,** 25–32 (1969).
301. O. A. Golikova, L. M. Ivanova, A. A. Pletyushkin, and V. P. Semenenko, (Russian Inventors Certificate No. 145 106) *Soviet Phys. Semiconductors*, **5**, No. 3, 366–369 (1971).
302. Peter T. B. Shaffer, *Mater. Res. Bull.*, **4,** 13–24 (1969).
303. G. K. Gaule, J. T. Breslin, J. R. Pastore, and R. A. Shuttleworth, *Boron Synthesis, Structure, Properties*, Proc. Conf. Asbury Park, N.J., 1959, pp. 159–174.
304. G. K. Gaule and R. L. Ross, *Boron Prepn., Properties, Appl.* Papers, 2nd Paris Internatl. Symp., 337–338 (1964).
305. U.S. 3,329,917, July 4, 1967.
306. B. B. Anisimov, Sh. Z. Dzhamagidze, and R. R. Shvangiradze, *Soviet Phys. Semiconductors*, **6,** (11), 1873–1874 (1973).
307. A. F. Konstantinova, K. K. Abrashev, *Geol. Petrog. Mineral Magmat. Obrazov. Sev. Vost. Chasti Sib. Platformy*, **197,** 308–322 (1970) (Russ.).
308. J. F. H. Custers, *Physica*, **18,** 489–496 (1952); *Nature*, **176,** 360 (1955).
309. J. J. Brophy, *Phys. Rev.*, **99,** 1336–1337 (1955).
310. W. J. Leivo and R. Smoluchowski, *Bull. Am. Phys. Soc.*, **30,** (2), 9 (1955).
311. I. G. Austin and R. Wolfe, *Proc. Phys. Soc. (London)*, **B69,** 329–338 (1956).
312. P. T. Wedepohl, *Proc. Phys. Soc. (London)*, **B70,** 177–184 (1957).
313. G. B. Rodgers and F. A. Raal, *Rev. Sci. Instr.*, **31,** 663–664 (1960).
314. R. M. Chrenko, *Phys. Rev. B*, **7,** (10), 4560–4567 (1973).
315. F. A. Raal, *Am. Mineralogist*, **42,** 354–361 (1957).
316. M. Seal, *Ind. Diamond Rev.*, **29,** 408–412 (1969).
317. R. H. Wentorf, Jr. and H. P. Bovenkerk, *J. Chem. Phys.*, **36,** 1987–1990 (1962).
318. G. N. Bezrukov, V. P. Butuzov, N. N. Gerasimenko, L. V. Lezheiko, and Yu A. Litvin, *Fiz. Tekh. Poluprov.*, **4,** (4), 693–696 (1970).
319. V. S. Vavilov, G. E. A. Konorova, V. F. Sergienko, and M. I. Guseva, *Soviet Phys. Semiconductors*, **4,** (1), 12–16 (1970).
320. V. S. Vavilov, M. A. Gukasyan, M. I. Guseva, and E. A. Konorova, *Soviet Phys. Semiconductors*, **6,** (5), 741–745 (1972).
321. U.S. 3,435,398, March 25, 1969.
322. Ger. Offen. 1,792,696, Nov. 25, 1971.
323. U.S. 3,735,321, May 22, 1973.
324. S. B. Lang, in *Temperature, Its Measurement and Control in Science and Industry*, Vol. 4, Part 2, Instrument Society of America, Pittsburgh, 1961, pp. 1015–1023, 1153–1167.
325. K. Takami, T. Komatsu, T. Horie, Hitachi Ltd., S. Yamasaki, A. Mochizuki, Japanese Natl. Railway, Tokyo, Japan, in *Temperature, Its Measurement and Control in Science and Industry*, Vol. 4, Part 2, Instrument Society of America, Pittsburgh, 1973, pp 1311–1316.

326. A. Smakula, N. Skribanowitz, and A. Szorc, *J. Appl. Phys.*, **43,** (2), 508–515 (1972).

327. W. N. Lawless, *Advances in Cryogenic Eng.*, **16,** 261 (1971).

328. W. N. Lawless, *Rev. Sci. Instr.*, **42,** (5), 561–566 (1971).

329. W. N. Lawless and E. A. Panchyk, *Cryogenics*, **12,** 196–200 (June 1972).

330. W. N. Lawless, R. Radebaugh, and R. J. Soulen, *Rev. Sci. Instr.*, **42,** (5), 567–570 (1971).

331. L. G. Rubin and W. N. Lawless, *Rev. Sci. Instr.*, **42,** (5), 571–573 (1971).

332. D. Bakalyar, R. Swinehart, W. Weyhmann, and W. N. Lawless, *Rev. Sci. Instr.*, **43,** (8), 1221–1223 (1972).

333. W. N. Lawless, in *Temperature, Its Measurement and Control in Science and Industry*, Vol. 4, Part 2, Instrument Society of America, Pittsburgh, 1973, pp. 1143–1151.

334. R. A. Brand, S. A. Letzring, H. S. Sack, and W. W. Webb, *Rev. Sci. Instr.*, **42,** (7), 927–930 (1971).

335. A. T. Fiory, *Rev. Sci. Instr.*, **42,** (7), 930–933 (1971).

336. H. S. Sack and M. C. Moriarty, *Solid State Commun.*, **3,** 93–96 (1965).

337 S. B. Lang, S. A. Shaw, L. H. Rice, and K. D. Timmerhaus, *Rev. Sci. Instr.*, **40,** 274–284 (1969).

338. S. B. Lang, L. H. Rice, and S. A. Shaw, *J. Appl. Phys.*, **40,** 4335–4340 (1969).

339. S. B. Lang, *Nature*, **212,** 704 (1966).

340. S. B. Lang and F. Steckel, *Rev. Sci. Instr.*, **36,** 929–932 (1965).

341. S. B. Lang M. D. Cohen, and F. Steckel, *J. Appl. Phys.*, **36,** 3171–3174 (1965).

342. S. T. Liu, J. D. Heaps, and O. N. Tufte, *Ferroelectrics*, **3,** 281–285 (1972).

343. H. P. Beerman, *IEEE Trans. Electron Devices*, **ED-16,** (6), 554 (1969).

344. A. M. Glass, J. H. McFee, and I. G. Bergman, Jr., *J. Appl. Phys.*, **42,** (13), 5219–5222 (1971).

345. E. C. Hirschkoff, O. G. Symko, and J. C. Wheatley, *J. Low Temp. Phys.*, **4,** 111–115 (1971).

346. R. E. Hamilton, *J. Can. Ceram. Soc.*, **37,** 62–64 (1968).

347. W. St. Clabaugh, E. M. Swiggard, and R. Gilchrist, "Preparation of Barium Titanyl Oxalate Tetrahydrate for Conversion to Barium Titanate of High Purity," *J. Res. Natl. Bur. Std.*, **56,** S.289 fr., Research Paper 2677 (1956).

348. W. R. Northover, "Tracer Determination of the Amount of Lathanum Coprecipitation in Barium-Strontium Titanyl Oxalate," *J. Am. Ceram. Soc.*, **48,** (4), 173–175 (1965).

349. K. S. Mazdiyasni, R. T. Dolloff, and J. S. Smith II, *J. Am. Ceram. Soc.*, **52,** (10), 523–526 (1969).

350. W. D. Kingery, *Introduction to Ceramics*, John Wiley & Sons, Inc., New York, 1960. J. T. Jones and M. F. Berard, *Ceramics: Industrial Processing and Testing*, Iowa State University Press, Ames, Iowa, 1972.

351. M. Sayer and A. Mansingh, *Phys. Rev. B*, **6,** (12), 4629–4643 (1972).

352. P. Petrov and T. Racheva, (Bulg.) *God. Vissh. Khimikotekhnol. Institutr, Sofia 1967*, **14,** (1), 91–101 (1971).

353. S. G. Bishop and W. J. Moore, *Appl. Opt.*, **12,** 80–83 (1973).

354. K. Yamamoto, W. Shimotsuma, H. Moriga, and T. Shimizu, *Nat. Tech. Report*, **15,** (2), 133–144 (April 1969).

355. U.S. 2,633,521, March 31, 1953, Bell. Tel. Lab.

356. Gulton Patent. 2,915,407 Dec. 1, 1959.

357. Jan Bekisz, *Przegl. Electron.*, **6,** (2), 84–89 (1965).

358. Senzo Miura, *Oyo Butsuri*, **28,** 117–119 (1959).

359. W. Becker and K. Lark-Horovitz, *Proc. of the Natl. Electronics Conference*, **8,** 506–508 (1953), Chicago, Ill.

360. J. W. Thornhill and K. Lark-Horovitz, *Phys. Rev.*, **82,** 762 (1951).

361. L. I. Mendelsohn, E. D. Orth, R. E. Curran, and E. D. Robie, *Am. Ceram. Bull.*, **45,** (9), 771–776 (1966).

362. L. I. Mendelsohn, E. D. Orth, and R. E. Curran, *J. Vac. Sci. Techn.*, **6,** (3), 363–367 (1968).

363. U.S. 3,642,527, Feb. 15, 1972.

364. T. N. Kennedy and F. M. Collins, US Clearinghouse Fed. Sci. Tech. Inf. AD 683,368 (1969).

365. Tara Matsumoto and Takeshi Shintani, U.S. 3,503,030, March 24, 1970.

366. D. S. Bersis and C. A. Karybakas, *J. Sci. Instr.*, **14,** 777–780 (1967).

367. German Offen. 2,117,446, April 8, 1971, Sprague Electric Co., North-Adams, Mass.

368. H. Fischmeister and E. Exner, *Metall.*, **18,** 932–940 (1964).

369. H. Fischmeister and E. Exner, *Metall.*, **19,** 113–119 (1965).

370. H. Fischmeister and E. Exner, *Metall.*, **19,** 914–945 (1965).

371. F. Thümmler and W. Thomma, *Metallurgical Rev.*, **115,** 69–108 (1967).

372. P. A. Marshall, Jr., D. P. Enright, and W. A. Weyl, "On the Mechanism of Sintering and Recrystallization of Oxides." *Proc. Internatl. Symp. on the Reactivity in Solids.* (*Gothenburg 1952*), 1954, pp. 273–284.

373. J. A. Hedvall, the pioneer in solid-state chemistry, deserves the merit to quote the original statement of Aristotle's: "Ta hygra meikta malista ton somaton."

374. J. A. Hedvall, *Solid State Chemistry*, Elsevier Publishing Co., New York, 1966, contains 186 references.

375. G. F. Hüttig, *Angew. Chemie*, **68,** 376 (1956).

376. J. A. Hedvall, *Z. Elektrochem.*, **45,** 83–93 (1939).

377. C. Wagner, *Z. phys. Chem. B*, **34,** 309–316 (1936).

378. H. Schmalzried, *Z. phys. Chem. neue Folge*, **33,** (1–4), 111–128 (1962).

379. P. Reijnen, *Proc. Internatl. Symp. on Reactivity of Solids, 1968*, Wiley-Interscience, New York, 1969, pp. 99–112.

380. D. W. Readey, *J. Am. Ceram. Soc.*, **47,** 366–369 (1966).

381. W. Shockley, *Electrons and Holes in Semiconductors*, Van Nostrand, New York, 1950.

382. C. Kittel, *Introduction to Solid State Physics*, 2nd ed., John Wiley & Sons, Inc., New York, 1967, pp. 383–401. See also ref. 16.

383. E. Spenke, *Elektronische Halbleiter*, Springer-Verlag, Berlin, 1955, pp. 320–368; 1st ed., 1955, p. 347.

384. K. Hirabayashi, *Phys. Rev. B.*, **3,** (12), 4023–4025 (1971).

385. H. Schweickert, *Ber. Phys. Ges.*, **3,** 99 (1939).

386. E. Kh. Enikeev, L. Ya. Margolis, and S. Z. Roginskii, *Dokl. Akad. Nauk SSSR*, **130,** (4), 807–809 (February 1960).

387. John B. Taylor and I. Langmuir, *Phys. Rev.*, **44,** 423–458 (1933).

388. Elias P. Gyftuponlos and Jules D. Levine, *J. Appl. Phys.*, **33,** 67–73 (1962).

389. L. Dobrochev and T. L. Matskevich, *Sov. Phys. Tech. Phys.*, **11,** (8), 1081–1087 (1967).

390. E. Kh. Enikeev, L. Ya. Margolis, and S. Z. Roginskii, *Dokl. Akad. Nauk SSSR*, **124,** (3), 606–608 (1959).

391. F. Wilessow and A. Terenin, *Naturwiss.*, **46,** (5), 167–168 (1958).

392. H. Farnsworth, *Proc. Phys. Soc.*, **71,** Pt. 4, 703–704 (1958).

393. H. Oettel, *Hermsdorfer Techn. Mitteilungen*, Heft 30, 951–958 (1970).

394. R. W. Atkinson, *Solid State Techn.*, **14,** (5), 51–54, 63 (1971).

395. R. E. Trease and R. L. Dietz, *Solid State Tech.*, **15,** 39–43 (January 1972).

396. S. H. Wemple, D. Kahng, C. N. Berglund, and L. G. Van Uitert, *J. Appl. Phys.*, **38,** 799–805 (1967).

397. D. Kahng and S. H. Wemple, *J. Appl. Phys.*, **36,** (9), 2925–2929 (1965).

398. N. Feldstein, *Solid State Tech.*, **16,** (12), 87–92 (1973).

399. D. R. Turner and H. A. Sauer, *J. Electrochem. Soc.*, **107,** 250–251 (1960).

400. Tsuneharu Nitta, Kaneomi Nagase, and shigeru Hayakawa, *J. Am. Ceram. Soc.*, **49,** (8), 457 (1966).

401. U.S. 3,586,534, March 2, 1971.

402. E. J. Nolte and R. F. Spurak, *Television Eng.*, **I,** (11), 14–18 (1930).

403. R. J. Bondley, Gen. Elec. Res. Lab. Report (April 1, 1947).

404. H. J. de Bruin, A. F. Moodie, and C. E. Warble, *J. Austral. Ceram. Soc.*, **7,** (2) 57 (1971).

405. H. J. de Bruin, A. F. Moodie, and C. E. Warble, *J. Mater. Sci.*, Gold Bulletin **5,** (3), 62–64 (1972). Also Australian Patent Appl. 25585/71.

406. U.S. 3,393,072, Manufactured by Vitta-Corporation, Wilton, Conn. 06897.

407. Kitty Ettre, *Ceram. Age*, 57–60 (1965).

408. German Offen. 2,055,657 (May 18, 1972).

409. U.S. 3,645,785 (Feb. 29, 1972).

410. F. Stöckmann, *Halbleiterprobleme*, **6,** 279–320 (1966), Verlag Vieweg und Sohn, Braunschweig.
411. L. J. van der Pauw, *Philips Res. Report*, **13,** 1–9 (1958); *Phys. Tech. Rev.*, **20,** 220–224 (1958/1959).
412. L. B. Valdes, *Proc. IRE*, **42,** 420–427 (1954).
413. Sadayenki Murashima and Fumio Ishibashi, *Japan. J. Appl. Phys.*, **11,** (5), 685–691 (1972).
414. R. G. Mazur and D. H. Dickey, *J. Electrochem. Soc.*, **113,** 255 (1966).
415. Private communication from Aly A. Mahmoud, Univ. of Missouri at Columbia, Elec. Eng. Dept.
416. E. V. George and G. Bekefi, *IEEE Electron Devices*, **ED-17,** (1), 27–30 (1970).
417. H. P. Wagner and K. H. Besocke, *J. Appl. Phys.*, **40,** 2916–2922 (1969).
418. I. R. Weingarten and M. Rothberg, *J. Electrochem. Soc.*, **108,** 167–71 (1961).
419. K. W. Keller, *Z. angew. Physik*, **11,** 346–350 (1959).
420. K. W. Keller, *Z. angew. Physik*, **11,** 351–352 (1959).
421. J. D. Holm and K. S. Champlin, *J. Appl. Phys.*, **39,** 275–284 (1968).
422. U.S. 2,970,411 (Feb. 7, 1961).
423. E. M. Rabinovich, *Izv. Akad. Nauk SSSR*, *Neorganicheskie Materialy*, **7,** (4), 545–560 (1971).
424. S. S. Flaschen, A. D. Pearson, and W. R. Northover, *J. Am. Ceram. Soc.*, **43,** 274–278 (1960).
425. J. E. Stanworth, *J. Soc. Glass Technol.*, **36,** 217 (1952).
426. G. Bosson, F. Gutman, and L. M. Simmons, *J. Appl. Phys.*, **21,** 1267–1268 (1950).
427. H. W. Trolander and R. W. Harruff, *Digest of the 6th International Conference on Medical Electronics and Biological Eng. 581, Tokyo (1965).*
428. I. S. Steinhart and S. R. Hart, *Deep Sea Res.*, **15,** 497 (1968).
429. H. W. Trolander, D. A. Case, and R. W. Harruff, in *Temperature, Its Measurement and Control in Science and Industry*, Vol. 4, Part 2, Instrument Society of America, Pittsburgh, 1973, pp. 997–1009.
430. R. H. Bube, *Proc. IRE*, **43,** 1836–1850 (1955).
431. G. Hoehler, *Ann. Phys.*, **4,** 371 (1949).
432. G. Hoehler, *Z. tech. Phys.*, **21,** 128 (1940).
433. A. Radkovsky, *Phys. Rev.*, **73,** 749–761 (1948).
434. W. R. Beakly, *J. Sci. Instr.*, **28,** 176–179 (1951).
435. E. Keonjian and I. S. Schaffner, *Trans. AIEE*, **73,** Pt. 1, 396–400 (1954).
436. A. B. Soble, *IRE Trans. Component Parts*, **CP4,** 96–101 (1957).
437. K. S. Cole, *Rev. Sci. Instr.*, **28,** 326–328 (1957).
438. I. S. Sokolnikoff and R. M. Redheffer, *Mathematics of Physics and Modern Engineering*, McGraw-Hill Book Co., New York, 1958, pp. 694–708.
439. Bull. 9612, Rosemount Eng. Co., Minneapolis, Minn. (1962).

440. R. C. Rose, in *Temperature, Its Measurement and Control in Science and Industry*, Vol. 4, Part 2, Instrument Society of America, Pittsburgh, 1973. Paper R33 (not published).

441. L. J. Eriksson, F. W. Kuether, and J. J. Glatzel, in *Temperature, Its Measurement and Control in Science and Industry*, Vol. 4, Part 2, Instrument Society of America, Pittsburgh, 1973, pp. 989–995.

442. C. A. Logan and W. E. Cook, *Control Eng.*, vol. 17, (December 1970).

443. D. Kimball and R. W. Haruff, in *Temperature, Its Measurement and Control in Science and Industry*, Conference paper L9 (unpublished).

444. I. M. Diamond, *Rev. Sci. Instr.*, **41,** 53–60 (1970).

445. H. W. Trolander and R. W. Harruff, "A Thermistor Temperature Sensor Suitable for Use as a Transfer Standard," *Proc. 5th Temp. Meas. Soc. Conference*, Hawthorne, California, 1967.

446. For details: Yellow Springs Instruments Co., Yellow Springs, Ohio 45387, Keystone Carbon Co., St. Marys, Pa. 15857.

447. J. A. Becker, C. B. Green, and G. L. Pearson, *Elec. Eng. Transactions*, **65,** 711–725 (1946).

448. R. E. Burgess, *Proc. Phys. Soc. B*, **68,** P+11, 908–917 (1955).

449. A. J. Friedman and M. R. Ody, *Rev. Sci. Instr.*, **43,** (4), 612–613 (1972).

450. J. W. Ekin and D. K. Wagner, *Rev. Sci. Instr.*, **41,** 1109–1110 (1970).

451. E. E. Swartzlander, Jr., in *Temperature, Its Measurement and Control in Science and Industry*, Vol. 4, Part 3, Instrument Society of America, Pittsburgh, 1973, pp. 2337–2351.

452. L. A. Frohbach, *Proc. of the 5th Temp. Measurement Soc. Conference*, Hawthorne, California, 1967.

453. A. E. Beck, *J. Sci. Instr.*, **33,** 16–17 (1956).

454. Cl. R. Droms, in *Temperature, Its Measurement and Control in Science and Industry*, Vol. 3, Part 2, Reinhold Publishing Corp., New York, 1962, pp. 339–346.

455. C. A. Pippin, AEC Report RTP, 649 UC 37 TID 4500 (December 1965).

456. G. Halverson and D. A. Johns, in *Temperature, Its Measurement and Control in Science and Industry*, Instrument Society of America, Pittsburgh, 1973, Vol. 4, Part 2, pp. 803–813.

457. G. van Rijn, Mrs. M. C. Nieuwenhuys-Smit, J. E. von Dijk, J. L. Tiggelman, and M. Durieux, in *Temperature, Its Measurement and Control in Science and Industry*, Instrument Society of America, Pittsburgh, 1973, Vol. 4, Part 2, pp. 815–826.

458. A. C. Anderson, in *Temperature, Its Measurement and Control in Science and Industry*, Vol. 4, Part 7, Instrument Society of America, Pittsburgh, 1973, pp. 773–784, 979–1009.

459. J. F. Cochran, C. A. Shiffman, and J. E. Neighbor, *Rev. Sci. Instr.*, **37,** 499–512 (1966).

460. P. A. Ainsworth, *Feinwerktechnik + Micronic*, **76,** Jahrgang Heft 3, 130–134 (1972).

461. German Patent 972,851, 20, October 1948.

462. W. Seith, *Diffusion in Metals*, Springer-Verlag, Berlin, 1955.

463. J. R. Black, "Metallization failures in integrated circuits," RADC Techn. Rept. TR-68-243 (October 1968).

464. J. R. Black, *IEEE Trans.*, **ED-16,** (4), 338–347 (April 1969).

465. A comprehensive study was made by R. Rosenberg and M. Ohring, *J. Appl. Phys.*, **42,** (13), 5671–5679 (1971).

466. A. D. LeClaire, *Phys. Rev.*, **93,** 344 (1954).

467. H. B. Huntington, *J. Phps. Chem. Solids*, **29,** 1641–1651 (1968).

468. R. A. Oriani, *J. Phys. Chem. Solids*, **30,** 339–351 (1969).

469. M. Sapoff, in *Temperature, Its Measurements and Control in Science and Industry*, Vol. 4, Part 3, Instrument Society of America, Pittsburgh, 1973, p. 2109.

470. K. W. Wagner, *A.I.E.E.*, **41,** 1034 (1922).

471. H. Lueder and Spenke, *Z. tech. Phys.*, **16,** 373 (1935).

472. H. Lueder, W. Schottky, and E. Spenke, *Naturwissenschaften*, **24,** 61 (1936).

473. K. Lark-Horovitz, E. Bleuler, R. F. Davis, and D. Tendam, *Phys. Rev.*, **73,** 1256 (1948).

474. K. Lark-Horovitz, E. Bleuler, R. F. Davis and D. Tendam, *Phys. Rev.*, **74,** 1255 (1948).

475. E. E. Klontz and K. Lark-Horovitz, *Phys. Rev.*, **82,** 763 (1951).

476. K. Lark-Horovitz, M. Becker, and H. Y. Fan, *Phys. Rev.*, **76,** 730 (1949).

477. M. Nachmann, L. Cojocaru, and L. Ribeo, *Nukleonik. (Germany)*, **10,** (C 1), 1–2 (1967).

478. N. N. Gerasimenko, A. V. Dvurechenskii, V. I. Panov, and L. S. Smirnov, *Soviet Phys. Semiconductors*, **5,** (8), 1439–1441 (1972).

479. P. Nagels, M. Ali, and M. Denayer, *Radiat. Eff. Semiconduction Proc. 1967*, edited by Frederick L. Vock, Plenum Press, New York, 1968, pp. 452–459.

480. F. F. Morehead, B. L. Crowder, and R. S. Title, *J. Appl. Phys.*, **43,** (3), 1112–1118 (1972).

481. R. Bäuerlein, *Raumfahrtforsch.*, **15,** 45–51 (1971).

482. W. R. Owen, *IRE Trans.*, **NS9**, 296 (1962).

483. J. K. D. Verma and P. S. Nair, *Naturwissenschaften*, **52,** (4), 82 (1965).

484. S. P. Solovev, I. I. Kuzmin, and V. V. Zakurkin, *Ferroelectrics*, **1,** 19–22 (1970).

485. H. A. Sauer, S. S. Flaschen, and D. C. Hoesterey, *J. Am. Ceram. Soc.*, **42,** 363–366 (1959).

486. R. D. Goodwin, *Proceedings of the Second Cryogenic Engineering Conference*, Boulder, 1957, pp. 254–268.

487. M. Böel and B. Erickson, *Rev. Sci. Instr.*, **36,** 904–908 (1965).

488. A. C. Anderson, *Rev. Sci. Instr.*, **44,** (10), 1475–1477 (1973).

489. C. Blake, C. E. Chase, and E. Maxwell, *Rev. Sci. Instr.*, **29,** 715–716 (1958).

490. R. C. La Force, S. F. Ravitz, and W. B. Kendall, *Rev. Sci. Instr.*, **35**, 729–732 (1964).

491. C. A. Mossman, J. Lundholm, Jr., and P. E. Brown, "Thermistor Temperature Recorder for Meteorologial Survey," Oak Ridge Natl. Laboratory, ORNL 556 (May 17, 1950).

492. C. M. Proctor, Dept. of Oceanography, A. and M. College of Texas, Techn. Rept. Ref. 55-15 T (March 28, 1955).

493. B. M. Zeffert and R. R. Witherspoon, *Anal. Chem.*, **28**, (11), 1701–1705 (1956).

494. W. F. Libby, *Science*, **171**, 499 (1971).

495. J. W. Hightower (private communication) Rice University, Texas.

496. F. I. Badgley, *Rev. Sci. Instr.*, **28**, 1979–1984 (1957).

497. E. D. Ney, R. W. Maas, and W. F. Huch, *J. Meteorol.*, **18**, 60–80 (1961).

498. G. H. Hall and R. M. Lucas, ISA 68-805 (1968); *Proceedings of the Fourth AFCRL Scientific Balloon Symposium*, Office of Aerospace Research, Washington, D.C., **4**, 279–293,(1967); **5**, 121–129 (1967).

499. F. L. Staffanson, *J. Appl. Meteorology*, **10**, 825–832 (1971).

500. J. Cooney, *J. Appl. Meteorology*, **11**, 108–112 (1972).

501. W. R. Blackmore, *Can. J. Phys.*, **37**, 1331–1338 (1959).

502. K. Kenyon, "Preparations for measuring horizontal thermal microstructure in the deep sea," MPL Tech. Memo 121, Univ. of California, San Diego, 1962.

503. J. C. Cook and K. E. Kenyon, *Rev. Sci. Instr.*, **34**, (5), 496–499 (1963).

504. M. C. Brewer, "Some results of geothermal investigations of permafrost in northern Alaska," *Am. Geophys. Union Trans.*, **39**, (1), 19–56 (1958).

505. E. C. Bullard, "Measurement of temperature gradient in the earth, in Runcorn." *Methods and techniques in Geophysics*, Vol. 1, Wiley-Interscience, New York, 1960, pp. 1–9.

506. P. Chadwick, "Heat flow from the earth at Cambridge. *Nature*, **178**, (4524), 105–106 (1956).

507. L. R. Cooper and C. Jones, "The determination of virgin strata temperatures from observations in deep survey boreholes," *Royal Astron. Soc. Geophys. J.*, **2**, 116–131 (1959).

508. W. H. Diment and J. D. Weaver, "Subsurface temperatures and heat flow in the AMSOC core hole near Mayaguez, Puerto Rico," Natl. Acad. Sci. Natl. Research Council publication 1188, 1964, pp. 75–91.

509. G. W. Greene, H. A. Lachenbruch, and M. C. Brewer, "Some thermal effects of a roadway on permafrost," U.S. Geol. Survey Prof. Paper 400-B, B141–B144 (1960).

510. A. H. Lachenbruch and M. C. Brewer, "Dissipation of the temperature effect of drilling a well in Arctic Alaska," *U.S. Geol. Survey Bull.*, **1083-C**, 73–109 (1959).

511. A. H. Lachenbruch, M. C. Brewer, G. W. Greene, and B. V. Marshall, "Temperature in Permafrost," Am. Inst. Physics, *Temperature, Its Measurement and Control in Science and Industry*, Vol. 3, Part 1, Reinhold Publishing Corp., New York, 1962, pp. 791–803.

512. A. D. Misener and A. E. Beck, "Thr measurement of heat flow over land," *Methods and Techniques in Geophysics*, Vol. 1, Wiley-Interscience, New York, 1960, pp. 10–61.

513. G. Newstead and A. E. Beck, "Borehole temperature measuring equipment and the geothermal flux in Tasmania," *Australian J. Phys.*, **6,** 480–489 (1953).

514. K. E. C. Nielson, "Temperature measurements with thermistors in concrete," *Swedish Cement and Concrete Res. Inst. Bull.*, **34,** 44 (1959).

515. R. Raspet, J. H. Swartz, M. E. Lillard, and E. C. Robertson, "Preparation of Thermistor Cables used in geothermal Investigations," *U.S. Geol. Survey Bull.*, **1203-C,** C1–C11 (1966).

516. J. H. Sass and A. E. LeMarne, "Heat flow at Broken Hill," *New South Wales, Royal Astron. Soc. Geophys. J.*, **7,** 477–489 (1963).

517. J. H. Swartz, "A geothermal measuring circuit," *Science*, **120,** (3119), 573–574 (1954).

518. J. H. Swartz, "Geothermal measurements on Eniwetok and Bikini Atolls," U.S. Geol. Survey Prof. Paper 260-U, pp. 711–739 (1958).

519. R. P. von Herzen, "Measurement of heat flow through the ocean floor," Am. Inst. Physics, *Temperature, Its Measurement and Control in Science and Industry*, Vol. 3, Part 1, Reinhold Publishing Corp., New York, 1962, pp. 769–777.

520. J. Krog, *Rev. Sci. Instr.*, **25,** 799–800 (1954).

521. J. Krog, *Rev. Sci. Instr.*, **27,** 408–409 (1956).

522. H. M. Whyte and S. R. Reader, *J. Appl. Physiol.*, **4,** 623 (1952).

523. L. Clark and H. W. Trolander, *J. Am. Med. Assoc.*, **155,** 251–252 (1954).

524. C. Clark, *Med. Biol. Eng.*, **6,** 133–142 (1968).

525. P. I. Hershberg, A. Kantrowitz, and E. J. Kass, *IEEE Trans. Bio-Med. Electron.*, **10,** 82–83 (1963).

526. J. P. Henry and C. D. Wheelwright, "Bioinstrumentation in MR-3 flight," *Proceedings on the Results of First U.S. Manned Suborbital Space Flight*, NASA, U.S. Govt. Printing Office, p. 37, June 1961.

527. B. Lyon, Memorial Res. Lab. Hospital Medical Center of North-Carolina, Oakland, North Carolina.

528. J. Gershon-Cohen, JoAnn D. Haberman-Brueschke, and E. E. Brueschke, *The Radiologic Clinics of North America*, **3,** (3), 403–431 (December 1965).

529. A. M. Lilienfeld, J. M. Barnes, R. B. Barnes, E. Brasfield, J. F. Connell, E. Diamond, J. Gershon-Cohen, J. Haberman, H. J. Isard, W. Z. Lane, R. Lattes, J. Miller, W. Seaman, and R. Sherman, "A cooperative Pilot Study," *Cancer*, **24,** (6), 1206–1211 (1969).

530. J. Villablanca and R. D. Myers, *Am. J. Physiol.*, **210,** 703–707 (1965).

531. O. Z. Roy and J. S. Hart, *Med. Biol. Eng.*, **4,** 457–466 (1966).
532. O. Z. Roy and J. S. Hart, *Am. J. Physiol.*, **213,** 1311–1316 (1967).
533. G. J. Deboo and T. B. Fryer, 17th Ann. Conference on Medical-Biological Eng. IEEE–ISA (1964).
534. C. D. Hull, J. Garcia, and E. Chracchiolo, *Science*, **149,** 89–90 (1965).
535. A. J. Rampone and M. E. Shirasu, *Science*, **144,** 317–319 (1964).
536. M. Ogata, *Kyushi J. Med. Sci.*, **10,** 61–79 (1959).
537. A. Slater and S. Bellet, *Med. Biol. Eng.*, **7,** 633–639 (1969).
538. B. N. Zeffert and S. Hormats, *Anal. Chem.*, **21,** 1420–1422 (1949).
539. J. R. Campbell, J. T. Pender, and R. P. Puri, *J. Roy. Tech. Coll. (Glasgow)*, **5,** 89–98 (1950).
540. S. B. V. Kulkarni, *Nature*, **171,** 219–220 (1953).
541. B. E. Ballard and F. M. Goyan, *J. Am. Pharm. Assoc.*, **47,** 40–42, (1958).
542. R. D. Johnson and F. M. Goyan, *J. Pharm. Sci.*, **56,** (6), 757–759 (1967).
543. A. V. Hill, *Proc. Roy. Soc. (London)*, **A127,** 9–19 (1930).
544. W. Simon and C. Tomlinson, *Chimica*, **14,** 301 (1960).
545. A. P. Brady, H. Huff, and J. W. McBain, *J. Phys. Coll. Chem.*, **55,** 304 (1951).
546. C. G. McGee and B. R. Y. Iyengar, *Anal. Chem.*, **23,** 1103–1106 (1953).
547. J. J. Neumayer, *Anal. Chim. Acta*, **20,** 519 (1959).
548. R. H. Müller and H, Stolten, *Anal. Chem.*, **25,** 1103–1106 (1955).
549. B. R. Y. Iyengar, *Rec. trav. chim. Pays. Ba.*, **73,** 789 (1959).
550. C. Tomlinson, *Microchem. Acta*, **63,** 457–466 (1961).
551. C. Tomlinson, Ch. Chylewski, and W. Simon, *Tetrahedron*, **14,** 949 (1963).
552. A. Adicoff and W. Murbach, *Anal. Chem.*, **39,** (3), 302–306 (1967).
553. W. Wagner, *Z. Phys. Chem. Leipzig*, **218,** Heft 516, 392–416 (1961).
554. A. Magnus and F. Becker, *Z. Phys. Chem.*, **196,** 378–396 (1951).
555. P. Conway, J. Mooi, and C. R. E. Harris, *J. Phys. Chem.*, **62,** 665–667 (1958).
556. M. Knoester, K. W. Taconis, and J. I. M. Beenakher, *Physica*, **33,** 389–407 (1967).
557. R. Bachmann, F. J. DiSalvo, Jr., T. H. Geballe, R. L. Greene, R. E. Howard, C. N. King, H. C. Kirsch, K. N. Lee, R. E. Schwall, H. U. Thomas, and R. B. Zubeck, *Rev. Sci. Instr.*, **43,** (2), 205–214 (1972).
558. F. E. Harris and L. K. Nash, *Anal. Chem.*, **23,** 736–739 (1951).
559. A. B. Hart and D. S. Weir, *Chem. and Ind. (London)*, 563–564 (1957).
560. C. H. Miyama, *J. Bull. Chem. Soc. (Japan)*, **29,** 711–715 (1956).
561. C. H. Miyami, *Bull. Chem. Soc. (Japan)*, **30,** 10–13 (1957).
562. G. A. Grant, M. Katz, and R, Riberdy, *Can. J. Techn.*, **29,** 511–519 (1952).
563. R. Belcher and R. Goulden, *Industrial Chemist*, **26,** 320–322 (1950).
564. A. Weissberger, *Technique of Organic Chemistry*, Vol. VIII, Wiley-Interscience Publishers, New York, 1953, pp. 669, 711.

565. E. Giladi, A. Lifshitz, and B. Perlmutter-Hayman, *Bull. Res. Council of Israel*, **A8,** G2, 75–80 (1959).

566. L. S. Bark and S. M. Bark, *Thermometric Titrimetry*, International Series of Monographs in Analytical Chemistry, Vol. 33, Pergamon Press, Elmsford, N.Y., 1969.

567. H. I. V. Tyrell aand A. E. Beezer, *Thermometric Titrimetry* Chapman and Hall, London, England, 1968.

568. J. Jordan, I. M. Kolthoff, and P. J. Elving, *Treatise on Analytical Chemistry*, Ed. Part I.4, Wiley-Interscience, New York, 1968, pp. 5175–5242.

569. L. D. Hansen and E. A. Lewis, *Anal. Chem.*, **43,** 1393–1397 (1971).

570. R. V. Mrazek and H. C. Van Ness, "Titration calorimetry used for measuring heats of mixing," *A.I.Ch.E. J.*, **7,** 190 (1961).

571. I. Grethe, "Titration calorimetry used for measuring heats of metalligand interaction," *Acta Chem. Scand.*, **17,** 2487 (1963).

572. R. M. Izatt, D. Eatough, J. J. Christensen, and R. L. Snow, *Phys. Chem.*, **72,** 1208 (1968).

573. J. J. Christensen, L. D. Hansen, and R. M. Izatt, "Titration calorimetry used for measuring heats of proton ionization," *J. Am. Chem. Soc.*, **89,** 213 (1967).

574. J. J. Christensen, R. M. Izatt, L. D. Hansen, and J. A. Partridge, "Titration calorimetry used for measuring equilibrium constants," *J. Phys. Chem.*, **70,** 2003 (1966).

575. J. J. Christensen, D. P. Wrathall, and R. M. Izatt, *Anal. Chem.*, **40,** 175 (1968).

576. T. R. Pattee, *Electronics*, **16,** 102, 105, 190, 192, 194, 196, 198, 200 (1943).

577. A. J. Wheeler, *Electronics*, **30,** 169–171 (1957).

578. P. Guené, *Ann. Radioelectr.* (*French*), **11,** 317–330 (1956).

579. Anon, *Wireless World*, **64,** 441 (1958).

580. T. R. Nisbet, *Electronic Design*, **31,** (September 1958).

581. G. Whitehaus, *Instrum. Control Sys.*, **45,** (9), 72–73 (1972).

582. W. H. Brattain and J. A. Becker, *J. Opt. Soc. Am.*, **36,** 354 (1946).

583. J. A. Becker et al., OSRD 5991, Final Report on Development and Operating Characteristics of Thermistor Bolometers (1945).

584. E. M. Wormser, *J. Opt. Soc. Am.*, **43,** (1), 15–21 (1953).

585. R. Hansen, *Proc. Opt. Soc. Am.*, **36,** 355 (1946).

586. E. Hesse and D. Mendez, *Z. Angew. Physik.*, **27,** (4), 289–291 (1969).

587. R. de Waard and S. Weiner, *Appl. Opt.*, **6,** (8), 1327–1330 (1967).

588. F. J. Low, *J. Opt. Soc. Am.*, **51,** 1300–1304 (1961).

589. S. Zwerdling, R. A. Smith, and J. P. Theriault, *Infrared Phys.*, **8,** (4), 271–336 (1968).

590. S. Zwerdling, J. P. Theriault, and H. S. Reichard, *Infrared Phys.*, **8,** (2), 135–142 (1968).

591. F. J. Low and A. R. Hoffman, *Appl. Optics*, **2,** (6), 649–650 (1963).

592. B. W. Kennedy, *Rev. Sci. Instr.*, **40,** 1169–1172 (1969).

593. W. F. Netusil, in *Temperature, Measurement Control in Science and Industry*, Vol. III, Part 2, Reinhold Publishing Corp., New York, 1962, pp. 449–453.

594. M. Cavallins, L. Menegheth, G. Scoles, and M. Yealland, *Rev. Sci. Instr.*, **42,** (12), 1759–1763 (1971).

595. T. G. Phillips and K. B. Jofferts, *Rev. Sci. Instr.*, **44,** 1009–1014 (1973).

596. E. K. Labartkava, *Akusticheskii Zh.*, **6,** (4), 468–471 (1960) [Soviet Phys., **6,** (4), 468–471 (1960)].

597. S. Morita, *J. Phys. Soc. (Japan)*, **7,** 214 (1952).

598. E. P. Manche and B. Carroll, *Rev. Sci. Instr.*, **35,** 1486–1488 (1964).

599. C. L. Lövborg, *J. Sci. Instr.*, **42,** 611–614 (1965).

600. G. S. Weaving, *J. Sci. Instr.*, **44,** 55–57 (1967).

601. J. C. Chambers and G. R. Bastedo, Navord Report 5950, NOLC Report 449 (April 1959).

602. H. Weiss, *Siemens Rev.*, **33,** (3), 103–107 (March 1966).

603. German Patent 1,490,498, Dec. 14, 1963.

604. F. R. MacDonald, *Instruments and Control Systems*, **42,** 101–103 (1969).

605. J. H. Bollman and J. G. Kreer, *Proc. IRE*, **38,** 20 (1950).

606. J. J. Gano and G. F. Sandy, *IRE Trans. Electron. Computers*, **EC-7,** 61–64 (1958).

607. H. Lineau and I. Seifert, *Matronics*, **12,** 216–222 (1957).

608. R. H. Barnes and P. J. Freud, *J. Appl. Phys.*, **43,** 3224–3225 (July 1972).

609. E. Weise, *Z. techn. Phys.*, **18,** 467–470 (1937).

610. K. Lämmchen, *Hochfrequenztechnik*, **40,** 119–133 (1933).

611. K. J. Schenfer and A. A. Ivanov, *Elektrichestvo*, **61,** 14 (1940), abstracted in *ETZ (Germany)* **29,** 642 (1941).

612. H. Straubel, *Z. techn. Phys.*, **18,** 464–467 (1941).

613. D. Baker Moore, *Rev. Sci. Instr.*, **37,** 1089 (1966).

614. S. B. Schwartz and A. E. Wilson, "Advance in Cryogenic Engineering," 1, *Proceedings of the 1954 Cryogenic Engineering Conference*, Plenum Press, New York, 1960.

615. Yvonne Lortie, *J. Phys. Rad.*, **16,** 317–320 (1953).

616. M. Varicak and B. Saftic, *Rev. Sci. Instr.*, **30,** 891–895 (1959).

617. A. V. Bulyga and A. G. Shashkov, *Inzhen. Fiz. Zh. USSR*, **6,** (12), 95–100 (1963).

618. A. V. Bulyga, *Inzhen. Fiz. Zhur. Akad. Nauk. Belorus. USSR*, **4,** (3), 46–52 (1961).

619. R. W. Roberts, P. E. McElligott, and G. Jernakoff, *J. Vac. Sci. Tech.*, **1,** (2), 62–64 (1964).

620. P. D. Zemany, *Rev. Sci. Instr.*, **23,** 176–177 (1952).

621. *Dushman's Scientific Foundations of Vacuum Technique*, John Wiley & Sons, Inc., New York, 1949.

622. R. A. Rasmussen, *Rev. Sci. Instr.*, **33,** 38–42 (1962).

623. M. T. Pigott and R. C. Strum, *Rev. Sci. Instr.*, **38,** 743–744 (1967).

624. K. Kraus, *Rev. Sci. Instr.*, **36,** 1191–1194 (1965).

625. J. W. Allen, M. M. Fulk, and M. M. Reynolds, Cryogenic Engineering 1, *Proceedings of the 1954 Cryogenic Engineering Conference*, Paper D-1, Plenum Press, New York, 1960.

626. P. Niiler, *Rev. Sci. Instr.*, **36,** 921–924 (1965).

627. R. B. Lambert, H. A. Snyder, and S. F. K. Karlsson, *Rev. Sci. Instr.*, **36,** 924–928 (1965).

628. R. K. Gould and W. L. Nyborg, *J. Acoust. Soc. Am.*, **31,** (2), 249–250 (1959).

629. C. T. Walker and C. E. Adams, *J. Acoust. Soc. Am.*, **31,** (6), 813–814 (1959).

630. L. C. Pharo, *Rev. Sci. Instr.*, **36,** 211–216 (1965).

631. H. H. Hausdorff, *Vapor Phase Chromatography*, Academic Press, Inc., New York, 1957.

632. D. Ambrose and R. R. Collerson, *J. Sci. Instr.*, **32,** 323 (1955).

633. D. Ambrose and R. R. Collerson, *Nature*, **177,** 84 (1956).

634. R. E. Walker and A. A. Westenberg, *Rev. Sci. Instr.*, **28,** 789–792 (1957).

635. A. D. Davis and G. A. Howard, *Chem. and Ind. BIF Rev.*, **R25–R26** (April 1956).

636. D. Buhl, *Anal. Chem.*, **40,** 715–726 (1968).

637. A. D. Davis and G. A. Howard, *J. Appl. Chem.*, **8,** 183–186 (1958).

638. C. E. Bennet, St. Dal Nogare, L. W. Safranski, and C. D. Lewis, *Anal. Chem.*, **30,**(5), 898–902 (1958).

639. D. A. Conlan and E. L. Szonntagh, *J. Gas Chromatogr.*, **6,** (8), 485–487 (1968).

640. R. Kieselbach, *Anal. Chem.*, **32,** (13), 1749–1754 (1960).

641. S. Dal Nogare and R. S. Juvet, *Gas and Liquid Chromatography*, John Wiley & Sons, Inc., New York, 1962.

642. R. Stock, and C. B. F. Rice, *Chromatographic Methods*, Chapman and Hall, London, 1967.

643. I. Smith, "Chromatographic and Electrophoretic Techniques," 1, *Chromatography*, 3rd Ed., John Wiley & Sons, Inc., New York, 1969.

644. T. Suzuki, K. Imagani, and A. Kanno, *Rev. Ind. Miner.*, **50,** (9), 679–682 (1968).

645. J. H. Bollman, *Bell Lab. Record*, **20,** 258–262 (1942).

646. J. E. Tweddale, *Western Electric Oscillator*, **3,** 34–37 (1945).

647. J. C. Johnson, *Electronic Industries*, **4,** 74–77 (1945).

648. L. A. Meecham, *Proc. IRE*, **26,** 1278–1294 (1938).

649. R. L. Shephard and R. O. Wise, *Proc. IRE*, **31,** 256–268 (1943).

650. H. V. Malmstadt and C. G. Enke, *Electronics for Scientists*, W. A. Benjamin, Inc., New York, 1962, pp. 280–282.

651. R. P. Turner, *ABC's of Thermistors*, Howard W. Sams and Co., Inc., The Bobbs-Merrill Co., Inc., Indianapolis, Kansas City, New York, 1970, pp. 75–77.

652. R. A. Rasmussen, M.S. Thesis, University of Texas (June 1962).

653. K. Kraus, *Rev. Sci. Instr.*, **39,** 216–220 (1968).

654. J. W. Motto, Jr., *Electro-technol.*, 96–99 (1962).

655. G. Budzyuski, *Przeglad. Elektron.*, **5,** (8), 383–387 (1964) (Polish).

656. A. L. Reenstra, *IEEE Trans. Electron Devices*, **16,** (6), 544–554 (1969).

657. R. Street, "The absolute measurement of lower power at 3000 Mc/s," *Proc. IEE*, **96,** Part 2, 391–396, 194–196 (1949).

658. M. Soldi, "Applications of thermistors to bridges for measuring low conductance at high frequencies," *Alta freq.*, **21,** 243–259 (1952).

659. H. B. Wood, "An R.M.S. milliammeter of noval design for the measurement of current from zero to video frequencies," *J. Sci. Instr.*, **31,** 124–125 (1954).

660. J. A. Lane, "The measurement of power at a wavelength of 3 cm by thermistors and bolometers," *Proc. IEEE*, **102,** part B, 819–824 (1955).

661. H. Reich, and F. Panninger, "A new thermistor power measuring head for λ = 9–20 cm," *Nachrichtentech.*, **7,** 101–104 (March 1957).

662. J. Collard, G. R. Nicole, and A. W. Lines, *Proc. Phys. Soc.* (*London*), **63B,** 215–216 (1950).

663. Em. Tataru, Eg. Tataru, and M. Tataru, *Acta Phys. Polon.*, **36,** Fasc. 5, (11), 501–502 (1969).

664. G. P. Gibson and J. J. Courtin, "The Westinghouse Solution," *Elec. Manufacturing*, 126–130 (November 1959).

665. W. H. Elliot, "The Cutler-Hammer Solution," "Thermistors with negative Temperature Coefficients of Resistance," *Elec. Manufacturing*, 130–132 (November 1959).

666. J. J. Courtin, *Westinghouse Engineer*, 116–118, (July–Sept. 1962).

667. J. Noble, *New Zealand Eng.*, **23,** N3, 104–109 (March 1968).

668. R. Einziger, *Siemens Zeitschrift*, **46** (4), 244–246 (1972).

669. E. Andrich and K. H. Haerdtl, *Philips Tech. Rev.*, **26,** 119–127 (1965).

670. K. Kupka, E. Ramisch, H. J. Reebs, *Int. Elektronische Rundschau*, **23,** (11), 284–286 (1969).

671. A. M. Hardie, *J. Sci. Instr.*, **34,** 58–62 (1957).

672. N. W. Bell, *Rev. Sci. Instr.*, **31,** 65 (1960).

673. C. R. Bertin and K. Rose, *J. Appl. Phys.*, **42,** (1) 163–166 (1971).

674. W. N. Lawless, *Rev. Sci. Instr.*, **43,** (12), 1743–1747 (1972).

675. R. C. DeVries and J. F. Fleischer, *Am. Ceram. Soc. Bull.*, **49,** (9), 782–788 (1970).

676. F. J. Kopp and T. Ashworth, *Rev. Sci. Instr.*, **43,** (2), 327–332 (1972).

677. J. Maserjian, *Appl. Optics*, **9,** (2), 307–315 (1970).

678. B. B. Graves, *Rev. Sci. Instr.*, **44,** (5), 571–572 (1973).

679. J. J. Christensen, J. W. Gardner, D. J. Eatough, R. M. Izatt, P. J. Watts, and R. M. Hart, *Rev. Sci. Instr.*, **44,** (4) 481–484 (1973).

680. Y. Alon and M. Jones, *Rev. Sci. Instr.*, **40,** (5), 646–647 (1969).

681. J. A. Weller, *Bell Lab. Record*, **23**, 72–75 (1945).

682. R. A. Rasmussen, *Rev. Sci. Instr.*, **31,** 747–751 (1960).

683. K. A. Sharifov, V. N. Chebotarev, and T. Kh. Azizov, *Teor. Eksp. Khim.*, **2,** (1), 137–141 (1966).

684. S. N. Gadzhiev and K. A. Sharifov, *Russ. J. Phys. Chem.*, **35,** 562–563 (1961).

685. B. Lawrence, *Can. J. Chem.*, **41,** (9), 2210–2218 (1963).

686. C. H. Griffiths and H. K. Eastwood, *J. Appl. Phys.*, **45,** (5), 2201–2206 (1974).

687. C. A. Goodwin, H. K. Bowen, and D. Adler, *J. Appl. Phys.*, **45,** (2), 626–632 (1974).

688. H. H. Sample, L. J. Neuringer, and L. G. Rubin, *Rev. Sci. Instr.*, **45,** (1), 64–73 (1974).

689. T. Vojnovich, *Dissertation Abstracts B* (*USA*), **28,** (1), (July 1967), Paper No. 67–8940, 90 pp.

690. G. Mesnard and B. Stenghel, *Compt. Rend.*, **267,** (2), 95–97 (July 1968) (French).

691. A. Z. Chaberski, *J. Appl. Phys.*, **42,** 940–947 (1971).

692. U. Strom and P. C. Taylor, *J. Appl. Phys.*, **45,** (3), 1246–1253 (1974).

693. R. J. Schutz, *Rev. Sci. Instr.*, **45,** (4), 548–551 (1974).

694. B. A. Rotenberg and Yu. I. Danilyuk, *Izv. Akad. Nauk SSSR, Ser. Fiz.*, 1824–1827 (1966) (Russ.).
[Translation: *Bull. Acad. Sci. USSR Phys. Ser.* (*U.S.A.*), **31,** (11), 1867–1870 (Nov. 1967)].

695. J. W. Dean and R. J. Richards, Advances in Cryogenic Engineering, **13,** 505 (1968).

696. A. H. Clauet, M. S. Seltzer, and B. A. Wilcox, *Materials Science Research*, **5,** 361–384 (1971).

697. A. J. Greenfield, D. Lieberman, E. Zair, and S. Greenwald, *Rev. Sci. Instr.*, **45,** (11), 1417–1422 (1974).

698. R. J. Janik, J. N. Lechevet, and W. D. Gregory, *Rev. Sci. Instr.*, **45,** (11), 1456–1457 (1974).

699. H. H. Sample and L. J. Neuringer, *Rev. Sci. Instr.*, **45,** (11), 1389–1391 (1974).

700. C. A. Hague and J. H. Fritz, *Proceedings of 1973 Holm Seminar on Electric Contact Phenomena*, Illinois Institute of Technology, Chicago, Illinois.

701. U.S. Patent 3,832,666 August 21, 1974.

702. V. S. VAVILOV, *Radiation Effects in Semiconductors*, English Translation, Consultants Bureau, New York, 1965, pp. 196–199.

703. N. LIOR, J. LEIBOVITZ, and A. D. K. LAIRD, *Rev. Sci. Instr.*, **45,** (11),1340–1343 (1974).

704. J. DRATLER, JR., *Rev. Sci. Instr.*, **45,** (11), 1435–1444 (1974).

705. L. G. RUBIN and Y. GOLAHNY, *Rev. Sci. Instr.*, **43,** (12), 1758–1762 (1972).

706. E. E. SWARTZLANDER, JR., *IEEE Transactions on Geoscience Electronics*, Vol. GE-7, 267–273 (1969).

707. E. E. SWARTZLANDER, JR., *J. Appl. Measurements*, **2,** 38–41 (1973).

708. A. L. REENSTRA, *IEEE Transactions on Industry Applications*, Vol. IA-9, (1), 58–62 January/February (1973).

709. L. HANKE and H. LÖBL, *VDI Zeitschrift*, 28 October, (I), (1968).

710. Japan Kokai 73 99,691, Dec. 17, 1973.

711. Japan Kokai 73 99,692, Dec. 17, 1973.

Author Index

Subject Index